D0558033

CRC
LIBRARY

About Island Press

Island Press is the only nonprofit organization in the United States whose principal purpose is the publication of books on environmental issues and natural resource management. We provide solutions-oriented information to professionals, public officials, business and community leaders, and concerned citizens who are shaping responses to environmental problems.

In 1994, Island Press celebrated its tenth anniversary as the leading provider of timely and practical books that take a multidisciplinary approach to critical environmental concerns. Our growing list of titles reflects our commitment to bringing the best of an expanding body of literature to the environmental community throughout North America and the world.

Support for Island Press is provided by Apple Computer, Inc., The Bullitt Foundation, The Geraldine R. Dodge Foundation, The Energy Foundation, The Ford Foundation, The W. Alton Jones Foundation, The Lyndhurst Foundation, The John D. and Catherine T. MacArthur Foundation, The Andrew W. Mellon Foundation, The Joyce Mertz-Gilmore Foundation, The National Fish and Wildlife Foundation, The Pew Charitable Trusts, The Pew Global Stewardship Initiative, The Rockefeller Philanthropic Collaborative, Inc., and individual donors.

Instream Flow Protection

Instream Flow Protection

Seeking a Balance in Western Water Use

David M. Gillilan and Thomas C. Brown

ISLAND PRESS
Washington, D.C. • Covelo, California

Library of Congress Cataloging-in-Publication Data

Gillilan, David M., 1960–
Instream flow protection: seeking a balance in western water use/ David M. Gillilan and Thomas C. Brown.
p. cm.
Includes bibliographical references and index.
ISBN 1-55963-523-1 — ISBN 1-55963-524-X (pbk.)
1. Rivers—West (U.S.)—Regulation. 2. Streamflow—West (U.S.) 3. Water use—West (U.S.) I. Brown, Thomas C. (Thomas Capnor), 1945– . II. Title.
TC423.6.G55 1997
333.91'00978—dc21 96-52477
CIP

Printed on recycled, acid-free paper ♻

Manufactured in the United States of America

10 9 8 7 6 5 4 3 2 1

Contents

Preface *ix*

1. What's the Big Deal, Anyway? 1
2. The Loss of Instream Flows: A Short History of Water Use and Water Law in the West 9
3. Instream Uses of Water 45
4. How Much Water Should Be Left in Streams? 95
5. Instream Flow Protection Issues in the States 111
6. Methods the States Use to Protect Instream Flows 137
7. Effect of Instream Flow Protection on Other Water Uses 165
8. Federal Authorities and Approaches for Protecting Instream Flows 177
9. Federal Water Development Programs Affecting Instream Flows 225
10. Federal Environmental Protection Legislation and Programs Affecting Instream Flows 255
11. Reaching a Balance in Water Allocation 297

Notes *307*

References *379*

Index *403*

Preface

We consider instream flow protection to be a critically important environmental, economic, and quality of life issue, but the reader will find herein neither an impassioned plea for its protection nor an easy path to an improved balance in western water use. Instead, we have attempted to describe what is happening in the exciting, ever-changing world of river management in the West and let the facts speak for themselves. We have tried to present the multifaceted issue of instream flow in its full complexity, presenting along the way some basic concepts and principles that can help in reaching a wise division between instream and offstream water use.

We are indebted to many knowledgeable people who generously answered our questions, helping us to better understand their areas of expertise. These individuals, too numerous to list here, are cited where appropriate and included in the references. In addition, the following persons graciously reviewed and offered helpful comments on portions of the text: John Bartholow, Lee Lamb, Robert Milhous, Kathleen Miller, John Potyondy, Larry Schmidt, Andrew Walch, and Owen Williams. Carol LoSapio and Doug Brown helped with the figures, and Joyce Hart provided clerical assistance. We would like to thank our colleagues at the Rocky Mountain Station's Craddock Building for their valuable insights, and most important, for their friendship and good humor and the folks at Island Press—Barbara Dean, Barbara Youngblood, Cecilia González, Christine McGowan, and others—for their encouragement and careful assistance. Finally, we would like to thank Wendy, Gale, and Suzanne for their understanding and support throughout the many hours of extra work required to complete this book.

While writing this book, David Gillilan was working for Colorado State University and Tom Brown was, and still is, working for the research branch of the U.S. Forest Service. We gratefully acknowledge the assistance of these two institutions, and note that the views expressed herein are our own.

CHAPTER ONE

What's the Big Deal, Anyway?

The obvious and most common reference of the term *instream flow* is to water flowing within a stream channel. But why would people be talking about that? How could instream flow be an "issue"? And why does instream flow need to be "protected"?

These questions have occurred even to people who make their living in the water resources management field. Harvey Doerksen, who later came to work extensively with instream flow issues, relates an anecdote about a meeting he attended in 1973 soon after he had been appointed the assistant director of Washington's Water Research Center.[1] Speaker after speaker at the meeting of water resources professionals from Idaho, Oregon, Washington, and the federal government referred to "instream flow" as one of the region's primary water research needs. The term brought to Doerksen's mind only a vague image of flowing water but seemed to mean much more to the other participants. As subsequent speakers continued to make references to instream flow, Doerksen found himself wondering, So what's the big deal, anyway?

What many people don't realize is that streamflows in a vast number of the West's rivers and streams have been severely diminished, even extinguished, as water has been diverted for offstream use. Rivers have been a source of water for drinking, cooking, cleaning, and other domestic purposes for as long as there have been people in the West, and water has been used for industrial purposes for at least one and a half centuries. The host of domestic, public, and commercial offstream water uses associated with municipalities are among the most important contemporary uses of rivers, and the use of water in industry has helped to promote varied and productive economies throughout the West. The diversion of water onto irrigated fields was one of the first major offstream uses of water and continues to account for 80–90 percent of all water diversions in the West. Irrigation enabled the production of food and fiber in quantities sufficient to sustain not only westerners, but also residents of other states and countries. The use of water has, in fact, been the foundation on which the West's economies, cities, and towns have been built.

The West, however, is marked—even defined—by the scarcity of water. Rain and snow fall in abundance in some mountain ranges and along the Pacific Northwest coast, but arid conditions prevail over most of the West. The West is often said to begin at the 100th meridian, the line running north to south through the states of North Dakota, South Dakota, Nebraska, Kansas, Oklahoma, and Texas. As one travels west across these states, rainfall becomes more sparse, the landscape more arid, and farming more difficult. The eastern portions of the 100th meridian states receive enough rainfall in normal years to support crops, but west of the 100th meridian it is often not possible to grow crops without irrigation. Some parts of the West, such as the High Plains of Texas, the Mojave and Sonoran deserts, the Great Basin (centered in Nevada and western Utah), the Snake River Valley Plain of Idaho, and the deserts of eastern Oregon and the Columbia Plateau are even drier, frequently receiving less than ten inches of precipitation in an entire year. Sparse rainfall totals are reflected in the reduced frequency and size of western watercourses. A quick look at any map shows fewer rivers and streams in the western part of the country than in the East, and many of the western river channels appearing on maps actually contain little, if any, water during at least part of the year.

The combination of scarce supplies and numerous demands has put heavy pressure on the West's rivers and streams. Each diversion to an offstream use has left less water for the variety of water uses that take place within stream channels. Streamflows support fish, wildlife, and streamside vegetation and are used for a variety of recreational purposes, including fishing, swimming, wading, rafting, kayaking, and canoeing, as well as to enhance land-based recreational activities such as hiking, camping, picnicking, and bird-watching. Streamflows enhance the aesthetics of riverine environments and landscapes, maintain the viability of stream channels, and are used to transport goods, generate electricity, and dilute contaminants. Rivers are an integral part of the environment, and their "use" to support environmental processes has become more popular as our understanding and appreciation of these processes has grown. So while water scarcity is a fact in the West, perhaps even the region's single most defining characteristic, there has been no shortage of purposes for which western water is used.

When water is scarce, choices must be made about how it will be used. Legal systems in the contemporary American West have typically protected the offstream water uses of cities, farms, and industry at the expense of water used instream for fish and wildlife, recreation, aesthetics, and other purposes. Water rights—legally protected property rights to use water—have traditionally been granted only for offstream uses of water, sometimes in amounts that leave little or no water instream. The diversion of water from natural channels has historically been a prerequisite to the procurement of a water right. The allocation priorities inherent in these policies reflect the dominant water uses and values at the time that western water allocation institutions were developed, in the mid-1800s, when the public sought to encourage the rapid development and settlement of the West.

But conditions, needs, and values change. It is hardly possible these days to read or talk about water policy in the West without coming across the topic of change. Enormous population growth, concentration of the West's residents and economic activity in urban areas, higher standards of living, and increased mobility have all contributed to shifts in the water needs of western state residents. Water to meet new needs has largely become unavailable as existing watercourses and aquifers have become fully allocated and the best dam sites have already been developed. Talk of water conservation, transfers, and marketing has become commonplace as the need to use existing water supplies more efficiently has become apparent. Values, too, have changed, now that the need to settle the West is no longer a national priority. The proportion of the West's residents that makes its living directly from the development of natural resources has dropped dramatically, and nearly everyone—newcomers and fifth-generation westerners alike—is more aware of the impacts that water development has had on the environment. Water for fish and wildlife, recreation, aesthetics, and protection of the environment is valued more highly. In the words of longtime water lawyer and scholar Frank Trelease, speaking at an instream flow conference in 1976:

> Perhaps our ancestors, our fathers, we ourselves not too long ago were willing to throw away as worthless some scenic, recreational, and environmental factors. Perhaps they were regarded as worthless because of their abundance, but now we realize what is left is far from abundant, that it is scarce, partly because we have already thrown away so much, partly because there are now so many of us that we compete with each other for what is left, and partly because the opportunities for enjoyment have been broadened by the automobile and highway.[2]

Laws can be changed to accommodate these new conditions and values. As Trelease also said, "The law is a mechanism for getting things done, for accomplishing the purposes of society, for requiring some things and forbidding others. If the people of the United States or of a state desire to keep water in a stream or to put it back in a stream a law can be framed that will do the job."[3] In recent years there have been numerous efforts to do just that, to change water allocation institutions to better protect instream uses of water. But the changes have been slow. Differences in values, the desire to protect existing property rights, and the fact that "water law is a human institution and just as resistant to change as other institutions"[4] have combined to place numerous obstacles in the path of those who would increase the level of protection given to instream flows. Protection advocates have even been severely limited in their attempts to protect instream flows through the purchase of water rights.

Markets have been heavily championed as a means of shifting water from old uses to new, but market imperfections and some of the special characteristics of instream flow make it unlikely that markets can supply an adequate level of protection even in the absence of artificial barriers. The level of protection afforded instream water uses continues to lag well behind that given to offstream water uses, and is likely to remain behind without more aggressive efforts to make institutions more compatible with the preservation of instream flows.

Efforts to change laws are accompanied by substantial debate. Much of the general population has probably never heard the term *instream flow*, or, like Harvey Doerksen in 1973, associates the term only vaguely with an image of flowing water. But a considerable and growing number of people have become ardently committed to the idea of preserving instream flows. There is growing awareness of the fact that the benefits of preserving instream flow extend well beyond the immediate benefits derived by anglers and other recreationists. Leaving water in streams to protect aquatic and riparian environments provides value even to people who rarely if ever visit streams themselves, and to society as a whole. Instream flow protection preserves the options of future generations to enjoy the same natural environment that we enjoy today.

Some people, however, continue to view instream flow protection as an unwelcome constraint on the development of additional water supplies. Debates about the desirability of protecting instream flows encompass much larger issues, such as the meaning of property, the relationship of states to the federal government and of citizens to the state, and the relationship between economic growth and ecological preservation.

Our goal in writing this book is to provide readers with a comprehensive understanding of the many issues surrounding instream flow, and to shed new light on a poorly understood but very important natural resource topic. Conflicts over instream flows sometimes occur because of a difference in values, but they also seem to arise because of misperceptions and confusion about the basic facts of instream flow protection. Much of the debate over instream flows has taken place outside the view of the general public. If the public is to participate more effectively in the policy-making process, it will be necessary to raise the level of awareness about instream flow protection. Policy makers and natural resource managers too often have only part of the information they need to make effective decisions and may be unaware of the ways in which similar issues have been addressed in other states and situations. New developments affecting instream flows occur almost daily, so it is also useful to provide a comprehensive overview of where things stand today, so that we can better see what direction we need to move in tomorrow.

The following chapter contains a brief history of water uses and values in the West and describes the legal systems that have been developed to protect them. Water law, like other law, changes over time in response to changes in dominant uses and values. In recent years there has been a movement away from viewing water solely as an engine of economic development and toward increased recognition of the ecological and recreational values of water. Our book both directly and indirectly constitutes a study of the ways in which laws and policies have changed in response to new values and conditions, and chapter 2 provides the context for such a study.

Chapter 3 describes in more detail some of the most common instream uses of water. Instream uses of water are numerous and varied, and flow needs—the quantity and timing of flows necessary to accommodate instream uses—can differ for each type of water use and water user. This chapter presents some of the many factors that must be considered when evaluating instream flow needs and describes some of the ways in which analysts attempt to quantify those needs.

Chapter 4 focuses on perhaps the most basic instream flow protection issue: How much water should be left in streams? There is no single answer to this question, but this chapter presents and analyzes many of the factors that should be considered during an evaluation of instream flow protection needs.

Chapters 5 and 6 describe many of the basic issues associated with instream flow protection and the ways in which those issues have been addressed by the states. Many of the same issues arise in virtually all of the western states, but they have not always been addressed in the same ways. Legal and administrative changes designed to accommodate new conditions and values have taken different forms and been given different names, and attempts to review and understand all of them can lead to confusion. Our goal in writing chapters 5 and 6 is to emphasize similarities and enhance understanding of the basic methods used to protect instream flows, and to provide the information necessary both to think more creatively about solutions to instream flow problems and to benefit from lessons already learned through implementation of protection measures in other states.

Chapter 7 examines the effects that instream flow protection measures have on other water users. An evaluation of those effects can be fairly complex. In many situations instream flow protection measures have little, if any, detrimental effect on other water users and may even provide positive benefits. But the effects on other water users vary with the circumstances unique to each river. In some cases there are negative impacts that need to be taken into account when considering or designing potential protection measures.

The federal government is heavily involved in water management and instream flow protection issues. Chapters 8–10 describe the many different ways in which the activities of the federal government directly or indirectly affect instream flows, including numerous and varied direct protection efforts. Instream flow protection efforts of the federal government vary widely by program, agency, and region of the country, but in any form and degree are viewed by many westerners as a threat to existing water users and to state sovereignty over water management. The underlying conflict between federal and state authorities is a consistent theme throughout these three chapters.

The final chapter briefly summarizes some of the major themes of the book and lists the various methods relying on state and federal law that are available for protecting instream flow. The chapter closes with some observations about the effort to achieve a reasonable balance between instream and offstream water uses.

Each chapter focuses on particular aspects of the instream flow debate. It is hoped that by presenting these chapters together the book will constitute both a comprehensive, integrated introduction to the subject of instream flow protection and a useful reference for further research and understanding. The book should be useful to interested citizens, environmentalists, farmers, ranchers, land and water managers, administrators, policy analysts, policy makers, students, researchers, and others interested in natural resources and environmental management. Those not already exposed to the topic might think that a book of this length would be more than adequate to cover all important issues and activities related to instream flow protection. That is definitely not the case. Instream flow is a subject so broad, and so heavily examined by others throughout the last three or four decades, that we can hope to do no more than present a comprehensive introduction to the subject. To accommodate those who wish to research a particular topic in more detail, we have included a substantial number of references to additional materials. These references, together with additional comments, are presented as endnotes following each chapter. A complete list of references appears at the end of the book.

The debate over instream flows is a debate about how western rivers are, can be, and should be used; about how much water should be diverted for offstream use and how much water should be left to flow in streams. To those who have experience with the subject, the terms *instream flow* and *instream flow protection* bring to mind myriad concepts, principles, activities, and conflicts. As noted by Harvey Doerksen almost two decades after the meeting in which he had first wondered why instream flow was such a big deal, "The mere physical image of water in the stream fails to capture the intensity of the emotions involved in the issue, the political importance of instream flow in the water resources arena, and the symbolic importance of the issue as part of larger environmental concerns."[5] Better understanding of the issues surrounding the topic of instream flow protection should lead to more productive discussions, fewer conflicts based on inaccurate information, and new and more creative thinking about solutions to water resource problems. These in turn should lead to policies and practices that better meet the needs of the West.

A Definition of Terms

People who work with water resources use a variety of terms and phrases that have been given specialized meanings, and we have used many of these terms in this book. Before proceeding further it may be helpful to review the meanings of some of the more common terms related to instream flow protection.

In general, a *river* is a large *stream*. But there really is no definitive size at which a stream becomes a river, and in practice the two terms are often used interchangeably, a practice that we follow in this book. Rivers flow in *channels,* which are composed of the river *bed* and *banks.* A particular length of river comprising less than the whole, as measured in the direction of flow, is often referred to as a river *reach* or *segment.* A line perpendicular to the direction of flow from one bank to the other is referred to as a *cross section* or *transect.*

Water volumes in the West are usually measured in *acre-feet.*[6] An acre-foot is the volume of water necessary to cover an acre of land to a depth of one foot, approximately 326,000 gallons, or 43,560 cubic feet. The term used to describe the quantity of water flowing in a river at a particular time is *discharge,* though people often use the word *streamflow* to mean the same thing. Discharge can change from day to day, or even hour to hour; thus it is measured in units of volume per unit of time. Discharge may be measured in units of acre-feet per day, month, or year, but the most common unit of measurement for discharge is *cubic feet per second* (cfs), sometimes referred to as *second-feet.* Stream discharge is a product of the stream's cross section and velocity; for example, if a stream cross section is five feet wide and two feet deep, and the average velocity of water through the cross section is three feet per second, the discharge will be 5 feet × 2 feet × 3 feet/sec = 30 cubic feet per second (30 cfs). A continuous flow of one cubic foot per second produces approximately 2 acre-feet of water per day.

Flow rates (discharges) vary through time; the temporal pattern of a particular river's flows—the frequency and magnitude of its flows—is referred to as the river's *flow regime.* Streamflows in much of the West follow a pattern in which they are highest in the late spring and summer due to snowmelt in the mountains, and lowest in the fall and winter. However, flow regimes vary significantly among regions, and even among rivers in the same region. For example, streamflow in some areas is highest in winter due to the prevalence of winter rains, while streamflows in other areas, particularly the Southwest, are highest in summer because of the occurrence of heavy monsoon rains.

Two general classes of flows are *natural flows,* which are entirely a product of hydrologic and topographic factors, and *regulated flows,* which are influenced at least in part by the operation of diversion dams and storage reservoirs upstream. To *divert* water is to remove water from its natural watercourse for use somewhere else. Diversions are usually achieved through placement of a diversion dam or other structure in the watercourse in such a fashion as to route all or a portion of the river's flow into a canal or pipeline leading away from the river. To *dewater* a stream is to take so much water out of a stream that very little, if any, remains within the channel.

Instream flow is often used to refer to the precise quantity and timing of water flows necessary to sustain one or more specified instream use of water. But the term is more commonly used to mean the same thing as *streamflow,* which is simply water flowing within a natural channel. We use the term in this more common meaning throughout the remainder of this book, but in other situations the reader needs to be aware of the context in which the term is being used. We use the term *instream flow protection* to refer to the legal, physical, contractual, and/or administrative methods that have been used to ensure that enough water remains in streams to sustain instream.

The term *minimum flow* has been used to mean several different things and hence can be quite confusing. Some people use the term to mean much the same as the first definition of instream flow given above, the precise quantity and timing of flows necessary to sustain one or more specified instream use of water. The term has also been used to refer to the reduced level of flow remaining in a stream after at least some of the flow has been diverted. Others use the term to refer to a particular method of protecting instream flows in which a certain amount of water is made unavailable for diversion. Yet others use the term to refer to the lowest possible flow level that is compatible with a particular instream use. The latter meaning, when applied to fish or other biotic resources, is sometimes referred to as *preservation flow,* the smallest amount of water that will allow biotic organisms to survive a limited period of time, as through the winter. Confusingly, *preservation flow* has also been used to mean the level of flow that will sustain existing levels of the instream use for an indefinite period of time. The term *base flow* has been used to describe the lowest possible flow level compatible with a particular use, and the typical seasonal flow in a regulated river, but is more properly used to describe that portion of streamflow attributable to groundwater inflow rather than to surface or near-surface runoff. For clarity, we will not use the terms *preservation flow* or *base flow* in the remainder of this book. We will, however, use the term *minimum flow* in two different ways depending on the context: (1) to describe a particular method of protecting instream flows, described in more detail in chapter 6; and (2) to describe the smallest possible flow that is compatible with a particular instream use when describing efforts to limit the amount of water that can be protected for instream use.

CHAPTER TWO

The Loss of Instream Flows: A Short History of Water Use and Water Law in the West

Virtually all of the first uses of western rivers and streams took place instream. The first Americans, people who crossed the Bering Land Bridge from Asia and migrated south, settled primarily along the banks of rivers that emptied into the Pacific, from what is now Alaska down through central California. Riverside locations provided ease of access to water supplies for drinking and other culinary uses, but these native Americans settled along rivers primarily because the rivers provided convenient transportation and supplied them with abundant quantities of food.

Fish plied the waters of the western coastal rivers in enormous numbers, particularly in the watersheds of such major watercourses as the Columbia and Sacramento rivers. Some of the fish lived year-round in the rivers, but most—anadromous species such as salmon and steelhead—were found only certain times of the year, during great migrations from the sea to upstream spawning grounds. Using spears, nets, weirs, and a variety of other tools, the first Americans caught these fish at places where the fish collected in great numbers, at sites where the river channel narrowed or provided other obstruction. Fish and fishing were central to the lives of these first westerners, not only as a source of food, but as the focus of entire cultures, economies, and religions. Life revolved around cycles of fish spawning and migration, and the people of the coastal rivers thrived for several thousand years in this manner. Rivers were also used for transportation, in dugout canoes and other watercraft, and as avenues for expanding civilizations and communication between cultures.

Rivers played a less prominent but still important role in the lives of people who lived in other parts of the West. On the Great Plains most people lived as nomadic or seminomadic hunters and gatherers, covering great distances over the course of the year as they followed seasonal migrations of bison and other land animals. Plains rivers also provided fish, though not in the same quantities as the coastal rivers, and riverside lands were prime sites for hunting and for collecting fruits, nuts, berries, and other plants. Many tribes located winter villages near rivers, where there was constant access to water and food, while a few, such as the Mandans on the Missouri River, were able to establish permanent settlements based on agriculture in the lush river bottomlands. Life was similar in the mountainous regions and in the Great Basin, where people were largely migratory but spent ample amounts of time near watercourses whenever possible.

First Diversions

The great exception to this early American way of life, and one that encompassed a much different relationship of people to rivers, occurred in the desert Southwest starting over two thousand years ago. There, in one of the driest regions of the continent, some of the desert people began to irrigate crops, diverting water from rivers and streams and applying it to fertile but previously dry lands located some distance from river channels.

An example of this new use of rivers took place within the Hohokam civilization, which thrived in the deserts of what is now central Arizona for over a dozen centuries. The Hohokam built an extensive system of irrigation canals along the Salt and Gila rivers, some of which date back to at least the year 300 B.C.[1] The canals, up to 30 feet wide and 10 feet deep and totaling hundreds of miles in length, were hand dug with stone and wood tools for the purpose of watering crops along the upper terraces of the river valleys. The delivery of water to upper terraces allowed crops such as corn, beans, and squash to be grown on much greater expanses of land. The security and leisure time resulting from establishment of a successful farming economy allowed the Hohokam to move from mere existence to a standard of living well beyond that of their contemporary neighbors.[2]

There is not enough evidence for researchers to determine what kinds of social institutions the Hohokam developed to build the massive irrigation works and manage the use of water. It is clear, however, that the Hohokam canal system constituted the West's earliest "public works" project, because the massive endeavor required to build the canals would only have been possible through the coordination of multitudes of people and the application of sophisticated engineering skills.[3] Given the necessity of cooperation, and the communal traditions of most contemporary Indian tribes in the Southwest, it is likely that water from the canals was owned and used communally rather than privately.[4]

The Hohokam civilization reached a peak population of several hundred thousand people sometime around A.D. 1200, but then disappeared over the course of the next two hundred years.[5] The sudden disappearance may have been caused by sustained drought. As desert farmers, the Hohokam were completely dependent on irrigation from rivers that are not necessarily perennial even during normal years, and they would have been extremely vulnerable to droughts lasting several years. Some have speculated that the irrigation system itself led to the demise of the Hohokam, because irrigation without adequate drainage causes salts to build up in the soil and reduce crop yields.[6] There is evidence that the Anasazi Indians, who practiced similar methods of irrigation in the San Juan River watershed,[7] particularly during the years A.D. 1000–1300, also disappeared around this time, probably because of sustained drought.[8]

Though the Hohokam and Anasazi civilizations largely disappeared, irrigation did not. Irrigation had been well established among Pueblo Indian cultures of the Southwest for centuries prior to European exploration in the 16th century. The Pueblo tribes may well have been descendants of the Anasazi, or of the Hohokam, and thus old hands at diverting rivers for irrigation. However, river development at the scale practiced by the Hohokam did not occur again anywhere in the West until European settlement.

Water in the Spanish West

The Spanish had centuries of experience with arid conditions before explorers arrived in the New World, and they knew well the value of water in making arid lands productive.[9] They put that experience to work, diverting water from natural watercourses after they arrived in the West.

Efforts to "civilize" native populations entailed the collection and settlement of the Indians into central communities, often built around Catholic missions. Mining the rich ore bodies of the New World was also one of the goals of Spanish settlement, and the collection of Indians at missions was in large part designed to assure adequate labor forces for the mines.[10] Production of food and fiber in quantities beyond those required by the farmers themselves was necessary to support the mines, as well as the *presidios* (military bases) that accompanied Spanish development. Irrigation was essential to the accomplishment of these tasks, and the diversion of water for irrigation became widespread around Spanish towns and missions. The construction of canals and other structures to divert water from streams to water fields was often the first act of any new community, usually preceding even the construction of shelter.

The mines themselves required substantial quantities of water. The Spanish used the water-intensive "patio" process in the processing of ores, and a medium-sized mine using this process often required more water than a half dozen towns or missions. The availability of water was often the primary determinant of mineral production levels, and water shortages were often cited by beleaguered managers as justification for low output from the mines.[11] Mining was a new water use in the West, and one that had significant potential to affect both water quantity and water quality for downstream users after mines became prevalent throughout New Spain by the 18th century.

The use of water for domestic purposes in missions, presidios, and towns was also important, though hardly new. A much larger and newer consumptive use of water in many Spanish settlements was for domesticated livestock, most of which was introduced to the New World by the Spanish. Indigenous peoples had domesticated little beyond dogs and turkeys, whereas the Spanish introduced horses, cows, mules, pigs, sheep, and other animals that have since become an integral part of the West. The Spaniards also built gristmills along rivers and streams, using streamflows to turn the wheels and produce flour from grain. The use of water to turn Spanish gristmills probably constituted the West's first use of rivers as an energy source.

The use of water in the Spanish West was governed by the same laws used in Spain. In Spain, running water was owned by the Crown and was available to all for purposes such as drinking, fishing, and navigation. Spanish law in the New World continued to protect public uses of water, giving first priority to the use of water by communities as a whole. Water allocations to individuals occurred only after sufficient quantities of water had been secured to meet the needs of the town itself.[12] However, this did not preclude the use of water by individuals, and the law included numerous protections for private uses of water. Grants of water for specific purposes were generally associated with grants of land but were issued separately.[13] A land grant by itself entitled the grantee to the use of water only for domestic purposes. In all cases the right was to *use* the water, while the state retained actual ownership.

Spanish officials used a variety of principles to resolve conflicts over water use, relying less on the existence of established rights than on a desire to determine how the water might best be divided for the enhancement of the common good. Spanish water law emphasized the need for fair divisions of the available water, seeking to apportion water in "such a way as to offend no one."[14] The water rights of individuals were to be protected, but in cases of conflict between individuals and communities, "there is no question that in the Spanish and Mexican judicial systems the rights of the corporate community weighed more heavily than those of the individual."[15]

Other factors were also considered in the resolution of conflicts. The concept of "first in time, first in right," a prominent feature of many property systems, dates back at least to the Romans and appears in Spanish law as far back as the 13th century. Demonstration of a legitimate need for water was considered and was sometimes the most important factor in resolving disputes. And a well-established principle of Spanish law protected parties not subject to the dispute but who may potentially have been adversely affected by its resolution.

Other Europeans in the West

Though the Spanish were the first Europeans to move into the West, they did not long maintain a monopoly on the resources of the West. The French were among the first to follow the Spanish, though by a different route. The French entered the West from the northeast—from Canada and the Great Lakes—and southeast and were more acclimated to forests and water than to deserts. Fur rather than minerals was their resource interest, and their search for new sources of fur-bearing animals led them to explore the West's rivers. The Spanish had originally laid claim to virtually the entire West, but by the late 18th and early 19th centuries the French had spent large periods of time in control of what later became the Louisiana Territory—a huge region that encompassed most of what are now the western states of Montana, Wyoming, Colorado, North Dakota, South Dakota, Nebraska, Kansas, and Oklahoma.

Other Europeans powers also entered the West in search of furs. The British controlled most of the Pacific Northwest by the middle of the 19th century, and the Russians had established settlements and trading posts along the west coast, primarily in Alaska but extending south nearly to San Francisco Bay. Western rivers were used by trappers and mountain men of all nationalities as optimal sites for trapping fur-bearing animals and for the transport of furs and supplies; the fur trade used water almost exclusively instream. However, the biggest change coming to the West, and to western rivers, came from the westward migration of the Americans.

Trappers and mountain men from the fledgling United States had competed with the British and French in the West almost from the beginning. The Americans' numbers swelled following acquisition of the Louisiana Territory by the United States in 1803, and after the first great exploration of the northwest quadrant of the country by Lewis and Clark. Lewis and Clark followed the Missouri, Snake, and Columbia rivers from St. Louis to the Pacific Ocean and back again, opening the way to an influx of American fur trappers and, eventually, settlers. Settlers followed the Oregon Trail along the Platte, Snake, and Columbia rivers to Oregon's well-watered Willamette Valley in the early 1800s. A few hardy settlers took a variety of routes to California, where they were outnumbered by the Spanish but still found room to settle. Many others started moving into the Southwest—first Texas, but then spreading throughout the region. By the middle of the 19th century Americans were showing up in isolated pockets all over the West, and the era of the Spanish, French, and British was close to an end.

Riparian Law

The Americans brought their own set of rules governing the use of water, rules derived from the common law[16] of England.[17] The American colonies were established under the auspices of the English Crown and were subject to English laws. English water law was relatively simple and undeveloped, having unfolded in a land where water was abundant and conflicts over the use of water were correspondingly rare. Following the precedent of Roman law, the navigable waters of England belonged to the Crown and were available to the public for the purposes of navigation and fishing.[18] Ownership by the Crown prevented these important economic activities from being monopolized by individuals, thereby reducing the potential for conflict. Rights to the use of waters not being used for navigation were held by those who owned the banks of the streams and were therefore known as riparian rights.[19]

Water resource conditions were similar in the American colonies, and there was not much incentive to adopt a different set of water-use rules after American independence from England. Water-use conflicts were so rare in England and in the original American states that the body of water law was not well developed. The rudimentary concepts used by the courts to govern disputes over water use were "so sparingly tested that they could not be fairly described as a comprehensive legal doctrine."[20]

The heart of the original riparian doctrine was the idea that rivers had value primarily as an amenity. Rivers enhanced the value of surrounding land, and each landowner along a river was entitled to receive the benefits of free-flowing water. As originally constructed, what came to be known as the "natural flow" interpretation of the riparian doctrine held that landowners were allowed to remove water from streams only for basic domestic purposes such as drinking, bathing, cooking, and the watering of limited numbers of livestock. Landowners were otherwise required to leave rivers in an undiminished and unpolluted condition. Essentially, the natural flow doctrine favored instream uses of water over all but the most minimal subsistence uses taking place outside natural river channels. This doctrine made sense where water was abundant and there were few out-of-stream uses of water. The natural flow doctrine often gave way, however, when advances in technology made rivers valuable as a source of energy for turning the wheels of industry.

The riparian doctrine was modified during the Industrial Revolution to allow riparian landowners to make a reasonable use of the waters flowing over their lands. The general principles of the "reasonable use" interpretation, and of the riparian doctrine in America up to 1827, were set out by Justice Story of the Rhode Island District of the Circuit Court of the United States in the case of *Tyler v. Wilkinson*.[21] Story said that the rights to use natural streams were annexed by law to the land bordering the streams. Each landowner had the right to the use of water flowing over the land without diminution or obstruction, though the landowner did not own the water itself—the right was solely to the use of the water. When water flows were insufficient to meet all uses, the deficiencies were borne as a common loss, with each user cutting back by the same proportion. No riparian landowner had a right to use water to the prejudice of any other riparian landowner, but "reasonable" uses were allowed. The extent to which any particular use was allowed was determined by the potential injury to other riparian landowners should the use occur.

Though definition of "reasonable" uses could be difficult, other features of the reasonable-use riparian doctrine are quite clear. Only riparian landowners have rights to the use of water. Owners of nonriparian lands and any others wishing to preserve free-flowing waters do not have any legal rights to the water.[22] As the water right is a consequence of land ownership rather than a separate piece of property, the right is not lost simply because it has not been exercised. The relationship among riparian landowners is one of parity rather than priority, and the doctrine allows the entry and accommodation of new water users.[23] Water rights are relative rather than absolute. The possibility that existing landowners may develop new water uses, or that new water users may be added to a stream, means that riparian rights do not attach to a fixed amount of water.[24] And uses that are reasonable under existing circumstances may become unreasonable as new uses are initiated or conditions change, so that riparian rights for specific water uses may not be secure in situations in which there is not enough water to accommodate all desired uses.[25]

Many of the Americans who moved west in the mid-19th century relied heavily on these rudiments of English-American common law to govern their use of water resources. Texans adopted the common law of England while still a republic, in 1840. The California state legislature adopted English common law during the year of its admission to the Union, in 1850. Riparian principles were also codified in civil law. An 1866 statute of the Dakota Territory recognized riparian principles, and the Territorial Code was carried over to both North Dakota and South Dakota when they were recognized as states in 1889. The Dakota Civil Code, itself copied verbatim from a proposed civil code for New York State, was copied by the Oklahoma Territory Legislative Assembly in 1890. Other states that adopted riparian principles at an early date include Kansas, Nebraska, Oregon, and Washington.

Though introduced early to the West, the riparian doctrine of water rights did not long go unchallenged. The riparian doctrine was born in lands with humid climates, where precipitation was sufficient to grow crops and plentiful water supplies made conflicts between water users infrequent. Most of the American West was different, enormously different, and this fact was apparent to the earliest settlers. Physically, the aridity of the West bore little resemblance to the eastern climates from which the first settlers had come. Politically, westerners were far from the national government in Washington, and other sources of governmental authority were rudimentary or nonexistent. By both inclination and necessity westerners were an independent people, accustomed to making do with what they had and making their own rules as needed. With respect to water, the new rules that they created were often quite different from those they had used in the East.

The California Gold Rush

Miners provided the primary impetus for changing the rules allocating water in the American West. The West's mineral resources attracted Anglo-Americans just as they had the Spaniards. The first American mines were few in number, small, and widely scattered, but the situation changed dramatically after 1848, when James Marshall discovered gold in the tailrace of the sawmill he was building for John Sutter on the American River in California. Marshall's discovery drew a flood of people looking to strike gold and get rich or to live off of those who had. The population of California and, later, the entire West swelled enormously. By the end of 1849 California's non-Indian population had grown from an estimated 20,000 pre–Gold Rush residents to almost 100,000, and just three years later the state's population was over 223,000. Mining soon became the principal industry in California and later in much of the rest of the West.

The production of gold was intimately tied to the use of water. Gold deposits in the foothills had been washed out of the mountains by the action of rivers and streams over the course of geologic time. Miners imitated those natural processes with a number of smaller but faster methods, all of them designed to use water to separate the heavier gold from its surrounding dirt and rock. Panning for gold was the simplest method, but it was soon followed by the more profitable method of washing gravel through long wooden sluices. Cleats along the bottom of the sluices retained the heavier gold as the lighter gravels were washed through.

The first gold deposits were found primarily along streams, and the first miners usually established claims along the banks, where they could pan for gold directly. Those arriving later, after the streamside locations had all been claimed, were forced to establish "dry diggings" some distance removed from the streams and then haul gravel in sacks or wheelbarrows to the water to be washed.[26] As mining operations grew in size and sophistication this process was often reversed. Instead of bringing gravel to the water, streams were diverted from their natural channels to bring water to the claims. Diverting a stream from its natural course required considerable effort. Dams to divert the rivers were built of rock and wood, and channels to convey the water were dug with picks and shovels through the surrounding terrain, often through granite rock.

Hydraulic mining, in which water pushed through hoses under great pressure was used to wash entire hillsides directly into wooden sluices, became widespread in the early 1850s. Hydraulic mining required the diversion and delivery of huge volumes of water to sites often far from natural channels.[27]

Streams were also diverted from their natural channels to provide access to gold-bearing sediments within the actual streambed. The first large projects designed to enable mining of river beds occurred on the American, Feather, and Yuba rivers in the spring of 1850. These projects diverted the entire flows of the rivers, completely exposing their beds.

The use of water was so basic to the production of gold that enterprising forty-niners discovered they could make more money providing water to the mines than they could from mining the gold itself. Private companies were organized to build dams, flumes, and canals. Many of the companies were partnerships and joint-stock ventures formed among men at the diggings, but private corporations supported by outside capital constituted the largest companies. The size of the companies, and the scale of their waterworks, was substantial. Some dams were over a hundred feet tall; reservoirs impounded tens of billions of gallons of water; individual canal systems included hundreds of miles of canals, tunnels, and flumes; and investments by individual companies ran into the millions of dollars. By 1870, almost 7,000 miles of primary and secondary ditches had been constructed in the hills of California to move water from rivers to the mines.[28]

Development of the Prior Appropriation Doctrine

Spanish colonists settled the West under the sponsorship of the state, which provided the Spanish with established systems of government and law. The same was not true of the forty-niners, who moved to California on their own initiative. Nor was government awaiting them in California, for Marshall's discovery of gold occurred near the end of the U.S.-Mexican War, subsequent to the expulsion of the Mexican government but prior to the official transfer of the region to the United States. The lack of effective government could have created a serious problem for the forty-niners. There were no rules to define property rights in the gold fields, either between individual miners or between miners and the federal government. The miners did not own the land they were occupying, the minerals they were seeking to remove, or the water they were using. It was not even clear what rules should eventually apply—those of the federal government, which owned the land, or those of the state government, which had not yet been created but was widely anticipated. Rather than waiting for clarification of the rules by some level of government, the miners treated the problem as an opportunity; as there were no existing rules to guide their use of the land and its resources, the miners created their own.

The miners' rules were created independently in each mining camp and administered by committee. Adjudication of disputes and enforcement of the rules were undertaken by the committees if not by the aggrieved individuals themselves. The miners' greatest need was to establish rules governing access to the gold; because they did not own the land or minerals, the usual rules of property ownership did not apply. Instead, the miners adopted the "first come, first served" principle already in wide use on the public domain, where rights were based on occupation rather than on ownership.[29]

Mineral claims in the California camps were established by posting notices along the boundaries of the discovery—known as "staking" the claim—and then recording the claim with officials designated by the mining committees. The first to record a claim was entitled to its exclusive use, so later arrivals were required to either purchase an existing claim or discover and stake their own. One other possibility was to take over a claim that was not being actively worked. The miners believed that the gold should be available to those who stood ready to take it and were firmly opposed to the staking of claims for speculative purposes. For the same reason, limits were often placed on the amount of land that an individual miner or group of miners could claim. Specific procedures for staking and recording claims, size limits on the claims, and requirements for keeping a claim active varied from camp to camp, as set by the individual mining committees.

The miners were also in need of rules to govern the allocation of water. The first to arrive at the gold fields, in the earliest months of the rush, often had their choice of land to claim and water to use. That situation didn't last long, however, as the promise of wealth attracted large numbers of fortune seekers. The later arrivals often were able to find promising, previously unclaimed land but discovered that there was not enough water available to work the claims. As the Spanish in the Southwest had discovered earlier, water was frequently the limiting factor in the production of the region's mineral wealth.

The riparian principles used to allocate water in the East would have been of little use to the miners even if they had been inclined to use them. The miners found that water allocation principles based on plentiful rainfall, numerous streams, and the need to leave water in the stream for downstream users made little sense in regions where rainfall and streams were less abundant. As a result, the basic principles of riparianism were substantially altered or completely rejected by the forty-niners. In their place, the miners applied the same rules that they used to govern access to mining claims. When applied to water, these rules became known as the prior appropriation doctrine.

The miners staked a claim to water by physically taking—"appropriating"—what they needed. Construction of the diversion works necessary to take the water served as notice to other miners, who could ascertain the status of the stream by making a physical survey of the channel. The water was used wherever it was needed, whether on riparian land or at some distance removed from the natural channel. The miners were guided by the same "first in time, first in right" principle applied to the mining claims, so the first miners to appropriate water had the best right to continue using it. Subsequent appropriators were required to make do with what was left, if anything. Even if located upstream from a prior user's diversion works, a subsequent "junior" water user was required to allow enough water to pass to meet the needs of the downstream "senior" appropriator.

The "use it or lose it" principle was also incorporated within the prior appropriation system, so that miners not making beneficial use of their water were forced to surrender it to those who would. Unlike the situation with mining claims, however, limits were seldom placed on the amount of water that an individual could use. A miner or company was free to appropriate as much water as could be put to use, even if that meant there wouldn't be any water left for those who arrived later, or to sustain the integrity of the stream and its biota. Even in the early years, before the arrival of large mining corporations, the first miners in a watershed were often able to divert the entire flow of a stream. Dry streambeds, or rivers with vastly reduced flows, became common for the first time.

California gold soon attracted investments from all over the world, and the gold fields became dominated by increasingly larger and more sophisticated mining and water supply operations. Because these operations often extended beyond the camp and watershed boundaries observed by the original miners, the need for a uniform set of water allocation rules soon became evident. In the absence of definitive guidance from federal or state legislatures, the task of defining uniform principles primarily fell to the California state courts.

The California courts faced a difficult task. The courts had been organized following California's admission to the Union as a state in 1850 and derived their jurisdiction and powers from the California state constitution. The mining camps, however, were located almost exclusively on federal land, and it was not clear that the state had jurisdiction over activities occurring there. The courts had also been given conflicting directives from the state legislature. California's first legislature, in 1850, had adopted the common law as the state's legal foundation, which meant that the allocation of water should be governed by riparian principles. But just one year later the legislature adopted a statute that sanctioned the use of "customs, usages, or regulations established and in force at the bar, or diggings,"[30] even though the miners had, in many cases, adopted rules substantially at odds with those of the riparian doctrine.[31]

The uncertainty of their jurisdiction and the conflicting guidance given by the state legislature made it difficult for the early courts to define a uniform set of water allocation principles. Some courts settled disputes using common law principles. Other courts, or the same courts when confronted by different sets of facts, settled disputes using the miners' priority principles. Occasionally the courts took advantage of the freedom resulting from lack of precedence and clear guidance to develop hybrid doctrines that merged aspects of both the competing doctrines. Uniformity increased over time, however, as the courts became familiar with other courts' reasoning and as lower court rulings were confirmed or rejected by higher courts. And as the courts gained more experience with the unique rules prevailing in the mining camps, their rulings increasingly reflected precepts of the prior appropriation doctrine.

The California Supreme Court first recognized the validity of the miners' priority principles in 1853,[32] but it was not until the court's decision in the 1855 case of *Irwin v. Phillips* that the court clearly set forth its justification for adopting priority principles to resolve water disputes on the public domain.[33] The conflict arose between Mathew Irwin, a miner who had diverted the entire flow of a small creek to supply water to his claim and was selling excess water to others, and a group of miners including Robert Phillips, who had later established claims along the banks of the stream and had cut Irwin's dam in order to supply those claims with water. Phillips argued that the law of the land embraced riparian principles, and that he and the others in his group were entitled, having located their claims on riparian land, to enough water to work their claims. Irwin argued that the legal and economic circumstances of mining in the West required new rules, including the precedence of prior use, and that his water right should be protected on the grounds that it preceded the others' use.

Interestingly, the miners' committee that first heard the case agreed with Phillips, the subsequent riparian water user, probably because many in the camps were opposed to the concept of private companies selling water, as Irwin was doing. Irwin appealed this decision to the state courts, and a district court ruled in his favor. Phillips then appealed, but the supreme court upheld the district court's decision in favor of Irwin. The supreme court justified the decision on the grounds that the state legislature, in its 1851 legislation, had explicitly endorsed use of the miners' prior appropriation rules. The court also reasoned that the federal government had implicitly validated the new legal system by failing to object to its development. The supreme court's reasoning and justification in this case clearly set forth the principle that the rights of prior appropriators would be protected, even if the water was used on nonriparian tracts of land, and the case is often cited as marking the birth of the prior appropriation doctrine.[34] The doctrine was not applied by the California courts in all water disputes, because, as will be explained in greater detail below, the California courts continued to apply the riparian doctrine in certain circumstances. But by the 1860s, use of the prior appropriation doctrine was firmly established as the mechanism by which the California courts would resolve water conflicts occurring on the public domain.

Minerals were also discovered in other parts of the West, and miners soon spread the new water allocation rules beyond the confines of California. Gold was discovered in Nevada in 1850 and in the Rogue River watershed of southwestern Oregon in 1851. A small strike was made in northeastern Washington in 1855, and the first of several large strikes discovered in Colorado started a gold rush there in 1858. In 1859 the massive Comstock silver lode was discovered in Nevada. Miners flocked to Orofino and to the Salmon and Boise rivers in Idaho in 1860 and to southwestern Montana in 1862. Gold was discovered near Prescott, Arizona, in 1863, the Black Hills gold rush in the Dakota Territory began in 1874, and southeastern Alaska, near Juneau, experienced a gold rush in 1880.

Everywhere they went the miners found conditions much like those in California—federal ownership of the land, a scarcity of water, and an absence of government. They employed the same rules they had known in California because the rules were well suited to their physical and legal surroundings. Opportunities to develop the West's rich mineral resources attracted large numbers of people, and mining dominated the early economies of several of the western states and territories. Political and judicial support for the miners' activities was strong, and the rules the miners employed to further their activities met with wide acceptance in state and territorial courts and legislatures.

The Federal Role in Defining Water Allocation Rules

Despite the rapid spread of prior appropriation principles, continued uncertainty about the status and jurisdiction of state courts and laws relative to those of the federal government made the validity of these first actions subject to doubt. Federal court cases in the 1840s had established the principle that the United States had title to the minerals found on the public domain and could eject trespassers engaged in unauthorized mining,[35] so the miners' activities in establishing mineral claims, and those of the state courts and legislatures in validating them, did not have a strong legal foundation. It seemed reasonable to assume that the United States also possessed title to the water resources found on federal lands, but the federal government's continued silence on the topic left the status of federal mineral and water rights very much in doubt.[36]

The United States Congress finally addressed the issue of mineral and water rights on the public lands in 1866, 18 years after gold was first discovered in California. Some members of the Congress proposed that the federal government assert its claims to the mineral lands of the West and then sell the rights in order to raise funds that could be used to pay off debts arising from the recently ended Civil War. Members of Congress from the western states were strongly opposed to the idea and successfully moved to prevent the assertion of public ownership by passing the Mining Act of 1866.[37]

The 1866 Mining Act gave the miners—and other users of the public domain—virtually everything they wished. Rather than asserting federal ownership claims, the Act confirmed existing possessory rights on the public domain and declared that federal mineral lands would be "free and open to exploration and occupation" in accordance with local rules and customs. The Act was primarily concerned with mining, but Section 9 of the Act sanctioned private rights to the use of water whenever "by priority of possession, rights to the use of water for mining, agriculture, manufacturing, or other purposes, have vested and accrued, and the same are recognized and acknowledged by the local customs, laws, and decisions of the courts. . . . "[38] Appropriative rules dominated the local customs, laws, and court decisions affecting water use on the public domain, and with passage of the 1866 Act the federal government had finally recognized the prior appropriation doctrine.[39]

The prior appropriation doctrine later received additional federal recognition. The 1866 Act had stated that the United States government would recognize rights to water arising through appropriation on the public domain; an 1870 amendment made it clear that holders of federal land patents, grants, and homestead rights would also have to acknowledge preexisting appropriative rights.[40] Passage of the Desert Lands Act in 1877 strengthened the prior appropriation doctrine.[41] This act encouraged settlement of the more arid portions of the West by selling large tracts of land at low prices and giving settlers the right to appropriate water to meet their irrigation needs.[42] The act took recognition of the prior appropriation doctrine a step further by providing that "all surplus water over and above such actual appropriation and use, together with the water of all lakes, rivers and other sources of water supply upon the public lands and not navigable, shall remain and be held free for the appropriation and use of the public for irrigation, mining and manufacturing purposes subject to existing rights."[43] It was not until almost 60 years later that these three acts were definitively interpreted by the United States Supreme Court as having given states rather than the federal government the authority to determine the method by which the right to use water would be allocated on the public domain, but even at the time of their passage the acts were widely viewed as having endorsed the desirability of the prior appropriation doctrine.[44]

Irrigation and the Settlement of the West

Miners weren't the only ones to divert water from rivers and streams. There was a massive infusion of settlers to the West throughout the latter half of the 19th century, and many of these settlers discovered the need to divert water out of natural channels to sustain their livelihoods. As the first settlers moved out onto the eastern Plains, they found sufficient rain to grow crops and rivers to water cattle, particularly given that the first to arrive often settled along river valleys. But as settlers moved farther west, precipitation and watercourses became more scarce. Those who were able to claim land near rivers and streams were able to raise crops with the aid of relatively primitive diversion and irrigation works, often planting their crops in soils moistened naturally by large flood flows in the spring. But more widespread settlement required more sophisticated irrigation methods.

Irrigation on a relatively small scale was still practiced in the Spanish Southwest, and in a few other widely scattered localities throughout the West, but by the middle of the 19th century irrigation was not being practiced on a large scale anywhere in the American West. This situation changed in 1847, when the Mormons established themselves in what became the Utah Territory. The Mormons moved into the Great Salt Lake Valley, on the eastern edge of the Great Basin, in 1847 after a long and grueling trek across the Plains and Rocky Mountains. They found the valley largely uninhabited, which was not surprising given that the climate was extremely dry. Previous European visitors had generally not been tempted to stay for long, traversing the area only in the course of their explorations or trapping expeditions, and most Indians in the region survived by hunting and gathering rather than farming.

The Mormons completely changed what they found, remaking the forbidding environment to meet the needs of their settlement. Though generally arid, the valley received the water of several streams that flowed out of the mountains and across the valley floor before emptying into the Great Salt Lake. The first act of the Mormons upon reaching the Salt Lake Valley was to begin diverting water from the streams in order to make more of the desert soils productive. During the following years they were successful in diverting the water out of streams in nearby valleys as well, and to grow not only enough to sustain themselves, but to accommodate further growth throughout the region.

The Mormons soon extended their settlements into parts of what are now the states of Idaho, Wyoming, Colorado, Arizona, Nevada, and California. Everywhere they went the Mormons built diversion dams and canals to irrigate the land, creating the largest irrigation systems the West had theretofore seen. In just the ten years from 1860 to 1870 the Mormons built 277 canals, totaling 1,043 miles in length and irrigating 153,949 new acres of land.[45] The Mormons came to think of the Utah Territory they had settled as "the cradle of American irrigation," and as the discovery of gold and other minerals in California, Nevada, Idaho, Montana, and Colorado led to the development of western mining towns, Utah became the granary of the entire intermountain region.[46]

The Mormons were very influential in the settlement of the West, particularly because of their use of water. Their large irrigation works were highly noticeable, in large part because of their location at a crossroads of the West. Many had occasion to inspect the "miracle of irrigation" firsthand when they stopped in Salt Lake City to reprovision themselves on the way to points farther west. Others learned of the Mormons' irrigation by word of mouth, or came to Utah specifically to study the irrigation works. It was not long before the Mormons' irrigation methods spread throughout the West.

The key to the Mormons' ability to develop large-scale irrigation works was their coherence as a community. The Mormons' irrigation works were developed cooperatively, by large numbers of people, and thus could be developed and operated on a large scale. The Mormons' irrigation methods were copied at first by other colonies,[47] and later by a variety of farmer cooperatives, mutual irrigation companies, and privately financed ditch companies. Irrigation soon became the dominant water use in the West, far outstripping mining in terms of both the number of locations in which it was practiced and the total volume of water used.

The Growing Dominance of the Prior Appropriation Doctrine

The circumstances of westerners varied greatly, as did their needs and preferences for water allocation systems. Most westerners, consumed by the day-to-day need to make a living in difficult circumstances, didn't spend much time debating the finer points of water allocation laws—they simply favored those that would be of most benefit to their own circumstances. Many of the first ranchers and farmers, like the first miners, settled in the choice areas bordering the West's rivers and streams. Those who settled near water generally expected that they would be free to use the water not only for the usual array of domestic and livestock needs, but also for irrigating crops and pasture.

Those who arrived later were often forced to choose sites on drier land, where farming was less productive. Many of these later arrivals reasoned that they should not be prevented from watering their lands if they were physically able to build structures to divert and transport water to where it was needed. To deny settlers the possibility of bringing water to distant lands would make it impossible for much of the West to be settled. Many settlers expected that the law, when it was finally established, would validate the appropriation and use of water on nonriparian lands.

There was a clear conflict between the two water allocation doctrines. The riparian doctrine emphasized parity and protected future uses, whereas the appropriation doctrine stressed priority and promoted immediate development and use. Both doctrines had strong support in the West. Advantages of the common law riparian doctrine included its familiarity, the fact that it worked well in some of the more humid areas of the West, and, as mentioned earlier, its history as the foundation on which most of the western states and territories had built their legal systems. Advantages of the prior appropriation doctrine included the fact that it was well suited for use in the more arid climates, suited the prevalent ethic of development of the public domain, and had already spread quickly throughout parts of the West. At first, western settlers used both doctrines, adopting riparian law in some areas and appropriative law in others. But over time, states and territories throughout the West shifted toward placing more reliance on the prior appropriation doctrine.

The Mormons provided an example of the shift to appropriative principles. The rules the Mormons used to allocate water when they first arrived in the Salt Lake Valley were much like those of the Spanish. Water allocations were controlled for the first few years by church authorities and then after 1852 by the county courts. There was a heavy emphasis on the communal development and use of water. Rights to the use of water were also granted to individuals, but the courts were allowed to reject applications for these individual water rights, or to place conditions on the rights in order to protect the public interest. Maximization of the good of the community rather than of the individual was the goal of the courts, and the courts often planned for the future welfare of the community rather than permitting a few individuals to appropriate all that they wanted.

Eventually, however, the Mormons adopted water allocation principles much like those of the miners. By the 1860s the immigration of tens of thousands of people to the Utah Territory and the continuing development of new lands led to situations in which there often was not enough water to irrigate all of the cultivated land. In response, the territorial legislature passed laws in 1864, 1865, and 1866 establishing the principle

> that those farmers who first made use of the water should ever afterwards be entitled to sufficient water to irrigate the amount of land originally cultivated by them, and that later comers, whenever scarcity occurred, should not take the water until those enjoying prior rights had satisfied their needs, the latest comers being the first to be deprived, and those settling before them losing their water supply in succession as it became less and less.[48]

By the end of the 1860s, it became increasingly clear that Utahans had adopted the prior appropriation doctrine. Much the same thing happened in other states and territories. The first efforts to adopt the prior appropriation doctrine had been somewhat tentative, inspired largely by the desire to accommodate mining on the public domain. But as the need to divert water from natural watercourses for other purposes became apparent, and irrigation more common, state and territorial legislatures embraced the prior appropriation doctrine more widely.

The shift to the prior appropriation doctrine was handled differently by individual states. Some states, particularly those where rainfall was more abundant, saw no reason to completely eliminate the riparian doctrine as they expanded the appropriation doctrine, and so made great efforts to accommodate both doctrines. The Pacific states of California, Oregon, and

Washington, and the 100th-meridian states of North and South Dakota, Nebraska, Kansas, Oklahoma, and Texas all tried to take advantage of the developmental benefits of the new prior appropriation doctrine without upsetting the expectations of citizens who based their water claims on the common law riparian doctrine.

The accommodation of both doctrines was largely accomplished by applying each within its own limited sphere of influence. For example, in California, the state best known for its adoption of both doctrines, the state supreme court decided in an 1886 case that common law riparian rights—authorized by the state's first legislation in 1850—would prevail on lands granted by the federal government to the state or to private individuals, whereas appropriative rights—as authorized by the Mining Act of 1866 and the state legislature's adoption of appropriative principles in 1872—would prevail on the public domain.[49] In conflicts between the two, the common law rights attached to riparian land would prevail in situations in which the land passed out of the public domain before appropriations were initiated. If the land passed out of the public domain after appropriations had already been made, then the appropriative rights would be superior. Other states adopted similar distinctions, though with some variation. For example, Texas first segregated the domains of the two doctrines through geography, passing legislation in 1889 and 1895 that authorized appropriations only in the arid western half of the state, leaving the riparian doctrine as the sole method of establishing water rights in the more humid eastern half of the state.[50]

Alaska and the intermountain states of Montana, Idaho, Wyoming, Nevada, Utah, Colorado, Arizona, and New Mexico all made a complete shift to the prior appropriation doctrine, abolishing all remnants of the riparian doctrine. Colorado provides a good example. Sentiment in favor of the prior appropriation doctrine was so strong at the time of statehood in 1876 that the Colorado constitution provided that "the right to divert the unappropriated waters of any natural stream to beneficial uses shall never be denied."[51] Four years earlier Colorado's territorial supreme court had decided that appropriation was a necessity if irrigation was to be successful,[52] a theme that was repeated by the state supreme court in 1878 when it held that the appropriative right was embraced by statute and was, more fundamentally, necessitated by the climate.[53] The fact that appropriative law would completely replace riparian law in Colorado was made clear in 1882, when the state supreme court handed down its decision in the case of *Coffin v. Left Hand Ditch Co.*:

> The right to water in this country, by priority of appropriation thereof, we think it is, and has always been the duty of the national and state governments to protect. The right itself, and the obligation to protect it, existed prior to legislation on the subject of irrigation. . . . We conclude, then, that

> the common law doctrine giving the riparian owner a right to the flow of water in its natural channel upon and over his lands, even though he makes no beneficial use thereof, is inapplicable in Colorado. Imperative necessity, unknown to the countries which gave it birth, compels the recognition of another doctrine in conflict therewith.[54]

The argument that prior appropriation was a necessity in arid regions and that conditions precluded application of common law riparian principles proved compelling to the courts in the other intermountain states. For example, in 1909 the supreme court of Idaho stated in no uncertain terms that "there is no such thing in this state as a riparian right to the use of waters as against an appropriator and user of such waters. . . . In order to acquire a prior or superior right to the use of such water, it is as essential that a riparian owner locate or appropriate the waters and divert the same as it is for any other user of water to do so."[55] The riparian doctrine never received much attention in Alaska, and with passage of the Alaska Water Use Act in 1966 even minimal riparian features were repealed and Alaska joined the intermountain states as pure prior appropriation states.

Even most of the mixed-doctrine states later took steps to ensure the supremacy of the appropriation doctrine. For example, the Oregon supreme court ruled in 1908 that though the riparian doctrine had prevailed throughout Oregon until 1877, the common law riparian doctrine had been abolished on public lands since that year as a result of the federal acts of 1866, 1870, and 1877.[56] The next near the state legislature passed an act declaring that the appropriation of water according to the provisions of the prior appropriation doctrine would thereafter be the sole means of establishing new water rights in Oregon.[57] Riparian rights created before 1909 were to be protected only if "vested," which meant that the riparian proprietor had applied a specific quantity of water to a beneficial use prior to 1909, and that the use had not been discontinued for a period exceeding two consecutive years. By eliminating the possibility of additional riparian rights, transforming existing riparian rights into a form very much resembling appropriative rights, and providing for the abandonment of riparian rights after two years of nonuse, Oregon severely limited the application of the riparian doctrine and tilted the balance strongly in favor of the appropriation doctrine.

Most of the mixed-doctrine states followed Oregon in barring the creation of new rights based on common law principles and limited existing rights to the amount of water already put to a beneficial use. The common law riparian doctrine is still strong in California and Oklahoma, Texas still protects large numbers of riparian rights, and riparian landowners never lost their ability to assert future needs for water in Nebraska. Nevertheless, the prior appropriation doctrine has become the primary means by which the western states allocate and administer property rights in water.

Conflicts Stemming from the Change of Laws

Changes in laws are common, for laws must change to meet the changing needs of society. But these changes can be difficult, as evidenced by the controversy and strife surrounding abolition of the riparian doctrine. The disappearance, or at least de-emphasis, of the riparian doctrine was not universally welcomed in the West, even within the pure appropriation states. The replacement of one legal system by another is difficult, no matter how many advantages are claimed for the new doctrine. Changes reallocate property rights, upset expectations, and lead to uncertainty. Despite the strong language and forceful decisions of the state supreme court cases discussed above, there was substantial disagreement in virtually all of the states as to which rules should be used to allocate water.

For example, when Colorado citizen George Coffin argued in 1872 that he and several other farmers should have rights to the use of water by virtue of their ownership of riparian land, notwithstanding the prior, upstream appropriations of the farmers of the Left Hand Ditch Company, he was arguing from legal principles well established in other parts of the country. Coffin's argument was also based in part on early statutes enacted by Colorado's territorial legislature, including 1861 legislation that stated:

> All persons who claim, own or hold a possessory right or title to any land or parcel of land within the boundary of Colorado territory . . . when those claims are on the bank, margin or neighborhood of any stream of water, creek or river, shall be entitled to the use of the water of said stream, creek or river for the purposes of irrigation and making said claims available to the full extent of the soil, for agricultural purposes.[58]

Legislation enacted the next year continued the theme:

> Nor shall the water of any stream be diverted from its original channel to the detriment of any miner, millmen or others along the line of said stream, and there shall be at all times left sufficient water in said stream for the use of miners and farmers along said stream.[59]

These two pieces of legislation also provided for the appointment of commissioners during periods of scarcity "whose duty it shall be to apportion, in a just and equitable proportion, a certain amount of . . . water upon certain or alternate weekly days to different localities, as they may, in their judgment, think best for the interests of all parties concerned, and with a due regard to the legal rights of all."[60] This language appears to be based on, and to endorse, standard riparian principles, and yet the Colorado supreme court, in the *Coffin* decision, ruled that the legislature had in fact intended to guarantee appropriative rights. This, according to the court, could be seen from the fact that the legislature had, in 1864, without eliminating any of the provisions quoted above, inserted a new phrase into the legislation:

> Nor shall the water of any stream be diverted from its original channel to the detriment of any miner, millmen, or other along the line of said stream, *who may have a priority of right,* and there shall be at all times left sufficient water in said stream for the use of miners and agriculturists along said stream. [1864 amendment in italics]

The court's interpretation of the 1864 amendment may indeed represent the intent of the legislature to abolish riparian rights, but the court's conclusion is certainly one over which reasonable people might disagree. By eliminating the riparian doctrine and proclaiming that rights to the use of water would thereafter be allocated and adjudicated solely according to the principles of the prior appropriation doctrine, the court's decision may well have ensured a reduction in the number of future conflicts over the use of the state's water resources. However, there is little doubt that the decision also upset the expectations of a number of people who had already settled in Colorado, including George Coffin and many other farmers who found their crops drying up in years when upstream, prior appropriators failed to leave sufficient water in the stream for downstream use.

The transition was also difficult in other states. For example, in Montana the state supreme court maintained in its 1921 *Mettler v. Ames* decision that riparian rights had not existed in that state since passage of the territorial Bannock Statutes in 1865. But, as in Colorado, a reading of the statutes referenced by the court also suggests that the 1865 legislature was actually endorsing riparian rather than appropriative principles, a fact noted in 1872 by Chief Justice Wade:

> The whole purpose of the [1865] statute was to utterly abolish and annihilate the doctrine of prior appropriation, and to establish an equal distribution of the waters of any given stream in the agricultural districts of the Territory. . . . If this section of the law does not mean that there shall be an equal distribution of the waters of a stream among all the parties concerned in such water, without any regard whatever to the date of location or appropriation, then we are utterly unable to comprehend the language used.[61]

Courts in some states found that the West's climatic conditions had made use of the appropriation doctrine "an imperative necessity," but Wade feared that adoption of prior appropriation principles would lead to monopolization of the state's water resources.[62] He wrote, "The climatic and physical conditions of this country cannot be such as to create a law so at variance with natural equity and so fatal to the improvement and prosperity of our best agricultural district."[63] Wade favored the riparian doctrine over the appropriation doctrine because of its emphasis on equity and felt that if conditions truly necessitated new laws that the best answer would be to modify existing common law principles rather than to adopt an entirely new system.

Critics of the new doctrine also had other concerns. Legal scholar John Norton Pomeroy held that it was "unnatural" for the right to use water to be separated from ownership of the land over which the water ran. Pomeroy maintained that granting appropriators rights of way to build canals across land owned by others—a standard feature of appropriation law—constituted a gross infringement of the property rights of those landowners.[64] Elwood Mead, a prominent figure in the history of western water development and administration, said of the prior appropriation doctrine:

> It assumes that the establishment of titles to the snows on the mountains and the rains falling on the public land and the water collected in the lakes and rivers, on the use of which the development of the state in a great measure depends, is a private matter. It ignores public interests in a resource upon which the enduring prosperity of the community must rest. It is like A suing B for control of property which belongs to C.[65]

Judge Melville Brown, a participant in Wyoming's constitutional convention in 1889, believed that priority rights were "pernicious and an outrage upon the people. . . . When we appoint a board of control to manage this water system,[66] that we say belongs to the state, let us give them the authority to control it for the highest and best uses of the people of the state, and don't fix that control by saying that priority of appropriation shall settle the matter."[67]

The passing of former institutions and the rise of the prior appropriation doctrine was also mourned in Utah. Though adoption of appropriative principles served some useful purposes,[68] the demise of the Mormons' distinctive community-based systems did not please everyone. For several years the changes were largely ignored by many Utahans, who continued to form groups for irrigation and to ignore chronological priority, choosing instead to share water equally during times of shortage.[69] Some of the deficiencies of the new system were expressed by George Thomas, an economics professor at the University of Utah, who wrote a book about local irrigation institutions in 1920.[70] Thomas thought that the establishment of a priority system was "contrary to the spirit of the settlement of Utah," where the nature of the physical conditions was such that only cooperative, community action could succeed. Not mincing his words, he wrote that the new priority-based system was "a marked step in retrogression" and "a great step backward in water jurisprudence for the territory." "It is doubtful, if in all the legislative history of irrigation a more retrograde piece of legislation was ever placed upon the statute books than the law of 1880," he wrote. The previous system had relied on the concept that water belonged to the public—"the only sound basis to act upon"—and didn't require the services of highly paid lawyers or lead to bankruptcy for the farmers. Thomas believed that the old system had been inexpensive, prompt, efficient, and fair, and, mirroring the comments of Justice Wade in Montana, lamented that "the public doubts the wisdom of the past and looks afar for some new institution when a slight change in an old one would meet the situation better."

The Status of Water Use, Water Law, and Rivers at the Turn of the Century

By the end of the 19th-century mining was still an important economic activity and water use in the West, but it was declining relative to the importance of other activities and uses. The economies of the western states at the turn of the century were based primarily on agriculture, and irrigation accounted for the vast bulk of all water use. Cities and towns became more numerous and larger as the West was settled and also became important users of water. Relative to irrigated agriculture, however, they consumed very little water.

The goal of virtually all 19th-century settlers was to develop natural resources for sustenance and to create wealth, and in the arid West the diversion and use of water outside of natural channels was instrumental to the accomplishment of this goal. River diversions to support farming, ranching, mining, and the West's growing cities took precedence, and there was very little consideration given to the impact of these water diversions on the environment, of the need to leave water in stream channels to support aquatic life, riparian areas, wildlife, recreation, and aesthetics. The prevailing attitude was that if other water uses were in competition with the use of water to develop natural resources and settle the West—the development of natural resources and settlement of the West were considered synonymous—then those other uses had to give way.

Despite the reservations of many in the West, the principles of the prior appropriation doctrine had been widely adopted throughout the West by the end of the 19th century. The doctrine was inspired by, and perpetuated, the settlers' development goals. The basic features of the prior appropriation doctrine continue to be the basis of water law in the West: (1) the right to use water is obtained by taking the water and putting it to a beneficial use; (2) the right is limited to the amount of water that is beneficially used; (3) first in time is first in right; and (4) the water must be used or the right is lost.

These rules had a major impact on the uses of western rivers and streams. For instance, to take water and put it to a beneficial use one had to exercise some form of physical control over the water. Control was exercised by building storage and diversion dams or otherwise "developing" rivers, thereby altering natural patterns of water flow. The allocation of water to those who took it first provided incentives for settlers to take and put to use all the water that they could possibly use as quickly as possible, rather than leaving it for instream uses or for potential out-of-stream uses by future settlers. Beneficial-use requirements were in some ways designed to prohibit taking more water than necessary, but enforcement of beneficial-use requirements has been limited and sporadic, at best, throughout the entire history of the American West. Further, beneficial-use requirements had the effect of excluding some water uses—such as many of those that take place instream—that were not considered beneficial at that time. Leaving water in streams was widely considered to be a waste of water. In most cases, the only effective legal limit to the amount of water that could be removed from a stream by an individual appropriator was the presence of a senior appropriator downstream. Together, water users were limited in their diversions only to the amount of water physically flowing in the river.

The result of these rules and attitudes was that by the year 1900 most small-to-medium-sized rivers and streams in the West that were located near arable land, working mines, or cities had already been developed to at least some degree. Legal claims to water in many of these rivers often exceeded the physical amount of water that was available, especially during dry years and seasons, and river channels often contained very little, if any, water during all or part of the year. Rivers with relatively undiminished flow were found primarily in areas where settlement had not yet occurred, usually where the river itself was largely inaccessible, or where settlers were forced to allow water to pass because water users with more senior water rights were located downstream. Water also continued to flow in most of the West's truly large rivers, not because of a lack of desire to develop them, but because of a lack of capacity to fully do so. Rivers still flow in the West where they are largely inaccessible to development or there are senior water rights downstream, but the capacity to develop the West's largest rivers has been in place, and excercised, since shortly after the start of the 20th century.

Continued Development of the West's Rivers and Streams

The West's larger rivers had been left undeveloped primarily because projects at the scale required to harness such rivers—storage reservoirs and extensive water distribution systems—were beyond the financial and organizational means of individuals, or even of privately organized companies. Some of the larger municipalities, with their concentrations of wealth and valuable uses of water, could sometimes afford to build their own moderately large water projects. For example, in 1913 San Francisco completed construction of O'Shaughnessy Dam, first proposed in 1908, which created the Hetch Hetchy Reservoir in Yosemite National Park as a source of municipal water supply.[71] Also in 1913 Los Angeles completed a project to capture water in the remote Owens Valley and transport it 223 miles back to the Los Angeles Valley.[72] But other water users, particularly irrigators, had much more difficulty financing and building large-scale projects. Even more modest irrigation projects often ran into financial difficulties, as evidenced by a virtual epidemic of bankruptcies among private ditch companies around the turn of the century.

These conditions soon made it clear that further development of the West's water resources, and of the West itself, would require the participation of the federal government. Perhaps the most significant of the federal water resources development programs is the federal reclamation program, which began with passage of the federal Reclamation Act in 1902.[73] It did not take long for the federal reclamation program to become a major factor in the growth and development of the West. The first five major federal water projects were authorized in 1903, and an additional 19 projects in 15 western states had been authorized by 1907.[74] Reclamation projects were originally designed solely for irrigation, but hydropower generation—if needed for irrigation purposes—became an authorized purpose of reclamation projects in 1906, and other purposes such as flood control and municipal water supply were authorized in later years.

Reclamation projects typically consisted of large dams to capture and store water, smaller facilities to divert water out of streams, and extensive water conveyance and distribution systems. Many of these facilities were huge, their scale dwarfing those of water development projects previously undertaken in the West or anywhere else. For example, Roosevelt Dam in Arizona, part of the Salt River Project authorized in 1903, was built to a height of 325 feet and stores almost 1.5 million acre-feet of water. Several projects built during the Great Depression, such as the Boulder Canyon (now Hoover) Dam on the Arizona-Nevada border, the Colorado–Big Thompson Project in northern Colorado, the Central Valley Project in California, and the Columbia Basin Project in Washington state, were considerably larger. Hoover Dam, authorized in 1928, is 726 feet high and is capable of storing 32 million acre-feet of Colorado River water in Lake Mead. A single Columbia Basin Project dam, Grand Coulee, provides enough water to irrigate almost 500,000 acres, which was more than the total irrigated acreage in Utah after the first 50 years of irrigation development. Bureau of Reclamation projects eventually delivered water to over 100,000 farms covering millions of acres, and encompassed hundreds of storage dams and tens of thousands of miles of canals and other conveyance and distribution facilities.[75]

The generation of electricity also became a major water use in the first half of the 20th century. The nation's first hydropower plants—facilities that used the force of falling water to turn turbines connected to generators—were developed late in the 19th century and had become widespread on western rivers by the 1930s. The General Dam Act of 1906 authorized private power companies to build hydroelectric facilities on navigable streams, subject to congressional approval.[76] Many hydropower facilities were developed by private enterprise, but the federal Bureau of Reclamation and Army

Corps of Engineers built dozens of the largest hydropower plants, on some of the West's largest rivers, including the Columbia, Snake, Sacramento, Colorado, and Missouri. Thermoelectric plants, which generate electricity by burning fossil fuels, use water for cooling and also became major water users.

Residents of the West continued to use water in ever-growing volumes for other purposes as well. Many of the large dams built for irrigation and hydropower were also used for flood control, industry, municipal water supply, recreation, and navigation, as were hundreds of smaller dams built by municipalities, irrigation districts, and power companies. During the first half of the 20th century water was developed not only in more locations, but on a much larger scale. Vast quantities of water were removed from natural watercourses, and that which remained—or was returned to rivers after use—was subject to new impoundments and diversions downstream. The attitude of the times was best summed up by the comment of Senator George Norris, who in 1933, referring to a newly authorized project on the Tennessee River, stated:

> It is emblematic of the dawning of that day when every rippling stream that flows down the mountain side and winds its way through the meadows to the sea shall be harnessed and made to work for the welfare and comfort of man.[77]

Administrative Institutions, Water-Use Preferences, and the Public Interest

The 20th century also saw changes in the prior appropriation doctrine. Under the doctrine, uses of water were determined in a manner consistent with the attitude of the times. Decisions were made by individuals in accordance with their independently determined goals rather than by the government or other collective authorities acting in pursuit of the social welfare. As noted by one historian, the allocation of water according to the principles of the doctrine "was consistent with the cherished American ideal that individuals, not society, should control their destiny."[78] Though the prior appropriation doctrine may well have met the individualistic ideals of the times, by the turn of the century it had become apparent that operation of the system was accompanied by a number of problems.

One of the greatest problems was the prevalence of claims for excessive amounts of water. Appropriators often asserted, and courts granted, rights to quantities of water greatly in excess of that which could reasonably have been put to use for the stated purposes of the appropriations. Sometimes people claimed too much water out of ignorance. Most westerners had no previous experience with irrigation and didn't know how much water their crops would need, how best to apply the water, or even how to measure it. Faced with a situation in which they had to claim water as soon as possible or run the risk of forever losing access to additional supplies, many erred on the side of claiming too much water rather than too little. Others claimed too much water for strategic purposes, to reserve water for future needs or to sell to subsequent settlers.

There were also problems associated with resolving water-use disputes in the courts. The courts were the sole recourse of water users under the prior appropriation system. The expense of hiring lawyers and gathering data, the risk of offending one's neighbors, and the difficulty of obtaining a favorable result discouraged many potential water users from filing suits. Water rights conflicts also involved as many—or more—technical questions as legal questions, and the courts were ill prepared to deal with such issues.

A general lack of adequate record keeping plagued the appropriation system.[79] The lack of credible information made it difficult for potential claimants to determine whether there was a reliable supply of water still available for additional uses, and for courts to determine the priorities and amounts of existing rights. Even after claims were verified the problems continued, because there were inadequate mechanisms for ensuring that water was actually delivered to water users in accordance with their decreed priorities. It was difficult for upstream junior appropriators to know when they could divert and when they needed to let water pass by to downstream seniors. It was difficult for downstream seniors to know whether they were not receiving their full allocation of water because of natural variations in streamflow or because of out-of-priority upstream diversions. Without a system for defining and enforcing diversion schedules all up and down the river, water users were able to determine whether they were diverting out of priority or receiving their decreed shares only by undertaking a physical survey of all watercourse diversions. On small streams this process was extremely time consuming. On larger rivers physical surveys were not even possible. These problems eventually led people to call for a more active role in the administration of water resources by the states, and in the late 1800s and early 1900s the states responded by adopting new administrative systems to control the allocation and distribution of water.

Administrative systems adopted by individual states varied, but the general thrust was to give state administrative agencies more control over the acquisition of water rights, the distribution of water according to legal priorities, and the adjudication of contested rights. States typically entrusted some or all of these responsibilities to the office of the state engineer, or to a department of natural resources, water resources, or similar agency.

Under these administrative systems, anyone wishing to obtain a water right must typically apply to a state agency for a water-use permit, rather than simply posting a notice and taking the water. Applications usually contain information about the source and amount of water to be used, the nature and location of proposed diversion works, the amount of time needed to complete the work, the purpose for which the water is to be used, and any other information deemed relevant. Agencies then have the authority to issue or deny the permit, though the criteria by which such decisions are made vary considerably from state to state.

State employees also ensure that water is distributed in accordance with the priorities of the rights. Any water user not receiving his or her legal share of a river's flow may place a "call" on the river. In response to the call, agents of the state require any water users with rights junior to that of the calling water user to curtail their diversions until the senior right is satisfied. Diversions of the most junior water rights on the watercourse are shut down first, then the next most junior, and so on, until enough water is left in the stream to fulfill the senior right.

In many states, administrative personnel also play a major role in the adjudication of water rights disputes. Most states that use agencies to adjudicate disputes make the decisions of those agencies reviewable in state court, and courts are still heavily involved in water rights issues in the western states. These changes did, however, shift much of the administration of the water rights system away from the courts and into the hands of executive branch agencies staffed by people with technical expertise.

A discussion of administrative systems is relevant to the issue of instream flow protection for a few reasons. First, it provides the context in which water resource allocation decisions are currently handled in most of the western states. Second, as will be discussed in much more detail in chapter 4, the switch to administrative systems opened up the possibility of using some new mechanisms for protecting instream flows. And third, the switch to administrative systems emphasized the role of the public in deciding how water resources would be allocated, and often introduced new criteria to be used in doing so. Many states, including some of those that had most ardently embraced the prior appropriation doctrine, had already declared that their water resources belonged to the public rather than to individuals.[80] A fundamental tenet of water law, widely ignored or misunderstood, is that a water right gives someone the right to *use* water rather than actual *ownership* of water; ownership resides in the public. Many states chose to clarify or restate this fact at the time that they adopted administrative systems. Combined with the actual shift to public control—as exercised by state agencies—of the day-to-day operations involved in the administration of water resources, many observers hoped that the rise of administrative systems would create a much stronger role for the public in determining how water would be used. As summarized by one scholar, in reference to the switch to an administrative system in Wyoming, the major change

> lay in the subordination of the appropriator to the welfare of the state. The interest of the state or the community came first, that of the individual ir-

> rigator second. Gone were the days in Wyoming when an appropriator, without anybody's leave, could post a notice, dig a ditch, install a dam, and divert the waters of a stream. Water was too limited a resource to be diverted and wastefully used without regard for the rights of others. Since it was the property of the state, rights to its use were to be granted by the state, adjudicated by the state, and protected by the state.[81]

The subordination of private interests to those of the public never occurred in the totality envisioned by these observers. In general, the creation of administrative systems, though supported by reformers, seems to have been more a result of efforts to make the existing priority system work better than to make substantive changes to the system. However, the shift to public administration of water rights did result in some changes in the way water was allocated. These changes were accomplished in large part through the use of public interest or public welfare requirements in state constitutions and statutes. For example, Wyoming's constitution states, "No appropriation shall be denied *except when such denial is demanded by the public interests*" (emphasis added).[82] Statements similar to this are now found in the constitutions or statutes of many of the other western states, emphasizing the fact that appropriations will no longer be valid just because they are of benefit to someone; rights will be granted only if proposed water uses are also consistent with the public interest.

The "public interest" is, however, very difficult to define. Legislators in at least two states, Alaska and Oregon, have provided state agencies with guidelines for determining what is, and what isn't, in the public interest, but most of the other states have left questions of the public interest to the discretion of administrative officials.[83] In some cases, the rhetoric about protecting the public welfare has been stronger than subsequent implementation by administrative agencies. Wyoming's approach—in which individuals are not denied the right to appropriate unless their proposed uses are specifically found to be opposed to the public interest—was adopted in several other states, including Arizona, Kansas, Nebraska, Nevada, and Texas. North Dakota and Oklahoma took a slightly different approach to safeguarding the public interest, by authorizing their attorneys general to intervene in private adjudications if the states' primary water management officials felt that such action was necessary to protect the public interest. California, South Dakota, Utah, and Washington attempted to promote the public interest more affirmatively, by authorizing the issuance of water use permits only for uses thought to be of the most benefit to the public. Alaska's criteria include: (1) the benefit to the applicant resulting from the proposed appropriation; (2) the effect of the economic activity resulting from the proposed application; (3) the effect on fish and game resources and on public recreational opportunities; (4) the effect on public health; (5) the effect of loss of alternate uses of water that might be made within a reasonable time if not precluded or hindered by the proposed appropriation; (6) harm to other persons resulting from the proposed appropriation; (7) the intent and ability of the applicant to complete the appropriation; and (8) the effect upon access to navigable or public waters.[84]

Many of the western states also established preferences among beneficial uses of water.[85] Preferences accomplish some of the same goals as public interest requirements, because the establishment of preferences promotes water uses thought to be of most benefit to the state. In some states the preferences are applied during the permitting process, so that among pending applications those for preferred uses receive the first permits, even if submitted later than the other applications. In other states the preferences are applied during times of water shortage, so that preferred uses have first access to scarce water supplies even if priority dates would indicate a different result. This usually happens through the process of condemnation, which requires that compensation be paid to the holder of the displaced right.

Water for domestic and/or municipal needs receives the highest priority in all of the states that have established preferences. Such uses typically include drinking, cooking, washing, irrigation of small gardens, and limited stock watering. Courts and legislatures have tried to make it easier for cities to obtain future supplies, to engage in long-term planning, and to raise and commit funds for large projects by exempting them from the same "use it or lose it" provisions that apply to other water uses. In most states administrative officials are required to approve permit applications for domestic and municipal purposes before approving any competing applications, and cities in many states are given wide latitude to condemn senior water rights during droughts.

There is considerable variation in other preferred uses among the states, depending in large measure on the distribution of political power at the time the preference statutes or constitutions were enacted. For instance, the use of water for agriculture is favored over all but domestic uses in most states, because agricultural interests usually dominated state legislatures in the early part of the century when preference statutes were written. Industrial, manufacturing, and electrical generation purposes are usually less preferred, and the use of water for recreation, fish, and wildlife purposes is usually at or near the bottom of preference lists, if listed at all.

Virtually all of the western states now place strong reliance on state agencies to manage their water resources, and most states use public interest criteria to at least some degree in deciding among competing water uses. Montana was the most recent of the western states to adopt an administrative system, doing so in 1973. The sole holdout is the state of Colorado, which has continued to rely almost exclusively on the courts to administer its water resources. Colorado's constitution does, however, state that water belongs to the public,[86] and Colorado's state engineer has been given the responsibility for distributing water in accordance with priorities and for maintaining certain records.

Rivers, Water Use, and Water Law in the Contemporary West

Uses of rivers in the West since the early part of this century have changed largely in response to rapid and sustained population growth, particularly since the end of World War II. Over 77 million (31 percent) of the nation's 252 million people were living in the 18 western United States in 1990,[87] and most of the nation's fastest-growing states are in the West. Population growth has changed the essential character of the West, because virtually all of the growth has taken place in urban areas. The West is home to such huge cities as Los Angeles, San Francisco, Seattle, Phoenix, Denver, Houston, and Dallas and is now the nation's most urbanized region. Even the Plains states, which have experienced more moderate growth, have become more urban.

These demographic changes have led to some changes in water use. Municipalities now use a larger and growing proportion of the West's water resources, and industrial and commercial water uses have become more diverse and numerous. The use of water to generate electricity has become prominent in many areas, particularly along the West's largest rivers. Water users in the western states withdrew an average of 106.6 billion gallons of fresh water from natural surface-water sources per day in 1990, virtually all of it from rivers, or from reservoirs impounding rivers. These withdrawals accounted for 41 percent of all fresh surface water withdrawals nationwide.[88]

Irrigation remains by far the West's largest user of water. In 1990, irrigation accounted for 76 percent of all water withdrawals in the West. Thermoelectric plants (13 percent) accounted for the next largest proportion of withdrawals. Eight percent of all withdrawals were for public supply, domestic, and commercial purposes, and the remaining 3 percent was withdrawn for industrial, livestock, and mining purposes.[89] Almost 15 times the total amount of water withdrawn for all of these purposes combined was run through hydroelectric turbines, though much of this water was not "withdrawn" from stream channels at all.

Much of the water withdrawn for the purposes listed above is used more than once, because it is returned to natural water bodies after use. However, consumptive water use—the amount of water that is not returned to natural water bodies—averaged 69.5 billion gallons per day in the western states in 1990, accounting for 74 percent of the entire nation's freshwater consumption. The high proportion of the nation's water consumption taking place in the West is explained by the prevalence of water used outdoors, primarily for the irrigation of crops, but also for lawns, trees, parks, and other uses in urban areas. Most of the water used outdoors is consumed rather than being returned to natural water channels. Fifty-six percent of all water withdrawn for irrigation is consumed by crops and evaporation, and another 20 percent is lost during conveyance. Twenty-three percent of water withdrawn for domestic purposes is consumed, as is 15 percent of water withdrawn for industrial purposes, 11 percent for commercial purposes, and 2 percent of water withdrawn by thermoelectric facilities. Virtually none of the water used for producing electricity at hydroelectric facilities is consumed.

Western rivers of all sizes have been heavily developed. The entire flow of many rivers has been legally claimed for offstream uses since the turn of the century, or even before. Some of the West's major watercourses that are now dry or virtually dry during substantial portions of the year include the Snake River below Milner Dam in Idaho;[90] the Gila River and, below Theodore Roosevelt Dam, the Salt River in Arizona; the Powder River in Oregon; the Arkansas River near the Colorado-Kansas border; the Rio Grande River below Elephant Butte Reservoir in New Mexico; and the San Joaquin River below Friant Dam in California. Virtually every substantial river contains at least a few dams, as do many of the smaller rivers. Figure 2.1 shows locations of the nearly 10,000 reservoirs in the West with at least 100 acre-feet of storage capacity. If stock ponds and smaller reservoirs are included, the western total is over 30,000 dams and reservoirs.[91]

Depleted or nonexistent flows in river channels are a fact of life throughout much of the West. The era of truly large water development projects seems to be over, however. Most of the best sites have already been developed, federal budget deficits have put an end to the extensive federal subsidies on which many of these projects relied, and other projects face opposition based on society's growing awareness of the environmental effects associated with water development. As figure 2.2 shows, the extravagant era of western dam construction appears to have ended abruptly.[92]

The prior appropriation doctrine, as modified by administrative systems and public interest criteria—and by the public trust and federal reserved water rights doctrines discussed in subsequent chapters—continues to be the foundation of western water law. Changing conditions and values have, however, continued to promote modifications of the law. As should be evident from the foregoing history, western water law is not static. The Hohokam and Anasazi, Spanish, Mormons, and others developed unique sets of rules to facilitate establishment of communal irrigation societies. The desire to develop natural resources and to settle the West led the Spanish, the forty-niners, and independent irrigators to implement rules that rewarded individual initiative in the conversion of raw materials into life-sustaining and marketable products. The public interest in preserving the amenity values of rivers, as evidenced in the riparian doctrine, shifted to encompass appropriative principles when development became a greater priority. With the prior appropriation doctrine, those who stood ready to apply water in the

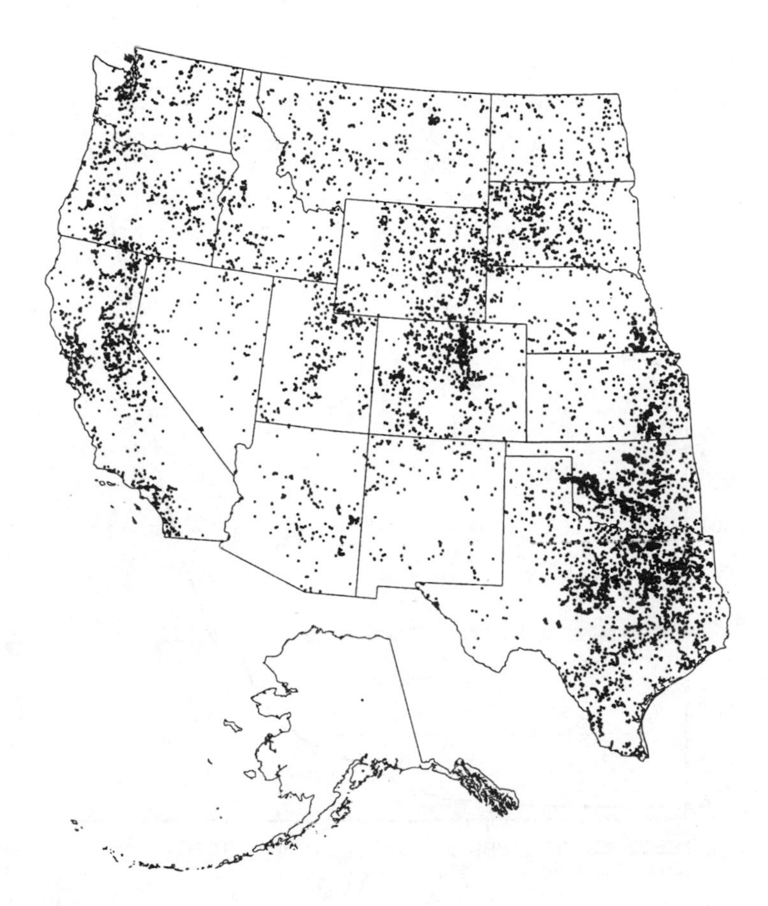

FIGURE 2.1
All Reservoirs Greater Than 100 Acre-Feet and Excluding Stock Ponds (one dot per reservoir)

pursuit of economic advancement received first priority to the use of the water. Now the public interest seems to be shifting toward the preservation of public values such as boating, swimming, fishing, hunting, and recreation.[93] The one constant in western water law has been that change, in response to new needs and conditions, has been frequent.

The clear need for change hasn't necessarily meant that changes have occurred immediately, or all at the same time. The history of the replacement of the common law riparian doctrine by the prior appropriation doctrine provides ample evidence to support this statement. Even the states that decided to completely abolish the riparian doctrine did so at different times and often in stages. Many of the states that adopted mixed-doctrine systems

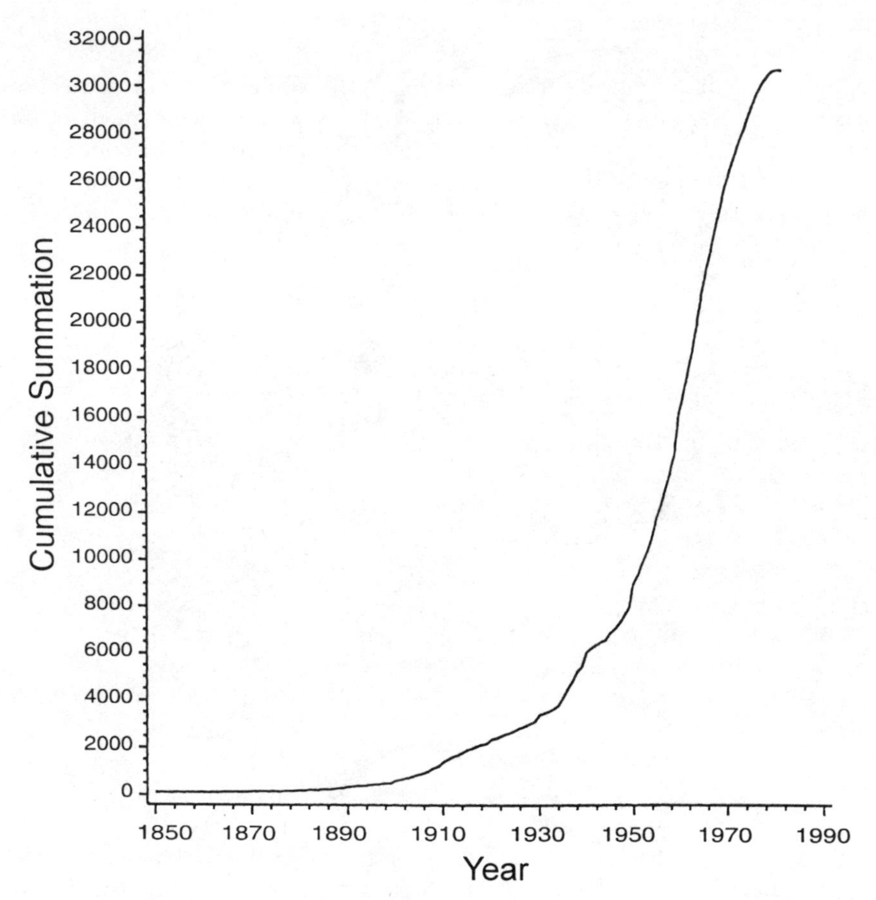

FIGURE 2.2
Cumulative Dam Construction for the 18 Western States

are still working to define the theoretical and operational relationships between the two doctrines.

One source of change in contemporary water law, and an issue very much at the forefront of contemporary water policy, has concerned the shift of water from previous uses to new uses, particularly from agriculture to municipalities. The desire to encourage water marketing while simultaneously protecting those dependent on previous uses of water has been a major source of innovation in water law in the last several years and will likely continue to be for the foreseeable future.

The desire to protect instream flows has also been a major inspiration for change in western water law. The population growth and economic development experienced by the West in the last several decades have been accompanied by shifting attitudes and values with respect to water use and the environment. Much of this may be attributable to levels of economic development and urbanization that have left western residents with much more leisure time than they had in the past. Westerners now spend much more time in recreational pursuits, much of it in the outdoors, than conditions permitted 100 or even 50 years ago, and western residents and visitors have been increasingly vocal in their demands for enhanced recreational opportunities. A new appreciation for the environment and environmental processes has also made itself felt in the last several decades. Through their activities, and sometimes through their advocacy and political action, westerners have expressed a growing desire to protect fish and wildlife habitats, riparian areas, the aesthetic qualities of the environment, and environmental sustainability.

Satisfying these desires requires that at least some water be left within stream channels. However, the water is often not there. Previous patterns of development, bolstered by laws developed in the mid-19th century to encourage and protect out-of-stream uses, have not favored instream uses. Rather, they have led to a substantial loss of instream flows throughout the West. Appropriative principles favor water development and diversion, but society is now expressing additional needs. In response, laws are beginning to change, but at this point changes to the law lag significantly behind the expression of needs. The laws seem not yet to have caught up with new conditions and values.

It should be apparent from an examination of the history of western water law that changing laws is a complicated business. Changing laws is difficult and controversial work. Just as the creation and spread of the prior appropriation doctrine upset expectations and generated controversy over a century ago, proposals to change the way that water resources are managed today have generated substantial argument and debate. Issues surrounding the protection of instream flows have become the focus of much of the current debate over water policy, and have attracted their share of controversy. Laws are changing, but as one would expect, the laws are not changing uniformly among the states, in either method or rate. The changes that are occurring in western water law with respect to instream flows are described in more detail in chapter 4. But before describing these changes, we turn to a closer look at the instream uses of water that westerners are seeking to protect.

CHAPTER THREE

Instream Uses of Water

Water flowing within stream channels is used for many purposes, each of which has its own flow-related needs. Most instream uses are environmental or recreational in nature, while others, primarily hydropower and navigation, have been developed solely for commercial purposes.

Fish and Other Aquatic Organisms

The conservation, restoration, and enhancement of fisheries are unquestionably among the most important instream uses of water. Protection of fish habitat is a primary goal of federal efforts to secure instream flows, and virtually every western state with an instream flow program includes the protection of fisheries as a valid purpose for seeking instream flows. In some states fisheries are the only instream use of water recognized by the law, and most of the methods used anywhere to quantify instream flow needs have been based on the needs of fish.[1] The following sections describe flow-related needs of fish.

Living Space

Water is the medium within which fish live. Fish take in the oxygen they need only through their gills, directly from the water, and are vulnerable to even short interruptions in the water supply. Some fish endure short-term interruptions in streamflow by finding refuge in pools, but others are lost when their medium for life is diverted for other needs, which is not uncommon in more arid portions of the West.

Food

Fish eat algae and small invertebrates such as insects, worms, mollusks, and crustaceans. Aside from terrestrial insects that fly or fall into the water, fish survive on organisms that themselves rely on streamflows for their existence. In addition to providing basic habitat for the organisms that fish eat, streamflows help dislodge algae and other benthic organisms—those found attached to or living on river beds—and carry them downstream to waiting fish. In essence, rivers provide fish with a constant flow of food as well as of water, so a reduction in streamflow may lead to a reduction in food supplies.

Cover

Cover offers fish protection from predators, temperature fluctuations, and the high energy of moving water. Cover is provided by submerged boulders and other rocks, submerged or overhanging vegetation, deep water (including pools), water turbulence, undercut banks, or large woody debris such as logs, branches, and rootwads.[2] The availability of low-velocity resting spots is of particular importance, as otherwise fish would be forced to expend a great deal of energy to avoid being swept downstream. Streams containing numerous resting spots, particularly if located next to high-velocity currents carrying food, constitute prime fish habitat for some species. Streamflow affects the availability of cover by influencing: (1) the density and distribution of woody debris in the stream channel; (2) the size and distribution of sand, gravel, and cobbles in the stream channel; (3) the presence, variety, and distribution of vegetation within and adjacent to the stream channel; (4) the creation and distribution of channel structures such as pools and overhanging banks; and (5) stream depth and turbulence.

Spawning, Incubation, and Rearing Habitat

Streamflow is important to fish reproduction. The necessary conditions vary by species, but all species require adequate space for spawning, incubation, and the growth of new fish. The life cycle of chinook salmon is illustrative.[3]

Chinook prefer spawning sites with swift current, porous gravels, cool temperatures, and water depths ranging from six inches to 3.5 feet. Eggs are laid in scooped-out depressions in the gravel, where they incubate for forty to sixty days. During this time the eggs must remain covered by water, and the gravels must remain porous and free of sediment so that water can circulate, bringing oxygen and removing wastes. Incubation is the most critical stage of the reproduction process, when mortality rates are highest, so adequate streamflows are crucial during this period. After hatching, the salmon *alevins* remain in the gravel for twenty to forty days, during which time the egg sack is absorbed. The salmon then emerge from the gravel as free-swimming *fry,* which require limited-velocity currents in which to feed. The fry feed on zooplankton and small insects until reaching *fingerling* length of two to three inches, at which time they ride high snowmelt-runoff flows to the ocean.

This process requires a succession of different flows, rather than a single "adequate" flow. Streamflow must first be conducive to the creation of porous gravel streambeds, must be sufficient to provide adequate space, and must be constant enough to sustain the eggs during incubation. Subsequent flows must be low enough to prevent fry from being washed downstream prematurely, but later must be high enough to facilitate migration to the ocean. The particular sequence, magnitude, and timing of flows necessary to facilitate fish reproduction vary enormously among species, but streamflow is of critical importance to all species.

Movement and Migration

Local food shortages, or undesirable environmental conditions such as excessive temperatures and turbidity, make it necessary for fish to change locations. To allow this relocation, streamflow must be sufficient to cover riffles and other obstructions.

Streamflows to facilitate movement are especially important for anadromous fish. Resident fish spend their entire lives in freshwater, often within a rather short reach of river, but anadromous fish spend most of their lives in the ocean and so must have clear passage to and from the sea.[4] As described above, the young fish typically migrate to the ocean with the aid of high flows resulting from snowmelt or a period of heavy rains. High flows enable the fish to complete the journey in relatively short periods of time, despite the sometimes enormous distances involved. After maturing in the ocean for several years, adult fish return to their stream of origin. Species of the salmon family (salmonidae) are the most well-known anadromous fish and include some varieties of trout[5] in addition to chinook (king), pink (humpbacked), chum (dog), sockeye (red or kokanee), coho (silver), and other species of salmon.[6] Other anadromous fish include sturgeon and, in California rivers, the introduced American shad and striped bass. Anadromous fish are of great ecological, cultural, commercial, and recreational importance in Alaska, California, Idaho, Oregon, and Washington.[7]

Temperature

Fish are cold blooded and therefore dependent on the maintenance of specific ranges of water temperature. Temperature affects internal metabolism rates, and all organisms have optimal temperature ranges for reproduction, growth, and survival. Water temperatures exceeding these ranges may lead to stunted growth, failure to reproduce, or fish mortality. Water temperature is so important to aquatic life of all kinds that the Federal Water Pollution Control Administration in 1967 described temperature as "a catalyst, a depressant, an activator, a restrictor, a stimulator, a controller, a killer, one of the most important and influential water quality characteristics to life in water."[8]

Fish are often categorized as either warm-water fish—thriving in water with temperatures ranging up to 80°F—or cold-water fish, which do not easily tolerate temperatures above 68°F. This dichotomy is approximate, as some species thrive in waters not clearly classified as either warm or cold, and some species that are well characterized as either cold- or warm-water fish can endure temperatures well outside their preferred range. Typical cold-water fishes in western rivers include salmon, trout, whitefish, and grayling. Perch, sunfish, and catfish are common residents of warmer waters, and various suckers, roach, and squawfish are found in waters of a variety of temperatures.[9]

Stream temperature is directly proportional to the amount of heat input, which comes primarily from solar radiation and air temperature. Factors such as the amount of riparian vegetation available for shading the water and the angle at which solar radiation strikes the water's surface are therefore important in determining water temperatures.[10] But temperatures are also inversely proportional to total streamflow volumes, so water temperatures tend to be lower when flow volumes are high and vice versa.

Turbidity and Dissolved Oxygen

Fish are also affected by stream turbidity and dissolved oxygen concentrations. Turbidity is a measure of the degree to which light is scattered or absorbed in water and depends on the amount of dissolved and, especially, suspended matter in the water. Suspended matter may directly injure fish by clogging and damaging gills, or by obscuring vision and thereby limiting the ability to find food. Suspended matter may indirectly affect health and survival by reducing growth rates and resistance to disease, modifying natural movement or migration patterns, reducing the availability of food, or by altering streambed composition, thereby affecting conditions necessary for spawning, incubation, and early growth.[11] Suspended matter may be either organic, such as algae, detritus, or fallen vegetation, or inorganic, such as sediment. In most streams, turbidity is primarily a result of suspended sediments, and in fact turbidity is commonly used as a measure of sediment concentration. The amount of material suspended in the water column is highly dependent on the velocity and turbulence of channel waters, and therefore on the amount of water flowing in streams.

Fish, like other animals, require oxygen, and streamflow rates influence the concentration of dissolved oxygen in several ways. Water contains oxygen due to its contact with air. Higher flow rates lead to greater turbulence, thereby bringing water from all depths of the stream channel into more frequent contact with air at the surface. Colder waters have greater capacity to hold oxygen in solution, so the effect of streamflow volume on temperature, as described above, is another way that streamflows affect dissolved oxygen levels.

Streamflow affects dissolved oxygen levels indirectly, via photosynthesis by algae, phytoplankton, and rooted plants within the stream, because streamflow affects the amount and type of habitat available for these biota. Oxygen concentrations are also affected by the amount of decomposing organic materials in the stream, because bacterial decomposition of organic materials requires large amounts of oxygen. The amount of oxygen consumed through decomposition and respiration is often measured and referred to as biochemical oxygen demand, or BOD. The presence and magnitude of photosynthetic and decompositional activities are discussed in more detail below.

Other Aquatic Organisms

Fish are often the focus of instream flow protection efforts, but it is important to realize that fish are but one component of complex aquatic ecosystems. Riverine biota include even bacteria and fungi, the smallest organisms, which decompose organic materials. Decomposition yields basic nutrients that are taken in by algae and phytoplankton, which, using energy from the sun, convert the nutrients into biomass. A wide variety of invertebrates, including insects, zooplankton, worms, mollusks, and crustaceans, feed on the algae and phytoplankton. Invertebrates, in turn, provide the primary source of food for many species of fish, as well as for amphibians and some nonaquatic animals such as birds. The invertebrates have the same general streamflow-related needs as fish—living space, food, cover, temperature, and water quality. All of these organisms are important components of the aquatic environment.

The Importance of Fish in Western Rivers

The existence of fish in western rivers is highly valued by both residents and visitors. As discussed at greater length below in the "Recreation and Aesthetics" section of this chapter, fishing is an enormously popular recreational activity throughout the West. Commercial fishing is also an important economic activity, particularly in Alaska and the other Pacific coast states. Most salmon used for commercial purposes are caught offshore, but their numbers are critically dependent on the maintenance of suitable habitat for spawning, incubation, and initial growth in freshwater streams.

Fisheries also have important cultural and subsistence value. Coastal Indian tribes in particular depend on annual salmon harvests not only as events of great cultural importance, but to supply themselves with food and generate income. Numerous other tribes throughout the western states have acknowledged the longstanding importance of fisheries and fishing to their cultures. Though some contemporary tribes are not tied as strongly to fisheries as was once the case,[12] it would be a mistake to assume that traditional cultural values related to fish have been lost; the rituals, symbolism, and rites associated with fish and fishing still play an important role in many Native American cultures.[13] Fisheries are culturally important to many other Americans as well. It is difficult to separate this aspect of fisheries from other longstanding commercial and recreational uses of fish. The existence of healthy, functioning fisheries would be missed by many westerners who do not themselves fish either recreationally or commercially.

Last but not least, fish and other aquatic organisms are important to the survival of other forms of wildlife. Many species of amphibians, reptiles, birds, and mammals rely on fish and other aquatic organisms for all or parts of their diets and, like people, would be significantly impaired without the presence of these functioning biotic communities. The needs of wildlife are discussed in greater detail below.

Instream Flows and Aquatic Habitat

The existence of suitable aquatic habitat depends on many things besides instream flow. Riparian vegetation, channel structure and form, patterns and quantity of deposited sediment, winter snow and ice accumulation, the presence and concentration of potentially toxic chemicals, nutrient and energy cycles, interactions between fish and invertebrates, competition with and predation by other fish, and predation by birds and mammals, including humans, all affect aquatic organisms. The importance of instream flows is evident even here, however, as all of these "other" factors are themselves affected to varying degrees by instream flow.

The primary flow-related threats to fish in the West today come from diversions, reservoirs, and adverse water-quality conditions. Intensive offstream water use has left many western streambeds dry or nearly so, at least during some parts of the year. Dams of any sort can be major impediments to fish movement and migration, both upstream and down, and are especially problematic for the anadromous fish species, which are in serious decline throughout their range. Migrating fish lose their way in the huge slack water pools of reservoirs, are injured or killed as they pass through electricity-generating turbines, or are entirely blocked from spawning habitat in rivers and tributaries upstream. In many areas acid mine drainage, agricultural and urban runoff, and other sources of pollution have produced water-quality conditions that are injurious or lethal to fish. Of particular importance in many rivers have been excess sediment loads resulting from a variety of land-use practices in surrounding watersheds. Instream flows alone cannot solve all of these problems, but they will be an important part of efforts to improve stream conditions for fish.

Conflicts among Species

One theme of this chapter is the actual and potential conflict not only between instream and offstream uses of water, but also between different instream uses of water. This conflict is well illustrated by the needs of fish. The timing, magnitude, and quality of instream flows necessary for fish vary enormously among species and among life stages of individual species. Flows that meet the needs of one species are not always sufficient for other species and may even be injurious. Attempts to establish, enhance, and protect fisheries using instream flows must therefore be preceded by determinations as to which species are going to be protected and at what level.

Riverine development has completely eradicated some fish species from western streams, while other species have thrived. Depending on the region, 30–60 percent of fish species currently found in the West have been introduced, often for the purpose of improving sport fisheries.[14] The questions of what species we wish to preserve, and at what cost in terms of species lost, will need to be addressed when considering the potential impacts of existing and future river development on instream flows and aquatic ecosystems. Those who attempt to protect fisheries will need to decide how much emphasis should be placed on protecting naturally occurring species and how much on introduced species. They will also need to decide on the principles to be used in guiding such decisions—should the selection of species be based on commercial value? Recreational value? Biodiversity? Preservation of the status quo? These decisions will be especially problematic in light of the inherent difficulty in knowing specifically how the health of particular species depends on changes in flow regimes.

Wildlife and Riparian Areas

Wildlife is heavily dependent on water. Some species, such as beaver, muskrats, and waterfowl, live their entire lives in or around water and cannot exist in the absence of continuous supplies. Amphibians such as frogs and toads spawn in water and live exclusively in aquatic environments during early stages of their life cycles. Other wildlife species require less but still need water to sustain basic life processes.

Rivers play an important role in meeting all of these water needs directly, but they may play an even larger role in meeting wildlife needs indirectly, through the creation and support of wildlife habitat on riverbanks and floodplains. Whether flowing within the channel or beyond it in flood stage, rivers create conditions favorable to the propagation and growth of vegetation by moistening soils, supplying nutrients, and keeping groundwater tables high. These riverside, or riparian,[15] areas provide a diversity and density of vegetation that often far exceeds that of immediately surrounding areas. This effect is often noticeable even in relatively humid regions, but it can be positively dramatic in more arid regions. It is not unusual when traveling through the West to find huge regions of sparse vegetation interrupted by long strands of tall trees and lush, green, dense foliage when approaching one of the West's relatively few watercourses.

Streamflow and Riparian Vegetation

Riparian areas host a variety of trees that would not ordinarily be found in the absence of water. For example, cottonwoods and willows cannot exist without access to water and are found almost exclusively along the banks of natural or humanmade water channels.[16] Species like sycamore, ash, walnut, alder, maple, ironwood, palo verde, and mesquite grow in much greater concentrations in riparian areas than in drier areas removed from watercourses, and species such as blue spruce, Douglas fir, pines, oaks, junipers, and cypress often grow in riparian areas at lower elevations than they would otherwise be able to exist. Riparian areas are also host to a variety of shrubs, grasses, and flowering plants.

Riparian vegetation is more than just diverse. Water availability, thick soils rich in organic matter, a constant influx of nutrients from the river and watershed, and the cooler temperatures and humidity resulting from shade and plant transpiration provide ideal growing conditions. Riparian vegetation is often taller, healthier, and more dense than vegetation in surrounding terrestrial environments.

Streamflow supports riparian vegetation in several ways. First, streams provide a continual supply of nutrients and water to plants, some of which require large amounts of water to survive.[17] Second, water flowing in stream channels helps maintain the groundwater table, keeping water accessible to trees and shrubs located some distance from the stream. Third, the variation in flow level—especially occasional high flood flows—that occurs under natural conditions helps establish the quite specific conditions needed for propagation of new plants and maintenance of healthy vegetation.

Occasional high flows are particularly important for successful propagation. For example, cottonwood and willow seeds regenerate only in moist soil that is relatively clear of other vegetation.[18] Bare, moist soil is created by flood flows that scour away existing bankside vegetation and, through the processes of erosion and deposition, create new sandbars and mudflats. If subsequent flows are low enough to leave these cleared areas undisturbed but high enough to keep the soils moist, then cottonwood and willow seedlings can grow. Otherwise, the new seedlings are either washed away by high water or don't find enough water to survive.

New trees remain vulnerable to being washed out by high flows in subsequent years but, once established, can tolerate occasional periods of inundation or drought. The trees are, however, largely dependent on the maintenance of a water table that does not undergo severe changes of elevation on a frequent basis. The absence of high flows over a long period of time, as often occurs during extended droughts, or on rivers that have been extensively diverted or regulated by dams, will result in an absence of the conditions necessary for the trees to reproduce.

The seasonal timing of high flood flows is important. Because scoured or newly deposited soils are not likely to remain free of other vegetation for long, high flows must occur shortly before the tree seeds are dropped if propagation is to be successful. Just the right combination of circumstances does not happen every year. Several years may pass between periods of successful cottonwood and willow replacement.[19]

Other riparian tree and plant species are also dependent on particular conditions, though not necessarily the same ones that are favorable to cottonwoods and willows. Just as with fish, flows at different magnitudes and times are constantly creating habitat favorable to some species or life stages and destroying habitat favorable to others. Variable streamflows are therefore a major source of dynamism in the riparian environment. Streamflows rework channels and banks, alter soil moisture levels throughout the floodplain, and regulate the age, density, and types of vegetation. The recruitment and maintenance of riparian forests are in large measure dependent on flow regimes, so streamflows are integrally tied to the health of riparian vegetation.

Riparian Areas and Wildlife

Riparian vegetation is in turn the foundation for enormously productive ecosystems. Some of the value of riparian vegetation to fish and other aquatic organisms has already been described—riparian forests create cover by providing shade and contributing large woody debris to the watercourse, contribute food directly in the form of organic matter and indirectly in the form of fallen terrestrial insects, and help to maintain water quality by intercepting sediments and excessive nutrients that otherwise would have ended up in the water. In addition, a variety of terrestrial biota, ranging from microbes to invertebrates, is able to feed on the profusion of organic materials lying on the ground or growing on trees, shrubs, and grasses. These consumers of vegetation in turn become food for an array of other aquatic, amphibian, reptilian, and avian species. Some birds and small mammals eat from the bounty of seeds and fruit produced by riparian vegetation. Larger mammals feed directly on leaves and grasses or find ample prey among the smaller species.

Riparian vegetation provides more than just food. Vegetation density and resultant shade regulate temperatures and climate, provide cover from predators, and supply sites for nesting, burrowing, and perching. The importance of riparian areas as corridors for wildlife movement and migration is hard to overstate. The food and cover available in riparian areas are of critical importance to migrating birds, and the long strands of habitat reaching from mountains to plains and across deserts allow wildlife to move in response to seasonal climate changes, food availability, and pressure from predators.[20]

Riparian areas support an impressive variety of wildlife. Some species prefer the treetop canopy, some the mid-height trees and shrubs, and some the grasses and leaf litter at the forest floor. Some species are seldom found more than a few feet from water, some prefer the still lush but somewhat drier conditions farther from the river channel, and yet others prefer the far edges where riparian conditions merge into the more arid conditions of the surrounding terrestrial environment. Examples of wildlife species that are dependent on the more aquatic portions of the habitat include frogs, turtles, beaver, muskrat, mink, river otter, moose, and birds such as bald eagles, osprey, waterfowl, herons, egrets, dippers, and kingfishers. Wildlife species that are not as directly dependent on water but are found in much heavier concentrations in riparian areas because of the abundance of food, water, and cover include bats, toads, lizards, porcupines, kangaroo and cotton rats, rabbits, raccoons, coatis, skunks, bobcats, grey and red foxes, coyotes, deer, and elk. Hawks, owls, and a variety of other birds including flycatchers, warblers, wrens, tanagers, sparrows, and swallows are also usually found in much greater concentrations in riparian areas than elsewhere.

Threats to Riparian Areas and the Importance of Instream Flows

Riparian vegetation affects more than just aquatic and riparian biota. Vegetation that extends into the stream channel or grows on the floodplain creates obstacles that reduce streamflow velocities, thereby diminishing the erosiveness of flows, particularly of high overbank flows. Lower stream velocities and the deep, uncompacted soils found in healthy riparian areas help to reduce flood peaks by facilitating the infiltration of water into the ground during high flows and slowly releasing water back into the channel as flows subside. Reduced streamflow velocity lowers the capacity of a stream to carry sediment, so that riparian zones become areas of sediment deposition. Streambank stability is enhanced by extensive root systems, which help to hold soils together, thereby reducing problems such as increased width, increased bank angles, and the elimination of undercut banks (undercut banks provide cover for fish).

Despite their great importance, riparian habitats are in decline throughout much of the West.[21] Channel lining, bank stabilization, and other river channel "improvements" undertaken for flood control or other purposes have eliminated riparian vegetation from the edges of formerly natural watercourses. In some areas riparian vegetation—which in the aggregate can consume large quantities of water—has been removed to reduce competition for scarce water supplies. Riparian areas have been logged for timber and cleared to expand farmland and create sites for new houses and other development. Livestock, and in some areas wildlife, has put extra grazing pressure on the rich vegetation found in riparian areas, consuming tree seedlings and preventing the recruitment of new trees. Grazing may also damage riparian areas by trampling vegetation, compacting soils, and eroding streambanks.[22]

Water diversions have severely depleted flows to many downstream riparian areas. Reservoirs have inundated thousands of miles of riparian habitat. Dams built for storage and flood control have eliminated natural flow regimes downstream, particularly the large flood flows necessary for the natural recruitment of new vegetation. Groundwater pumping has resulted in the eradication or shrinkage of many riparian areas, either by lowering groundwater tables in the immediate vicinity or by intercepting subsurface water flows that otherwise would have ended up in stream channels. In the short run, reductions in streamflow expose and dry out amphibian eggs, create open spaces between the water source and the channel edge where wildlife is more vulnerable to predators, expose birds and animals feeding and resting in the stream channel to predators because of diminished water depth, and lead to the loss of protection for the nests of waterfowl and other ground-nesters. In the long run, reduced flows may also lead to a loss of vegetation, thereby depriving wildlife of food, nesting sites, and thermal and protective cover.

The protection of instream flows will be a key factor in the maintenance or restoration of remaining riparian areas. The quantity, variety, and distribution of species, both plant and animal, originally found in riparian habitats were those best adapted to existing streamflow conditions. Any changes from those conditions, whether in magnitude or timing of flows, will have some impact on the habitat; the greater the deviation, the greater the impact. Minimal flows will be necessary to enable species that spend much of their lives in the water to survive and to provide drinking water for other species. Higher flows will be necessary to sustain vegetation and wildlife on a longer-term basis. A variety of flows, particularly the occurrence of occasional high flood flows, will be necessary to create dynamism in the riparian habitat and to establish conditions necessary to the propagation of many tree and plant species.

Recreation and Aesthetics

Not the least of the species using rivers and riparian areas are humans. People have been drawn to rivers for many of the same reasons as other animals, including the abundance of food, water, and shelter found in riverine environments. But people have also been drawn to rivers for recreation and renewal, activities that have received growing attention since the middle of this century, when the rapidly expanding national economy dramatically increased both household incomes and the availability of leisure time. The

West's expanding population has not only swelled the quantity of water diverted for offstream uses; it has also increased the demand for stream-based recreation.[23]

Recreation plays an important role in maintaining physical and emotional health, and in enhancing the quality of life. Outdoor recreation plays an especially important role in the lives of the West's growing urban and suburban population, as many city dwellers crave temporary escape from their urban surroundings. Recreation also generates significant economic activity, to the point of being the leading source of income in many communities.

Some forms of river-based recreation are entirely dependent on streamflow. Others are enhanced by the presence of flowing water. The different forms of riverine recreation, and their relationship to streamflow, are discussed below.

Streamflow-Dependent Recreation

Boating and Floating. Boating is among the most popular of riverine activities. Powerboats, sailboats, jet boats, jet skis, canoes, kayaks, dories, rafts, tubes, and a variety of other engine-, wind-, oar-, self-, and river-powered buoyant devices are used to float rivers in the name of fun and recreation. Beyond enabling the existence of such activities, streamflow affects the quality of a recreational boating experience through its effects on boatability, whitewater, and rate of travel.[24]

Boatability—the capacity of the river to support boating free of excessive hits, stops, drags, and portages—is an important determinant of the quality of a boating experience. Running into rocks, river bottoms, trees, aquatic vegetation, or other obstacles has the potential to damage vessels and injure passengers. Delays and effort expended in prying boats off of obstacles, stopping to disembark, portaging, or taking other evasive action can likewise decrease the quality of the recreational experience. Boatability is maximized by streamflow levels that create clear channels. Clear channels are produced by flows that provide water depth sufficient to clear the river bottom and other obstacles and enough energy to remove obstacles such as downed trees, sandbars, and vegetation.

Whitewater affects the level of thrills, exhilaration, excitement, risk, and danger offered by a river boat trip. Whitewater is caused by a combination of streamflow velocity, channel form, and channel obstacles. Both streamflow velocity and stream channel form—including the presence or absence of various obstacles—are influenced by streamflow volumes, though the relationship is not always a simple one and varies considerably among individual rivers and river reaches. Larger streamflow volumes yield higher velocities and are capable of washing more debris into the river, which increases the potential for whitewater, but can also increase water depth, thus covering many obstacles. Lower volumes may expose more obstacles but at the same time reduce the power and velocity of the flow. Whether increased whitewater is "good" or "bad" depends entirely on the preferences of those who wish to boat the river, and preferences vary enormously between users of different types of crafts, between boaters with different levels of skill and experience, between individuals, and within the same individual at different times. Some boaters look forward to the challenge of whitewater, some prefer a more relaxing float, and some prefer a bit of both.

The *rate of travel* on a given reach of river is an important consideration for many boaters, especially on rivers with a limited number of access and exit points, or for boaters who are embarking on full-day or multi-day trips. Rate of travel is influenced by streamflow velocity, as well as by the boatability and whitewater characteristics discussed above. Larger streamflow volumes yield higher streamflow velocities, so—everything else being equal—an increase in streamflow will be associated with a decrease in travel time.

Water Contact Recreation. Virtually everyone has participated in water contact recreation. Not everyone has been water skiing or scuba diving, but most people have enjoyed swimming, wading, or just playing in the water. Streamflow volume affects the quality of these recreational experiences because of its influence on water depth, velocity, and temperature.

The most "desirable" *depth* depends on the particular activity and level of skill. Swimmers need enough depth to avoid touching the bottom; streamflow just sufficient to fill river pools and create "swimming holes" is ideal. Diving requires substantially greater depths. Water skiers need enough water to safely cover rocks and other obstacles. Rivers can't be deep enough to suit some scuba divers, whereas those who are wading or playing in the water usually prefer, and are safer in, shallower water. Safe, easy access to and exit from the water require water levels that are not too far below the top of the bank.

Streamflow *velocity* must be low enough to prevent recreationists from being swept downstream or into obstacles such as overhanging trees and submerged or protruding rocks. Lower velocities are particularly appropriate for children and those who have less skill or experience. But velocity should be high enough to keep water quality fresh and to provide beneficial challenges for those who wish to test their skills.

Fishing. Healthy fish populations are a prerequisite for a high-quality fishing experience, and good conditions for fish often mean good conditions for fishing. Both the variety and the abundance of fish species are important to the recreationist and depend on such streamflow-dependent variables as temperature, cover, food supplies, and turbidity, as described above.

Other flow-dependent characteristics also affect fishing quality. Flows of wadable depths and velocities allow fishing from within the stream channel.

Occasional high flows help keep riverbanks free of vegetation, thus providing ease of movement up and down the river and adequate space for casting. Sustained high flows adversely affect fishing quality, however, by increasing turbidity. Though fish activity tends to increase with higher flows, activity tapers off at very high flows.[25] Safety is threatened if flow levels rise too quickly, as recreationists could be left stranded or faced with life-threatening situations.

The most desirable streamflow regime will depend on the type of fishing that is done. Ideal conditions for bait fishing, spin casting, and fly fishing may vary substantially, and a variety of flows may be necessary in the long run to create the most desirable mix of fishing conditions.

Streamflow-Enhanced Recreation

Flowing water enriches many activities, including hiking, camping, walking, biking, picnicking, observing birds and wildlife, sightseeing, nature study, collecting rocks, photography, and just sitting or relaxing. All of these activities benefit from the aesthetic quality of flowing water, an effect that is discussed in greater depth below. But streamflow enhances recreation in other ways. For example, hikers, walkers, bird watchers, and others who move along the river all benefit from vegetation-scouring high flows that create open banks for traveling. The clearing out of vegetation affords bird and wildlife watchers more unobstructed views and enables those who are boating the river to gain access to off-river hiking areas. Flows that create gravel and sandbars in or next to the river give boaters and others a place to camp. These river-scoured, vegetation-free areas are often the only flat areas available for pitching tents and tend to be well drained and relatively free of insects.[26]

Watching birds and wildlife—two of the most popular outdoor recreation activities—is more productive around streamflow-dependent riparian areas than in most other kinds of habitat. The best streamflows for these activities are often those that create the best habitat—flows that lead to the greatest abundance and variety of species create the best opportunities for seeing wildlife. Streamflows that are conducive to the creation and maintenance of wildlife habitat also create good conditions for hunters, especially with respect to the support of habitat suitable for waterfowl.

Aesthetics

Nearly everyone enjoys the sight and sound of a free-flowing stream. A running stream can contribute to the pleasure, excitement, enjoyment, and relaxation experienced by participants in all of the recreation activities listed above. For example, the beauty of the surrounding environment was found to be the single most important factor in providing a good experience for those fishing in California rivers.[27] And not only recreationists enjoy flowing water; anyone who lives or works by, or happens to travel through or visit, a riverine area may be touched by its beauty. This preference for streamflow extends to cities, as evidenced by the profusion of parks, walkways, museums, nature centers, and other developments that have become popular features of the urban streamside environment.

Running water expresses a sense of life and vitality. This energy may vary from the vigor of rapids and waterfalls to the more steady and soothing movement of the ordinary stretch of river.[28] Within this picture of movement, attention is also drawn to areas of stillness, and the contrast between pools and riffles or between currents and eddies is visually pleasing.[29] The water surface creates different textures that people find attractive; it sparkles, reflects images, and ripples with the wind.[30] From farther away, rivers and riparian areas are important components of scenic vistas, providing elements of unity, vividness, and variety,[31] drawing the viewer in and affording a sense of mystery, involvement, and adventure.[32]

The sound of water falling over rocks, gurgling through streamside vegetation, or crashing into the banks, the calls of birds and wildlife, the splash of a fish feeding at the water surface, the smell of trees, flowers, plants, and moist soil all contribute to a pleasing environment. Often it is the mixture of several senses that creates a pleasurable experience. Sight, sound, and the feel of water spray on the skin are all important in creating feelings of power, beauty, and awe at large waterfalls and rapids.

The entire riparian environment contributes to the aesthetic experience. One recreationist described the importance of such aesthetics this way:

> Most of us who gravitate to rivers—whether boaters, anglers, birdwatchers, or just people who like to mess around on rivers—seek something more subtle than simply lots of big fish, good whitewater, or a new notch for our life list. In my frame of reference, that includes things like some sense of escape, a sense of well-being, and some feeling of isolation from the everyday things that bind me to the civilized world. Intrinsic in those sensations is the riparian zone. For me, that band of water-dependent vegetation, with its accompanying wildlife, diversity, and lushness, is the protective barrier that sets me up to achieve those more ethereal enjoyments. Without that enveloping membrane, I might as well simply float down a gutter or an irrigation ditch.[33]

These aesthetic qualities of the riverine environment provide equal enjoyment and tranquility for homeowners, farmers, ranchers, and others who live or work near rivers.

Aesthetic quality varies with the level of streamflow. Several researchers have observed that aesthetic values of most river stretches tend to be maximized at moderate flow levels.[34] Very high flows can drown out the contrasts between riffles and pools, mask apparent differences of velocity with the impression of a single kind of movement, and make islands and point bars disappear.[35] High flows can be turbid and may frighten people with their power or create an unwelcome sense that events are out of control. Very low flows may reduce aesthetic quality by limiting or eliminating white-water; causing an acute loss of visual and aural qualities at waterfalls, out-crops, and boulder dams; suggesting a feeling of abandonment because of the river features left stranded out of the water; decreasing the vividness of the contrast between pools and riffles; flowing at a level too low to be seen in river channels composed of massive rocks or boulders; decreasing river width, thereby leaving river bars and bed material rather than the river itself as the dominant visual impression; and causing islands and central bars to lose their identities and become mere extensions of the shore.[36] Sudden declines from higher flows to lower flows may reduce aesthetic quality, as people tend to find the resulting "bathtub rings" unsightly.

Aesthetic quality is dependent on both short-term and long-term flow regimes. There needs to be enough water present at the time of a visit to produce the features detailed above, but there also needs to be enough water flowing through the channel over longer periods of time to create the channel forms and maintain the riparian vegetation that play such large roles in the aesthetics of the river environment.

Economic Significance of River Recreation

The economic significance of an activity (recreation or any other) can be gauged by two quite different measures: the activity's economic *impact* and its economic *value*. Economic impact refers to the sales, income, and employment that result from participation in an activity. Economic value refers to the value recreationists receive from the activity—sometimes referred to as "willingness to pay"—over and above the recreationist's actual expenses.[37] In other words, economic impact focuses on the recreationist's expenditures, and economic value on the net benefit the recreationist receives. Measures of economic impact are typically of most concern to local businesses and to city or county governments, whereas measures of economic value are of most interest to public resource managers. Of course, both are of interest to the recreationist.

The total value of recreation—measured in terms of both economic impact and economic value—depends on participation rates. Participation rates for stream-based recreation are substantial. The 1991 National Survey of Fishing, Hunting, and Wildlife-Associated Recreation reports that, among U.S. residents age 16 and over, 13.7 million anglers fished a total of 126 million days in rivers and streams.[38] Participation rates ranged from 15 to 27 percent among residents of the western states.[39] Rates are similar for participation in nonconsumptive wildlife activities, many of which focus on riparian areas. For example, 17 percent of Pacific region residents age 16 and over traveled at least a mile from home for the primary purpose of a non-consumptive wildlife activity in 1991, as did 22 percent of Mountain state residents, 20 percent of West North Central state residents, and 12 percent of West South Central state residents.

Recreationists spend money on many kinds of goods and services. Trip-related expenditures include food and beverages, lodging, and transportation. Equipment expenditures include the purchase or rental of boats and other watercraft, rods, reels, tackle boxes, depth finders, artificial lures and flies, binoculars, cameras and lenses, film, field guides, special clothing, tents, and backpacking equipment. Other common expenditures include guide fees, land use rentals, magazines, memberships, contributions, licenses, stamps, tags, and permits. The total number of dollars spent on travel, equipment, and other items is substantial. Total expenditures for non-consumptive wildlife activities in the United States in 1991 were $18.1 billion, and anglers spent an additional $11.8 billion. Much of this spending is dispersed throughout the national economy, but many expenditures are concentrated in areas that offer good recreational opportunities and have a substantial impact on local economies.[40]

Economic value measures are used to help decide how best to spend scarce public resource management funds; thus, they are often used to compare two management alternatives. For example, economists measured the benefit received by participants under two different streamflow levels and found that a significant potential loss of visitor benefits would occur if streamflows through the riparian area next to Arizona's Hassayampa River were reduced to the point that flows became intermittent rather than perennial.[41] Another example of this approach was provided by a 1990 study that compared the marginal value of water left in streams to enhance downstream fisheries with the marginal value of water used off-stream for irrigation.[42] The study authors looked at all 99 major river basins of the contiguous United States and concluded that instream water values are highest in populated, arid areas, and that the *marginal* value of water for recreational fishing exceeds water's marginal value in irrigation in 52 of the 67 watersheds in which irrigation takes place.[43] These conclusions are especially interesting relative to the oft-heard comment that water is too scarce and valuable in the arid West to be left in stream channels; this study indicated that instream uses too are more valuable in arid areas.[44]

Environmental Protection

Recreation is not the only motivation behind people's interest in maintaining streamflows. Many people desire to protect the natural environment for its own sake, or for the enjoyment of others. There is little doubt that much of the support for protecting instream flows in the West comes from people who have never visited, and may not expect to visit, the rivers that are being protected. Both users and nonusers of the rivers seem to value maintaining rivers in a healthy, functioning condition and knowing they will be left in that condition for succeeding generations.

The environment and environmental protection activities have experienced a surge in popularity in the last few decades. The National Water Commission noted in 1973, "The people of the United States give far greater weight to environmental and aesthetic values than they did when the nation was young and less settled."[45] Preservation of the natural environment has since proven to be one of the most enduring goals of contemporary society.

Our society's renewed environmental consciousness has been inspired in part by advances in our understanding of environmental processes. We know more about the environment and how it works than we used to—about how seemingly independent pieces of the environment are actually related, and about how human activities affect the environment. Our renewed consciousness of the environment is probably also a result of an advanced economy that has provided us with relatively inexpensive food and material goods as well as more leisure time, and consequently greater opportunity to concern ourselves with, and enjoy, natural settings. Even those who don't consider themselves "environmentalists" often value maintaining more natural environmental conditions. Rivers and streams are an integral part of the environment and of environmental processes, tied inextricably to climate, geology, and biota, both plant and animal. When people express a desire to protect the environment, healthy rivers and streams are clearly a part of what they want to protect.

Existence and Bequest Values

The desire to protect the environment for its own sake or for enjoyment by others—a desire that exists apart from active personal use and may exist even if the person has never visited the particular area—has been studied by economists, who have coined the terms "existence value" and "bequest value" to capture at least part of this "nonuse" value.[46] Simply stated, existence value is the value a person derives merely from knowing that a certain amenity or state of affairs exists, quite apart from any personal use of that amenity that the person has previously made or intends to make in the future. Bequest value is tied closely to the ideas of sustainability and stewardship; it is the value that a person derives from knowing that an amenity will be passed intact to future generations for their enjoyment, use, and pleasure. Estimates of these values in economic terms—while admittedly rough—indicate that nonuse values are often substantial, comprising the principal source of value of efforts to preserve the environment.[47] Thus the primary value of keeping water in rivers may come not from the direct uses that are subsequently made of those instream flows, but from the knowledge that such live and healthy rivers exist and have been preserved for future generations.

Natural versus Regulated Flows

Preservation of the "natural" environment is a good point at which to introduce a discussion of natural and regulated flows. Natural flow conditions—the magnitude and timing of flows resulting from hydrologic and geologic conditions in the absence of development—have been replaced in many western rivers and streams by regulated flows, flows determined to some extent by the management of upstream dams and reservoirs. It is important to realize that, for some instream uses, it is not always clear that natural flows are superior to regulated flows or vice versa. Preservation of native riparian and aqueous species and the protection of ecosystems are most easily accomplished with natural flows. As mentioned earlier, the quantity, variety, and distribution of species originally found in riparian and aqueous habitats were those that were best adapted to existing conditions, including the magnitude and timing of streamflows, so leaving streamflows unaltered is probably the best way to protect existing biota. However, regulated flows may be more favorable to other species. For example, several of the West's most highly prized trout fisheries occur in the tailwaters below large dams.[48] Some types of recreation can also be enhanced by regulated flows, as natural flows may be either too low, too high, or too variable to provide optimum, safe conditions.[49] Whether a natural or regulated flow is best depends on the particular conditions and the particular goals established for the river or river segment in question.

Dams and reservoirs serve a variety of useful purposes and are a fact of life. On most rivers a return of truly "natural" flows is not feasible. But this does not mean that instream uses must be totally frustrated, even for those instream uses harmed by dams. Flow releases from dams have not always been designed with particular instream goals in mind, but they could be; regulated flows can be altered to provide for certain instream uses just as they have been used to provide for offstream consumptive uses. Where dams and reservoirs exist, we need to ask what uses are possible given existing conditions and knowledge, what are the trade-offs, and what uses should have priority. Regulated flows can be, and have been, used to sustain a variety of "natural" uses, including habitat for native aquatic and riparian species, in addition to other instream and offstream uses.

Some of the support for instream flow protection has been aimed at preserving natural hydrologic and ecologic conditions, and there is no doubt that these goals will continue to be an important part of the instream flow protection movement. But it is important to realize that for many—perhaps most—rivers, the protection of instream uses and values will be accomplished through the use of regulated flows from dams and reservoirs. Both natural and regulated flows have important roles to play in the protection of instream values.

Hydropower

The production of electricity using hydropower may be either an instream or out-of-stream use.[50] When production facilities are located within the stream channel, there is no need to divert water out of the river and hydropower is an instream use. When hydroelectric production facilities are located outside the stream channel, water must be diverted to the facility and streamflow within the channel is correspondingly reduced over at least some length of the river. In these cases hydropower is an offstream water use.[51] But even when generated instream, some do not consider hydropower an instream use of water, probably because many people view hydropower facilities and their operations as direct threats to other instream uses and values, and therefore as something different. Nonetheless, hydropower generation is a common, nonconsumptive, and economically significant use of streamflow in many western rivers.

Hydroelectric and Thermoelectric Water Uses

Virtually all electrical power generation facilities are dependent on the use of water. Water is obviously essential to hydropower, but thermoelectric plants also use huge quantities of water. In fact, more water is withdrawn in the U.S. for use by thermoelectric plants than for any other single use of water, including irrigation.[52] Most of this water is withdrawn to cool condensers and reactors.[53] Because only 2 percent of the water is consumed (evaporates), 98 percent returns to the surface water source from which it was withdrawn.[54] Even though the use of water by thermoelectric power plants is largely nonconsumptive, thermoelectric power generation is an offstream water use, as the power plants are located on land and require the diversion and transport of water from rivers and lakes to intended sites of use. By contrast, hydroelectric production facilities use the force of moving water to generate electricity. Water is thus the "fuel" of the hydroelectric plant, and the total quantity of that fuel is constrained by hydrologic conditions.

There are at least three different types of hydropower production facilities: storage, run-of-the-river, and pumped-storage. A storage facility uses a dam in the river channel to raise the level of water on the upstream side of the dam higher than that existing on the downstream side. "Penstocks," or pipes, are then used to direct the water falling from the upstream reservoir into the turbine blades, which are housed, along with the generators, in a power plant located at the foot of the dam. After being run through the turbines, the water is discharged from the plant into the river channel, from whence it flows downstream. If the dam is large enough to create a substantial reservoir of water, the timing of electricity production can be controlled by when water is released into the penstocks.

Hydropower plants without substantial reservoirs are run-of-the-river facilities. Without significant storage capacity with which to regulate flows, the timing of the electricity produced depends almost entirely on the timing of flows in the river.[55] Pumped-storage facilities reuse water, pumping it into an upper storage reservoir when demand for electricity is low and then allowing the water to flow back down through the power plant to generate electricity when power demand is high.[56]

Storage plant facilities are virtually always located within stream channels and constitute an instream water use. Pumped-storage plants are virtually always located outside the river channel; as offstream water users, pumped-storage facilities will not be considered further in this chapter.

The location of run-of-the-river facilities—despite the implication of the name—varies. Many run-of-the-river facilities are located within stream channels and are instream water users. But other run-of-the-river facilities are located offstream, whether due to the infeasibility of locating a power facility within the stream channel or due to the possibility of taking advantage of a greater drop in elevation elsewhere. Water is diverted out of stream channels and transported to some other location from which it can be dropped through penstocks to a power plant. The power plant for such a "diversion canal" facility may be located next to the same river channel at a downstream location, so that the water can be returned directly to the channel after it has been run through the turbines. Or, the plant may be located some distance away, so that the water must be transported through canals before being returned to the river channel. The power plant may even be located in a separate watershed, so that water is permanently removed from the original channel. Whether located just offstream or at some distance away, diversion canal facilities are properly considered to be offstream water users. Like other offstream uses, diversion canal hydropower plants reduce, or—depending on the magnitude of the diversion—completely eliminate flows through at least some stretch of the original channel.

The Significance of Hydropower to the West

Hydropower is used to generate less than 9 percent of all electricity produced in the United States. Hydropower is much more significant in the West, however, where 66 percent of the nation's hydropower is generated. Hydropower accounts for 18 percent of all electricity generated in the 18 western states, a total of some 158,152 million kilowatt hours (kWh) in 1992. The proportion of electric power generated by hydropower in each state varies widely, as shown in table 3.1.[57] Washington, Oregon, and California, in that order, are the largest hydropower producers, together accounting for virtually half (49.5 percent) of all the hydropower generated in the United States.

Hydropower has several advantages. The equipment used to generate electricity with hydropower is relatively simple, and the force of falling water has been harnessed to produce electricity since 1872, so the technology is highly developed. These factors make hydropower very dependable.[58] Hydropower is a renewable energy source, is not dependent on the fluctuating prices and availability of fossil fuels, is nonpolluting, is relatively cheap to operate because of low labor and maintenance requirements, and has low outage rates. Hydropower facilities also have long life expectancies.[59] The sale of electricity generated by hydropower accounted for approximately $10 billion in revenues for electric utilities in the 18 western states in 1992.[60] Some 1,725 million acre-feet of water passed through the turbines of hydropower facilities in the western states in 1990, the equivalent of 1,539 billion gallons of water every day of the year.[61] Not all of this was "new" water, be-

TABLE 3.1
Generation of Electricity in the Western States, 1992

State	Electricity generated by hydroelectric facilities (million kWh)	Electricity generated by all sources (million kWh)	Percent of all electricity from hydroelectric facilities
Alaska	918	4,167	22.0
Arizona	6,911	70,109	9.9
California	19,205	119,310	16.1
Colorado	1,505	31,899	4.7
Idaho	6,260	6,260	100.0
Kansas	0	31,764	0.0
Montana	8,223	25,468	32.3
Nebraska	1,075	22,387	4.8
Nevada	1,982	20,963	9.5
New Mexico	255	27,708	0.9
North Dakota	1,699	28,592	5.9
Oklahoma	3,210	45,943	7.0
Oregon	31,476	41,220	76.4
South Dakota	3,612	6,246	57.8
Texas	2,638	239,964	1.1
Utah	580	32,921	1.8
Washington	67,967	84,115	80.8
Wyoming	636	41,852	1.5
TOTAL—WEST	158,152	880,888	18.0
TOTAL—U.S.	239,559	2,797,219	8.6

cause the nonconsumptive nature of the water use in hydroelectric generation allowed the same unit of water to be used several times over.[62]

Hydropower and Instream Flows

The relationship between hydropower and streamflow starts with a simple fact: The amount of electricity produced by a hydroelectric facility is directly tied to the amount of water that passes through the plant's turbines. Specifically, the quantity of electricity produced is equivalent to the quantity of water passing through the turbines multiplied by the "head"—the difference in elevation between the power plant turbines and the water surface behind the dam—and by a coefficient representing the efficiency of the turbine-generator unit in converting water force to electrical power.[63] This is true for both storage plants and run-of-the-river facilities, the primary difference being that water flowing into the reservoir of a storage plant may be used at some later time, whereas water running through a run-of-the-river plant must be used immediately or not at all. So, all other factors being equal, greater flow will produce more electricity.[64] To produce more electricity, either more water must pass through the turbines, or the water that passes through the turbines must do so at greater force, as a result of an increased difference in elevation between the upstream water surface and the power plant turbine blades.[65]

The relationship between hydropower generation and the timing and magnitude of flows downstream from hydropower facilities is what leads many to view hydropower as a threat to other instream uses. Instream run-of-the-river facilities by their very nature do not substantially alter the timing or volume of water flows.[66] The same cannot be said of storage facilities, which are designed to replace existing flow regimes with regimes more closely tied to the needs of electric power consumers (and perhaps other water users). Tying flow releases to the needs of electric power consumers leads to extremely variable downstream flows, because electric power demand varies widely through the course of the day, week, and year. Unlike many other goods, electricity is consumed instantaneously, at virtually the same time that it is produced.

Power demands are lowest in the middle of the night when most people are sleeping, and highest during working hours when commercial and industrial establishments are in operation and people are using electricity at home. Power demand is higher on weekdays than on weekends and changes with the seasons. In colder regions, power demand is highest during the winter months when electricity is used for heating. In warmer regions, power demand is highest during the summer months when electricity is used for air conditioning and other forms of cooling.

If changing power demands were to be met solely by hydropower facilities, flows would be released in strict accordance with fluctuations in power demand. In reality, individual hydropower facilities are connected to other power generators by an extensive electrical transmission grid, and power demand is met through the conjunctive, integrated use of multiple production facilities. Generating facilities are independently owned and operated, but in many ways constitute a single integrated system.

Within this system hydropower fills some special needs. Individual production units are used to meet different types of power demand, referred to as "base load," "intermediate load," and "peak load." Base load is the minimum amount of electricity that must be continuously available in the system. Peak load is the amount of electricity that must be available to meet the highest daily, weekly, and yearly demand. Intermediate load, as the name implies, is the level of demand that occurs more often than peak but less often than base, and refers to the amount of electricity that must be supplied at levels falling between the two extremes.

Base load demand is usually produced by large thermoelectric facilities, which operate most efficiently and economically at relatively constant levels of output. Base loads may also be met by other facilities, including hydroelectric generators, so long as the level of power output may be sustained indefinitely. For run-of-the-river facilities, this output is the amount of power produced from the minimum foreseeable level of streamflow through the facility.[67] For storage plants—which can store high flows for use during otherwise low-flow periods—the sustainable level is that which can be produced using the lowest *average* level of flow through the generating facility.

Peak demands are met by generating facilities that have the capability to come on-line very quickly, in response to sudden increases in demand for electricity. Gas-turbine and pumped-storage hydroelectric plants are often used for this purpose, but so too are other kinds of hydroelectric facilities. A major advantage of hydroelectric plants is that their generators can be stopped or started in a matter of minutes, or even less time. Most thermoelectric facilities take much longer and do not produce electricity as efficiently under fluctuating loads. Because of this responsiveness, hydroelectric storage plants are ideal for the production of peak power needs. For the same reason, hydroelectric facilities are often used to meet intermediate load requirements. Another advantage of hydroelectric facilities in meeting intermediate load demands is that hydropower facilities are able to maintain high efficiencies over a wide range of partial load conditions.[68]

Some electrical generating capacity needs to be kept on reserve in order to meet unexpected needs, such as the sudden failure of some other generating unit or a sudden surge in electrical demand. Typically, the generating capacity that is kept on standby represents approximately 15 percent of total generating capacity. Hydropower storage facilities, again because of their responsiveness, are often used for this purpose.[69]

Peaking power cannot be supplied by every generating facility and so commands a premium in the power market. Hydropower operators therefore prefer to use their facilities for this purpose.[70] As a result, streamflows beneath hydropower storage plants are often diminished during periods of off-peak electric demand and then surge to potentially very high levels during periods of peak demand.[71] Rather than responding to changes in natural events such as precipitation, the timing and magnitude of downstream flows reflect fluctuations in demand for electricity.

Rapidly fluctuating flows may not be compatible with other instream uses of water. For example, fluctuating flows can negatively affect fish and wildlife by alternately exposing and submerging habitat, including incubation areas. Rapidly fluctuating flows may also produce hazardous conditions for anglers, boaters, and other recreationists and have adverse effects on aesthetics. Accommodation of other instream uses is possible, but, given that existing operating procedures are designed to maximize production efficiency and net revenues, it may come at the price of diminished energy efficiency and utility revenues. The effects of accommodation would extend well beyond the immediate river and generating facility, for, barring a concurrent reduction in power demand, local reductions in power production would have to be made up by increased production at some other facility. Efforts to alter hydropower production in order to meet other instream needs on the Columbia and Colorado Rivers, and the associated costs, are described in later chapters.

Power pricing has the potential to increase minimum streamflow levels. Power customers value supply reliability and offer higher prices for "firm" power than for power supplies that are only intermittently available. Unreliable supplies—such as those produced using streamflows greater than minimum historical levels—cannot be sold as firm and thus bring lower prices.[72] This gives hydropower facility operators the incentive to increase the minimum level of flows moving through hydroelectric facilities. This might be accomplished by purchasing the water rights of those using water for consumptive purposes upstream, protesting new upstream diversions, or applying for the legal right to keep higher minimum levels of streamflow in the river.[73] To the extent that existing streamflows have already been diminished, the reintroduction of water into the stream channel for hydropower purposes might coincidentally result in the furtherance of more "natural" conditions. Additional water may also be of benefit to other instream uses such as fish, wildlife, and recreation. This, of course, will depend in large part on the type of hydropower facility through which the additional water will pass and the reservoir operating rules in place at those facilities.[74]

In summary, the relationship between hydropower and instream flows is very complex. Depending on the type of facility, hydroelectric generation can be an instream water use. Like other instream uses, hydroelectric generation could benefit from efforts to enhance instream flows. Also like other instream uses, hydropower has the potential to be in conflict with other water uses taking place within the stream channel—and often is, to a degree exceeding that typically found in conflicts among other types of instream uses. Hydropower generation is economically a highly significant use of water with ramifications that extend well beyond the actual stream channel, and efforts to change operating procedures to enhance instream values must take into account the impacts that will occur in other areas.

Navigation

Navigation was one of the earliest instream uses of water. Today, most commercial navigation occurs along the coasts, but some rivers, referred to as inland waterways, continue to be used for this purpose. Although a number of rivers scattered throughout the West are still occasionally used for commercial navigation, navigation is a major use on only a few of the largest western rivers.

Commercial Navigation

Riverine transport became a primary method of transporting goods during the 19th century. Use of rivers as public highways was so important to the nation that the instream right of navigation was protected and held in trust by the federal government for the use of all Americans. The use of rivers for moving goods declined enormously with the advent of railroads and, later, highways, and riverine navigation no longer plays such a prominent role in the nation's economy. It is still important, however, in some localities.

Commercial navigation includes the movement of goods or people in vessels and also the floating downstream of wood products. Floating of fuelwood, ties, and timbers was once common on some western rivers, but this practice largely ended by the time of the Second World War. Passenger transport is still common on some rivers, though primarily as an adjunct of commercial transport, or as a tourist operation that might better be categorized as recreation. Thus, commercial navigation today primarily refers to the transport of bulk goods. The vessels used to transport goods on the inland waterways are almost exclusively barges and tow boats.

The primary western rivers now used for commercial navigation are listed in table 3.2.[75] There are other rivers in the West used for commercial navigation but only on a much smaller scale.

Instream Flows and Commercial Shipping

The basic flow-related needs of navigation are relatively simple: There must be enough water to fill the channel to sufficient width and depth to accommodate the passage and maneuvering of commercial vessels; current velocity must not be so high as to impede travel in the upstream direction; and the channel must be kept free of obstacles. Low flows decrease depth and

TABLE 3.2
The West's Primary Navigable Rivers

River	Navigable length	Goods transported in 1986 (millions of tons)
Columbia/Snake	470 miles	42.2 Columbia 5.0 Snake
Missouri	734 miles	7.0[a]
San Joaquin	125 miles	2.4
Willamette	118 miles	3.7
Sacramento	248 miles	1.2

[a]Over 6.5 million tons of this total was transported between Kansas City and St. Louis, a reach of the Missouri that does not border any of the western states.

width and expose obstacles. High flows not only create problems in running against the current, but also are likely to cause undesirable changes to channel morphology, such as shifting channels and the creation or movement of sandbars. High flows may also wash new obstacles, such as trees, into the river. Very high or very low flows may make it difficult or impossible for vessels to dock.

A characteristic peculiar to shipping is the need for relatively constant flow levels. For example, flows in the Missouri River must be maintained between eight and nine feet of depth throughout the navigation season, which, under current management plans, extends from April 1 to December 1 in a normal year. Flows may be varied outside that range only during the remaining months of the year, or in response to flooding or drought conditions beyond the control of upstream reservoirs. The shipping season may be shortened, or temporarily suspended, in the absence of sufficient water supplies to maintain necessary conditions in the navigable portion of the river.

Most of the West's navigable rivers have been developed to resemble canals or a long series of slackwater pools. For example, the Columbia and Snake rivers between Bonneville Dam and Lewiston, Idaho, are actually just a series of slackwater pools formed by Bonneville and seven other large dams. Each reservoir extends to the foot of the next dam. Only the 145-mile stretch of the Columbia from Bonneville Dam to the river's mouth is free flowing. Barge traffic moves between the pools using locks at the dams. The locks provide a total lift of 730 feet from Bonneville Dam to Lewiston. As on most other rivers developed for navigation, the locks were constructed, and are currently operated and maintained, by the U.S. Army Corps of Engineers.

Commercial shipping can require ample quantities of water. For example, the Missouri River Navigation and Bank Stabilization Project (managed by the Corps of Engineers) requires a flow of about 25,000 cubic feet per second (cfs) to maintain a depth of eight feet, and that provides only "minimal service" for navigation throughout the 300-foot-wide channel.[76] A flow of 25,000 cfs for one month is equivalent to nearly 1.5 million acre-feet.

If navigation takes place solely on reservoirs, as on the Columbia and Snake rivers above Bonneville Dam, it doesn't require any flow beyond that necessary to operate the locks, so long as reservoir levels are maintained. Reservoir levels necessary for navigation are not always maintained, however. For example, the drawdown of reservoirs on the lower Snake River for the purpose of aiding salmon migration has the potential to reduce reservoir pools below the minimal operating level necessary for navigation, thereby eliminating the possibility of barge transportation on lower reaches of the Snake and Columbia during at least some parts of the year.[77] Where there are reaches of free-flowing river between reservoirs, navigational water needs in those reaches are the same as for free-flowing rivers.

The Value of Navigation

For shippers, river navigation is cheaper than transport by rail or truck. This is primarily because the federal government constructs, maintains, and operates inland waterway navigation projects at no cost to users. But the value of water used for navigation tends to be quite low, such that navigation is not usually an economically efficient use when there are alternative uses for the water.[78] For example, researchers determined that the long-run net value (benefits minus costs) of water used for navigation on the Missouri and Columbia rivers was negative.[79] Navigational uses of water can also be in competition with other instream uses. For example, it takes 43 million gallons of water to fill a lock at one of the Columbia River dams; each time a lock is used, a quantity of water sufficient to generate the electricity needs of a Northwest household for an entire year bypasses the hydroelectric turbines.[80] Commercial barge traffic may crowd out or create unsafe conditions for instream recreational uses. Channelization for navigation purposes eliminates substantial amounts of aquatic and riparian habitat for fish and wildlife and can have undesirable impacts on riverine aesthetics.

Riverine navigation receives less priority at federal and state levels than it once did, and there have been no new large expenditures for inland waterway improvements in recent years. Despite the large volume of commodities moved on some rivers, commercial navigation constitutes only a marginal use of rivers across the West.

Channel Maintenance

Instream flows maintain stream channel conditions conducive to the efficient transport of water. By keeping channels free of excess sediment and vegetation, and by preventing unnecessary erosion of those channels, instream flows maintain the capacity of rivers to transport high water flows and to minimize damage to riverside development.

Sediment Transport

Rivers transport sediment as well as water.[81] Much of the sediment introduced into rivers is a result of natural erosional processes that cause sediment to be washed from hillsides and surrounding ground into river channels. Sediment is also introduced into rivers and streams as a result of human activities in surrounding watersheds. Rivers occasionally receive large volumes of sediment through more dramatic events such as landslides or mudflows, and rivers may themselves be the source of sediment, through erosion of riverbeds or banks.

Sediments are transported downstream when streamflow energy is sufficient to move them. Streamflow energy is mostly a result of velocity and turbulence, the magnitude of which relies primarily on channel slope and the volume and rate of water flowing in the channel (stream discharge). Once in motion, sediments continue to move downstream, sliding, rolling, or bouncing along the riverbed or carried in suspension until stream energy decreases. When streamflow energy decreases, larger materials stop moving along streambeds, and smaller particles drop out of suspension. Streamflow energies may decrease at particular places in the river, such as the inside edge of river bends or in river reaches where channel slopes become less steep. Energies may also decrease throughout an entire river or river reach, as happens when stream discharge is diminished.

The processes of river channel formation and maintenance are largely determined by the balance between a river's sediment supply and transport capacity. If sediment supply exceeds transport capacity, the river or river reach is "capacity-" or "energy-limited," and sediments accumulate along the riverbed and banks in a process known as aggradation. If transport capacity exceeds sediment supply, the river reach is "supply limited," and—if river channels are erodible—rivers downcut their beds and/or erode their banks in a process known as degradation. In aggrading reaches, sediments are deposited faster than they are removed; in degrading reaches, sediments are removed faster than they are being supplied. River reaches in which sediment deposition and erosion are essentially in balance are said to be in equilibrium.[82]

Impact of Human Water Use on River Channels

Aggradation and degradation occur naturally as rivers adjust to changes in prevailing conditions. The processes can also be stimulated by human activities. Foremost among the activities that accelerate aggradation are the diversion of water, alteration of flow regimes, and introduction of additional sediment. Diversions cause aggradation and stream channel narrowing by reducing stream discharge, and hence the stream energy necessary to move sediments. Storage dams—which are designed to replace natural flow patterns with regimes better suited to other purposes—reduce the magnitude and frequency of high flows that have the most sediment transport capacity.[83] Increased sediment inflows can cause aggradation by overwhelming the transport capacity of rivers. Additional sediment is introduced to rivers by agriculture, silvicultural practices, sand and gravel mining in or near rivers, and the building of roads, bridges, and dams, all of which loosen soils and increase the potential for sediments to be washed into the channel.

No matter what the source, sediment accumulations on river edges, bars, and banks can cause problems. Channels become narrower and shallower as they aggrade. This process is reinforced if accompanied by a reduction in peak flows, because the absence of high, scouring flows allows vegetation to encroach into the channel on the newly deposited sediments. Over time, vegetation stabilizes the sediments, causing a more permanent narrowing of the channel. Narrowed, shallower channels have less capacity to transport high water flows. When higher flows finally do arrive, they cannot be contained within the channel and can cause substantial damage. Overbank flows erode streambanks, roadbeds, bridge abutments, diversion structures, and fields; inundate and wash away houses, barns, and other structures; create swamps; increase water loss through evaporation from wide, shallow water surfaces; damage surrounding wildlife habitat; and reduce the aesthetics of surrounding areas. Aside from these flood-related problems, narrow, shallow channels do not provide as much habitat for aquatic organisms or as many opportunities for recreation.

Human activities can also accelerate degradation of river channels. Two such activities are gravel mining and dam construction. Current and former riverbeds and banks are prime sites for gravel mining. But gravel mining in riverbeds increases erosion by exposing softer sediments beneath the gravel and by reducing the river's supply of bed-lining materials. Dams are a primary cause of increased erosion. In the short run, when large quantities of dirt and other materials are being moved around, dam construction actually increases sediment loads. But in the long run dams and their associated reservoirs cause degradation downstream by trapping sediments needed to form channel beds. The lower-velocity waters in reservoirs cannot transport much, if any, sediment, so sediments entering the reservoir drop out of suspension and collect on the reservoir bottom. By the time water reaches the dam, where it is eventually released, it contains little of the sediment needed to maintain the downstream channel bed.

Also, the "sediment-starved" waters released from dams have excess transport capacity and are therefore highly erosive. Rapid fluctuations in flow rate increase the susceptibility of riverbed and bank material to erosion, so erosion is further accelerated when flows are released from dams at rapidly fluctuating rates, as occurs when hydroelectric dams are operated to produce peaking power.

Degradation often results in "channel incision," in which rivers erode their beds and cut down further into the ground. It is not unusual for channel beds to show these effects for miles, even hundreds of miles, downstream from large reservoirs. In addition to causing damage to streambanks, roadbeds, bridge abutments, water diversions, and other structures located along river channels, river channel degradation can impair environmental and recreational values. Incised channels damage riparian vegetation by lowering local groundwater tables and by keeping vegetation isolated from occasional large flood flows. Degradation may also reduce the number and size of sandbars and islands available for wildlife habitat, camping, hiking, and other recreational activities. For example, the disappearance of sandy riverbanks and islands has been a major concern on reaches of the Colorado River downstream from Glen Canyon Dam, particularly within the Grand Canyon. The dam has caused the river's ample sediment loads to be deposited in Lake Powell. Together with operation of the dam to produce peaking power, sediment deprivation has produced highly degrading flows below the dam.

Channel Maintenance Flows

River and land managers have increasingly relied on instream flow manipulation to mitigate damages resulting from altered channels. For energy-limited systems the goal is to identify a flow regime that has sufficient capacity to move sediments through the system and prevent the encroachment of vegetation into the channel. For supply-limited systems the goal is to identify a flow regime that has less potential for erosion, such as one with high releases scheduled only when sediments from downstream tributaries and other sources are available, or one with constraints on the rapidity with which flow rates are changed from one level to another. River or watershed development that causes any change to flow regimes will have some impact on river channels; the idea behind "channel maintenance" flows is to minimize undesirable changes and ensure that the effective function of stream channels is maintained.

Determination of the magnitude, duration, and timing of flows necessary to produce such channels is not easy.[84] Extremely large flows have the capacity to affect stream location and alter the gross form of channels. However, such flows occur infrequently. By contrast, very small flows occur very frequently, but they don't have much energy to move sediments. Researchers have concluded that for most rivers intermediate flows, with discharge and frequency intervals between those of the extreme events, do most of the work in moving sediments and forming channels.[85] Research efforts originally focused on identification of "effective" or "dominant" discharges, which are flow levels that—given both their energy and their frequency of occurrence—have the greatest capacity to move sediments and maintain stream channels in the long run. More recent research has focused on the identification of a range of flows mimicking natural hydrographs that have the capacity to transport sediment and maintain river channels in functional condition.

Some Limitations in the Study of Channel Maintenance Flows

There are two caveats that should be included in any discussion of channel maintenance flows. First, the study of stream channel formation and maintenance—usually referred to as fluvial geomorphology—is a relatively young science.[86] So far it is a more qualitative than quantitative science, which means that its practitioners are better at identifying and describing trends than at making precise predictions.[87] Though the study of river channel creation and maintenance has made possible more informed conclusions about the consequences of particular courses of action, it is still difficult to precisely identify flow regimes that will accomplish desired objectives. The basic concepts of fluvial geomorphology are sound, but the identification of flow regimes that will adequately maintain stream channels remains controversial.

Second, the importance of sediment transport processes in forming and maintaining channels depends in part on the characteristics of the particular river channel at issue. For example, channels formed of bedrock are not easily eroded, so channel form in bedrock is predominantly controlled by geology rather than hydrology. The processes described above are more obvious—at least within human time scales—in alluvial terrain than in bedrock terrain. This bedrock-alluvium distinction is overly simplistic when compared to the actual circumstances of any given river, but it illustrates the need to avoid making generalizations about the effects of instream flows on channel maintenance and form. The flows needed to maintain the channel of any given river need to be determined on an individual basis.

Water Quality

Instream flow protection is primarily about water quantity—about finding ways to leave a specified volume of water within stream channels at specified times to accomplish designated purposes. However, instream flows also affect water quality. Instream flows improve water quality by limiting the volume of pollutants introduced into rivers and by decomposing and diluting wastes that do reach the river.

Pollution Avoidance

Very few offstream water uses are 100 percent consumptive. Some portion of water withdrawn for offstream purposes is usually returned to the environment as "return flow" after use. Agricultural return flows have higher concentrations of salts (from contact with soils), fertilizers, and animal wastes and may contain pesticides. Return flows from municipal sewage treatment plants have higher concentrations of sediments and organic and biological wastes. Industrial return flows may be the source of heavy metals, petroleum products, and volatile organic compounds such as solvents and thinners in addition to large volumes of organic waste. Even relatively non-consumptive uses of water, such as the cooling of thermoelectric power plants, may adversely affect water quality, by raising the temperature of return flows. Increasing instream flows by reducing offstream use therefore has the potential to reduce the pollution of rivers and streams.

The actual relationship between instream flow and water quality can be quite complex. An example is the quality of Colorado River water in the Grand Valley, an irrigated area near the confluence of the Gunnison and Colorado Rivers in west central Colorado. Colorado River water entering the Grand Valley has a salt concentration of roughly 300 milligrams per liter (mg/l). The salt concentration of irrigation return flows in the valley range from 1,500 mg/l to over 10,000 mg/l.[88] The return flows are more saline in part because evaporation and crop transpiration remove only pure water, leaving the salts behind. However, the major cause is that water leaches salts from the highly saline soils and the underground rock formation that underlies the irrigated lands as the return flow makes its way back to the river.

Several solutions to this problem have been proposed, including lining canals to reduce seepage, using more efficient irrigation to lessen water application requirements, diverting return flows to evaporation ponds, and stopping irrigation altogether.[89] Lining canals and improving irrigation efficiency would reduce diversion requirements. Provided that the conserved water would not subsequently be used for other offstream purposes, these measures would leave more water in the river for instream purposes—an example where efforts to improve water quality would increase instream flows. Routing saline return flows to evaporation ponds would improve the river's water quality but decrease river quantity because there would be no return flow to the river—water quality would improve, but instream flows would decline. The last solution, purchase and retirement of irrigated land, would have the most dramatic effect. By eliminating an offstream use, this proposal would improve both water quality and instream flows. However, even though sale would be voluntary, this solution faces stiff opposition from communities dependent on irrigated agriculture.

Waste Decomposition

Water quality is also affected by the capacity of rivers and streams to decompose pollutants. Pollutants introduced into rivers and streams meet a variety of fates; some are absorbed into riverbed and bank sediments, evaporate into the atmosphere, infiltrate into groundwater, or are transported to the ocean. Other wastes, particularly organic wastes, decompose while in the water. Organic wastes are broken down through physical processes facilitated by river turbulence and velocity, and decomposed by bacteria. The capacity of rivers to decompose waste products is of great benefit to society. Municipal sewage treatment plants and other generators of organic waste, such as pulp and paper mills, are among the heaviest users of rivers' capacities to decompose wastes.

The amount of waste that can be decomposed depends on many factors, including the types and amounts of bacteria that are present and the temperature and dissolved oxygen concentrations of the water. Because (as explained with regard to the needs of fish) higher streamflows increase dissolved oxygen concentrations, instream flows can aid decomposition processes. And, all other factors held constant, more water can decompose more waste.

Dilution

Instream flows also improve ambient water quality through dilution. Dilution improves water quality not by breaking down pollutants, but by lowering concentration levels. Water quality is usually dependent on the concentration of pollutants, rather than the absolute quantity of pollutants. For any given quantity of pollutant introduced to a stream, the concentration of that pollutant will be half as high if there is twice as much water. Lower concentrations may be within acceptable limits for certain uses of the water, such as the maintenance of cold-water fisheries or water contact recreation, when higher concentrations are unacceptable.

The use of instream flows to dilute wastes has a long history. For example, the dilution of waste and sewage was probably the first instream use for which releases from federal reservoirs were authorized.[90] More recently, instream flows have been used or advocated as a means of diluting pollutants, such as effluent from placer mining operations in Alaska,[91] and heavy metals draining from a Superfund mine site in California's Trinity River.[92]

Other Links between Instream Flow and Water Quality

Instream flows have been used in coastal areas to prevent saline intrusion. For example, instream flows have been at the heart of recent efforts to prevent saline intrusion into the delta formed by the San Joaquin and Sacramento rivers where they flow into the San Francisco Bay,[93] and they have been proposed as a means of protecting water quality in estuaries along the Texas coast.

The Clean Water Act requires states to set water-quality standards for all rivers and streams within their borders. Water-quality standards, as defined by the Act and interpreted by the Environmental Protection Agency, include both water-quality "criteria," which are maximum or minimum concentrations of various constituents that will be allowed, and "designated uses." The maintenance of designated uses may require water of a certain quantity as well as a certain quality; for example, upstream diversions might reduce streamflow volume below that necessary to sustain a cold-water fishery. Under the Clean Water Act, such withdrawals could be construed as violations of water *quality* standards. In this way, water "quantity" issues have sometimes been cast as water "quality" issues. As discussed in more detail in chapter 10, recent court decisions have increased the likelihood that instream flows will be linked to water-quality issues under the Clean Water Act.

Methods of Quantifying Instream Flow Needs

Each instream use depends on a particular range of flows, a range that varies for each river. Calculation of the appropriate flows, even if goals are clearly identified, is complex. There is no one widely accepted method of calculating the required flows, and a fully validated and accepted comprehensive methodology is not likely to be available in the near future.[94] There are, however, a variety of methods that have been developed and used to approximate the necessary flows, as well as an extensive literature describing and evaluating these methods.[95] A basic understanding of the variety and nature of these methods is fundamental to an understanding of instream flow protection. Presented below is a representative sample of some of the more popular methods.

Fish

The earliest efforts to quantify instream flow needs focused on the needs of fish. Five methods of determining fish flow needs will be summarized here: the Tennant method, the wetted perimeter method, empirical observation, PHABSIM, and IFIM.

The Tennant Method. The earliest and simplest of the methods to determine the flow needs of fish used generic indices correlating different levels of flow with subjective evaluations of the quality of fish habitat. The most widely applied of these index methods was developed by Donald Tennant and is known as the Tennant or Montana method. Tennant, a fisheries biologist, developed the method through personal observation of rivers and streams in the 1960s and 1970s. During his travels throughout Montana and several other western states, he stopped to investigate the fish habitat in rivers and streams that he encountered, evaluating the quality of the habitat in each stream using terms such as "excellent," "good," and "fair." Later, he checked hydrologic data to determine the flow level, expressed as a percentage of mean annual flow, in the river or stream at the time of his investigation. By compiling the results of these personal investigations, he established an index relating percentage of mean annual flow to the suitability of aquatic habitat, as reproduced in table 3.3.

Note that the index includes Tennant's estimate of the size of an annual flushing or maximum flow necessary to preserve desirable habitat characteristics over the long run. The Tennant method, being simple and easy to use, is often used as a guiding rule of thumb for initial survey or planning purposes. The method is also used to establish quantifications when the costs of performing more sophisticated analyses are not justified by the benefits of their use.

The flow levels required to provide habitat of a given quality vary by stream type and species of interest, so the Tennant method is often applied in modified form. The simplest modification is to adjust the numbers based on either stream type or region of the country in which the stream is located. A more extensive modification is to assemble the relevant data from scratch, using only the particular stream and species of interest—essentially retracing Tennant's procedure by evaluating fish habitat at a variety of flow levels, but only for the species and stream of interest.

Wetted Perimeter. The wetted perimeter method is another fairly simple method of quantifying the flow needs of fish. With this method a "critical" area of the stream—often a riffle, where flow is shallow—is selected by the analyst to serve as an index for the entire stream. The analyst then determines the minimum flow level necessary to provide adequate food production, spawning, and migration habitat for fish at that site. The method assumes that if the critical area is chosen appropriately, this level of flow will be sufficient to maintain fish habitat in other parts of the stream.

TABLE 3.3
Flow Necessary to Establish Habitat of Given Quality, Expressed as a Percentage of Mean Annual Discharge (Tennant/Montana Method)

Habitat quality/health	October–March	April–September
Annual flushing/max. flow	200	200
Optimum	60–100	60–100
Outstanding	40	60
Excellent	30	50
Good	20	40
Fair	10	30
Poor	10	10
Severe degradation	<10	<10

The necessary flow is based on measurements of wetted perimeter. For any given transect (cross section or "slice" of the river measured at right angles to the direction of flow, from bank to bank) and level of discharge, the "wetted perimeter" is the total length of riverbed and bank that is in contact with water. Wetted perimeter varies with discharge; higher discharge leads to deeper, wider flows, thereby wetting a greater width of bed and banks, whereas lower discharge covers a smaller width of bed and banks.[96]

Wetted perimeter rarely increases uniformly with discharge. In the typical channel, wetted perimeter increases fairly rapidly with discharge at low flow levels and then increases more slowly at higher flow levels. The level at which there is a noticeable change in the rate of increase in wetted perimeter (the inflection point on a graph of wetted perimeter vs. discharge) is usually chosen as the desired minimum flow level.

PHABSIM. The complexity of the relationship between flow level and fish habitat often leads analysts to use more sophisticated computer-based methods to identify relevant flows. One of the most widely used computer methods is the Physical Habitat Simulation System, more commonly known as PHABSIM.[97] PHABSIM is based on the relation of desirable fish habitat to four critical hydraulic variables: water depth, flow velocity, substrate,[98] and cover. PHABSIM calculates the values of these hydraulic variables at different levels of flow and compares them to the known preferences of particular fish species and life stages. Hydraulic values are based on data obtained through field measurements taken at specific river transects, and fish species and life-stage preferences—"species suitability criteria"—are established through biological studies.

The result of a PHABSIM analysis is a series of graphs depicting the relationship between discharge level and usable habitat for each species/life-stage combination at each transect. For example, for one particular fish species with suitability criteria for four identifiable life stages, there will be four graphs of usable habitat vs. discharge for each transect. The number of analyses required will be a multiple of the number of fish species, life stages, and transects being studied. An example is presented in figure 3.1.

Empirical Observation. Empirical observation is an alternative to the use of index and simulation methods. The relationship between flow levels and fish status can be determined by observing the health and size of fish populations under differing levels of flow. Though relatively simple in concept, this process requires substantial effort and time—possibly decades—to assemble the number of observations needed to make statistically valid predictions about the relation of discharge to fish population.

A variation uses multiple regression, a statistical method, to extrapolate from a smaller data set. With this approach, data on estimates of population size are regressed on several habitat variables to quantify the discharge-habi-

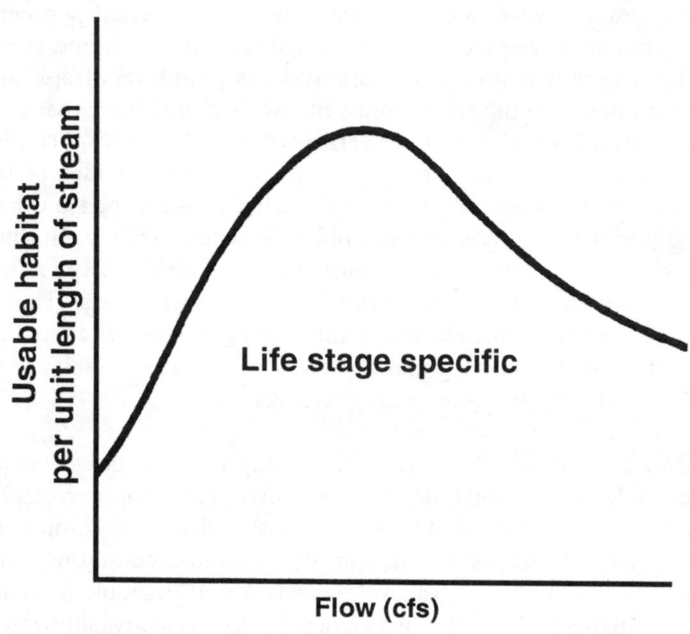

FIGURE 3.1
Output of Microhabitat Model (reach specific, life stage specific)

tat relationship. Again, data may need to be collected over considerable variation in time or space.

Instream Flow Incremental Methodology. The Instream Flow Incremental Methodology (IFIM) is the most sophisticated and comprehensive method of quantifying instream flow needs in widespread use today.[99] The term *incremental,* as applied to this and some other quantification methods, refers to the ability to evaluate impacts resulting from incremental changes in flow. The IFIM process uses computer software to combine the evaluation of "microhabitat" and "macrohabitat" variables and project the results through time. Microhabitat variables such as depth, velocity, substrate, and cover come from specific transects and are analyzed using tools such as PHABSIM. Macrohabitat values such as water quality, channel morphology, discharge, and temperature vary longitudinally along the stream. The result of an IFIM analysis is a "Habitat Time Series" that shows total usable habitat in a river segment through time, as a function of discharge. IFIM can be used, for example, to predict the amount of habitat that would be available during the year in the defined river segment under different flow regimes. In this case, IFIM output might specify the amount of habitat available 90 percent or more of the summer, or 60 percent or less of the year.

IFIM is state-of-the-art technology capable of supporting detailed analysis of streamflow quantification problems. But IFIM is more than just that; as used by stream analysts, the term incorporates the whole series of more or less specified steps taken to solve flow-related problems, of which the computer analysis is just a part. At the core of an IFIM analysis is the ability to evaluate incremental, rather than all or nothing, changes in streamflow magnitude and timing. Depending on circumstances that vary from situation to situation, a complete analysis using IFIM may take as long as a few years to complete.

Riparian Vegetation

A variety of methods have been used to quantify the flows needed to sustain riparian vegetation.[100] Perhaps the simplest of these methods is to assume that riparian vegetation requires a flow regime that mimics the range of natural flow magnitudes, including infrequent flows of very high magnitude. Rather than specifying flows at some fraction of natural magnitudes, the goal is to maintain the full range of flow magnitudes but to reduce flow duration and/or frequency as necessary to accommodate other water uses.

Some analysts have used evapotranspiration studies to quantify flows needed for riparian vegetation. By calculating the amount of water transpired by the desired abundance of trees and plants, one can then calculate the flow level required to supply that amount of water.

Others have used empirical observation and statistical regression techniques to determine the relationship between streamflow and riparian vegetation at both site-specific and regional scales. Flows necessary to establish and maintain riparian vegetation have been quantified based on species-specific criteria such as the magnitude and timing of flows necessary to promote recruitment and growth of particular cottonwood species. Adaptations of the wetted perimeter and PHABSIM methods have been used to estimate the impacts of various flow levels on riparian vegetation, and sophisticated vegetation succession models have been used to model rates of vegetation change resulting from different levels of flow on a dynamic basis.

Recreation and Aesthetics

Methods of quantifying flow needs for recreation and aesthetics are similarly diverse.[101] One researcher defined the concept of "canoeing zero" flow as the minimum flow necessary for passage of a canoe. He then used personal observation to determine canoeing zero flow for a series of rivers and regressed those flow needs on mean annual streamflow, producing an equation that shows how canoeing flow needs increase with rising mean annual flow.[102] Others have used the wetted perimeter technique, or have used PHABSIM with "recreational" rather than "species" suitability parameters.

And the IFIM approach has been used to determine "weighted usable surface area" for recreational activities through time.[103] However, to date these methods have not met with the same acceptance when applied to recreation as when applied to aquatic habitat.

A variety of other methods, most relying to at least some extent on collecting data from actual participants in recreational activities, have been developed to quantify the relation of recreation quality to instream flow. In perhaps the simplest of these, analysts have correlated historical recreational-use-intensity data to flow levels at time of use. This method has a number of limitations, primary among them that it assumes users know flow levels before they decide on making a trip and that other factors (such as timing of vacations) do not overwhelm any flow-dependent participation effect that might otherwise exist.

Another widely used method is to rely on the judgment of experts, such as professional river and fishing guides or landscape architects. Experts may be asked to define the minimum flow level necessary to sustain the given use, or may attempt to define optimal flow levels or an acceptable range of flows. Because sample sizes tend to be small, this method is not likely to yield a flow quantification that is as exact as some of the other quantitative methods, but it is efficient and can be very useful. The judgment of professionals may stand alone as a quantification method or may be used to establish initial parameters, develop guidelines for future study, or provide feedback on results obtained from other methods.

User surveys are probably the most effective method for quantifying recreational and aesthetic flows.[104] User-survey methods rely on statistical procedures to correlate responses to different levels of flow. The analyst may choose to survey all who participated in an activity or may survey only users who have substantial experience. River users may be surveyed on site about the particular level of flow they just experienced, or a smaller group of individuals may be gathered to systematically assess a range of flows. For example, boaters representing a range of skill levels may repeatedly run the same reach of river using a variety of watercraft at different flow levels and then rate beach, water, and boating conditions for each combination of craft and flow. The assessment of multiple flows by the same group of users is particularly effective where flow levels can be systematically varied over a relatively short period of time, generally with the cooperation of upstream reservoir managers.

Analysts may have users evaluate photographs, videotapes, or verbal descriptions rather than surveying participants on site. These indirect methods simplify the survey task, for participants do not have to be contacted on site at the particular time that the relevant flow levels exist. If the survey is well prepared and executed, this indirect approach can be quite successful. Figure 3.2 shows the result of one such effort, a mail survey of professional raft-trip guides and private trip leaders familiar with floating through the Grand

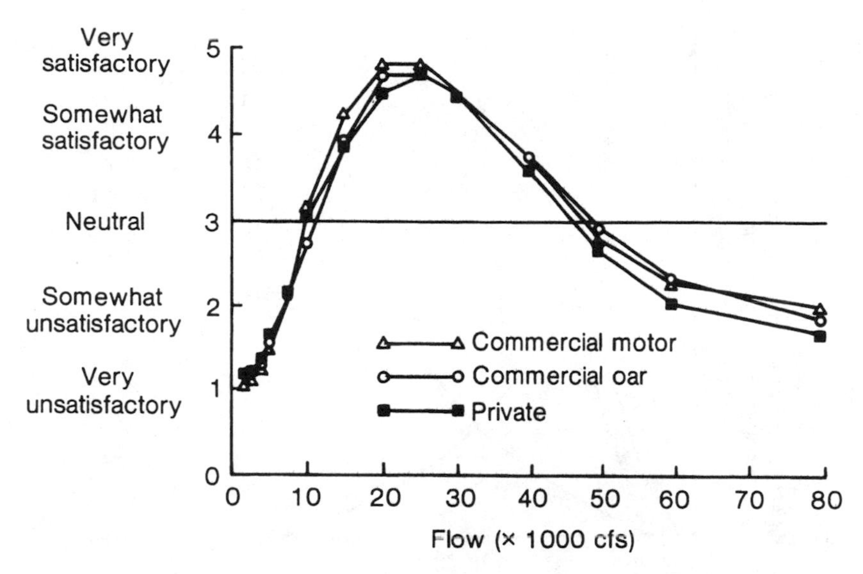

FIGURE 3.2
Guides' Constant Flow Level Preference Ratings, Colorado River through Grand Canyon

Canyon of the Colorado.[105] Respondents were asked to evaluate the acceptability of various flow rates, and they tended to agree that, regardless of trip type, the most favorable flow levels were in the 15,000 to 35,000 cfs range.

Channel Maintenance Flows

Channel maintenance flow quantification methods—of which there are several—are relatively new and still undergoing modification.[106] One frequently used method is based on hydrologic records. Historical flow data are analyzed to determine the frequency with which flows of various magnitudes occur. In general, flows close to the average discharge have higher frequencies than do either high or low flows, as in curve B of figure 3.3. The frequency of each flow level is then multiplied by the rate of bedload sediment transport at the respective flow level (curve A) to yield a curve that depicts sediment transport throughout the full range of discharges experienced by the stream (curve C). The discharge at which the product reaches a peak is identified as the effective discharge, which is the discharge that moves the most sediment over the long term.

A second method is based on the theory that a river's channel shape and form will have been configured by, and can therefore just accommodate, the dominant discharge. The discharge necessary to just fill a river's banks with-

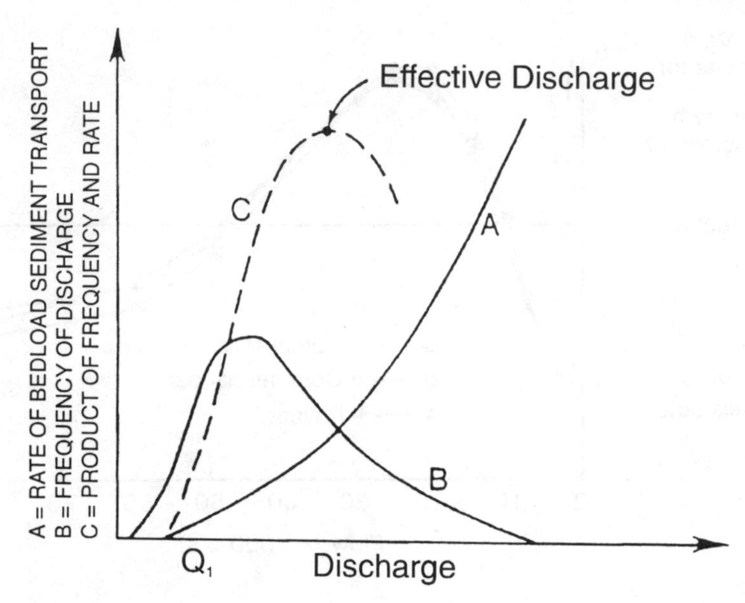

FIGURE 3.3
Computation of Effective Discharge

out flooding—the "bankfull" discharge—should therefore be the dominant or effective discharge. In practice the bankfull discharge is not easy to determine, because channel morphology is neither simple nor consistent throughout an entire reach. Several methods of calculating bankfull discharge are in use, and, depending on the particular river and reach, the choice of method can affect the result.

The hydrologic and bankfull methods frequently yield comparable results. For many rivers, the identified flow is the discharge exceeded on average just once every one to two years. The 1.5-year discharge is often used as a quick rule of thumb to identify the probable effective discharge. However, it is important to realize that this discharge is only a general average; the figure is subject to much variation and may not be appropriate for particular rivers.[107]

There are at least two other types of methods in widespread use for identifying effective discharge, one relying on modeling techniques and one on empirical observation. The modeling method uses sediment-transport equations to calculate the flow necessary to move sediments of particular size, shape, density, and quantity. Modeling methods are used when practical considerations make it difficult to obtain results through empirical observation. Empirical methods are facilitated where discharge levels are controlled by upstream dams. Water is released from dams in steadily increased increments, and researchers downstream identify the flow at which sediments actually begin to move in the desired quantities. Empirical methods yield more accurate results than theoretical methods, but physical, financial, and institutional constraints often prevent their use. In recent years efforts have been made to combine the two methods by using empirical data to calibrate model results.[108]

Using Multiple Methods: The Rio Chama

Each quantification method has its own advantages and disadvantages. The choice of method may depend on a variety of considerations, including the complexity of the problem, the amount of time and money available for the study, the level and availability of expertise, the degree of accuracy required, and the uniqueness of the resource for which flows are being quantified. Institutional factors may be important, as laws and regulations sometimes require particular measurements, or specify allowable methods. The choice of method may also depend on the degree to which study results are likely to be opposed, because a more controversial quantification will need to withstand more intense scrutiny. When faced with a difficult choice of method, analysts often choose to employ more than one method. And when flows must be recommended for multiple resources, the use of multiple methods becomes very likely. An instream flow quantification study for the Rio Chama is illustrative.

The Rio Chama, in northern New Mexico, is a large tributary of the Rio Grande. Rio Chama water is used extensively for agricultural and municipal purposes, and natural flows have been augmented by transmountain diversions from the San Juan River, in the neighboring Colorado River watershed. The Rio Chama is one of New Mexico's most popular recreational sites and supports a variety of instream recreational and ecological uses.

A 24.6-mile segment of the river between El Vado and Abiquiu reservoirs was designated by Congress in 1988 as a national Wild and Scenic River.[109] Pursuant to this designation, the federal agencies with management responsibility under the Wild and Scenic Rivers Act developed a management plan for the river. This plan identified water deliveries to downstream water rights holders as the highest management priority, because the numerous laws, compacts, and agreements allocating the river's water predated the Wild and Scenic designation. However, the existence of numerous dams on the Rio Chama and Rio Grande allows some flexibility in releasing water from El Vado Reservoir, making it possible to concurrently meet the needs of several water users. Accordingly, the management plan called for a comprehensive instream flow assessment to evaluate and quantify flows needed to support the variety of instream uses for which the river had been designated. This assessment was subsequently performed by a federal interagency group that included the Bureau of Land Management, Forest Service, Army Corps of Engineers, Bureau of Reclamation, and Geological Survey.[110]

The management plan called for an assessment of the instream flows necessary to support the following uses: (1) fisheries, including various species and/or life stages; (2) recreational boating, including 16-foot rafts, 12-foot rafts, kayaks, and canoes; (3) fishing, including boating access and the effect of variable flows on fishing success; (4) riparian habitat, including both maintenance and regeneration of riparian vegetation; (5) scenic and aesthetic qualities of the river, including water quality; and (6) endangered species, primarily the maintenance of foraging areas for wintering bald eagles and suitable habitat for the eagles' prey species.

The interagency group used the IFIM process to quantify the needs of fish. For reasons described in the following section of this chapter, the group focused its efforts on the brown trout. The group used data from past studies at four sites on the river and also developed a new, fifth site. Habitat suitability curves for different life stages of the brown trout were derived from data already on file with the U.S. Fish and Wildlife Service. Hydraulics were simulated using a computer model known as IFG4, a part of PHABSIM. Weighted usable area estimates were produced by comparing the hydraulic and habitat suitability data using a software program known as HABTAV. Analysts made recommendations for both "optimal" and "acceptable" flow levels. Because desired flow levels varied across the five sites, the flow providing the most total habitat at the five study sites combined was chosen as the recommended optimum flow.

The group studied benthic macroinvertebrate populations (the brown trout's primary food source) to determine if there was any flow dependency. Macroinvertebrates were collected in riffle areas at four different sites and then identified and categorized as to their sensitivity to environmental perturbations using EPA protocols. Plots of wetted perimeter versus discharge were used to identify flows below which wetted perimeter, and hence macroinvertebrate habitat, decreased sharply with further reductions in flow.

Riparian vegetation was inventoried and classified by type, extent, and age class. Analysts then described and photographed the effects of different flow levels on the vegetation. Several channel cross sections were surveyed and analyzed to determine the relationship between flow levels and channel morphological characteristics. Plots of wetted perimeter versus discharge were developed to determine the flow levels at which the edge of the water started to recede from rooting zones. Channel cross-section data were used to analyze the effects of high flows on areas containing riparian vegetation.

Experienced fishing guides were taken down the river at different flow levels and asked to evaluate the potential for fishing success. For recreational boating, past recreation studies on the river were reviewed and two empirical studies were performed. First an on-site user-preference survey was performed on weekends during which releases of 800–1,000 cfs were purposefully made from the upstream El Vado Reservoir. Boaters in different craft were asked to evaluate the flow level they had just experienced, using categories including "more than adequate," "adequate," "minimum," and "below minimum." Second, an on-river navigability assessment was conducted of 14 river segments identified during a preliminary survey as being the most critical for navigation. Participants, all of whom were experienced boaters, used a variety of craft to run the river at different flow levels, which, again, were requested in advance from operators of El Vado Reservoir. Each boater filled out a navigability assessment form after floating through each of the critical segments. In addition, videos were taken of the craft moving through the identified critical segments at different flow levels, for later assessment.

Minimum flows necessary to sustain aesthetic characteristics were ascertained using plots of submerged river bottom (wetted perimeter) versus discharge. The selected flow level was the flow below which wetted perimeter decreased sharply with further decreases in flow.

Three separate methods were employed to study the flow needs of bald eagles. First, analysts compared historical census data of bald eagle populations with streamflow data to determine whether there were any significant trends. Second, they analyzed the flow dependency of bald eagle prey species (fish) at riffle-run complexes at three different flow levels using IFIM. Third, because the visibility and accessibility of prey fish species depend in part on the degree of ice cover and clarity of the water, analysts used models to estimate winter ice cover and turbidity at different flow levels.

The flow levels recommended by study participants for each instream use are presented in table 3.4.[111]

Additional Comments on Quantification Methods

There are many more methods used to quantify instream flows than are described here, and more will undoubtedly be developed in the future. All of the quantification methods rely heavily on professional judgment—to choose representative transects, species, and river segments, to evaluate the relevance of particular stream features or resources, and to choose the methods that will be employed. Reliance on professional judgment has not been eliminated by the development of sophisticated computer models. If anything, the use of these elaborate quantification tools places even more of a premium on professional training and experience, for proper application of the tools depends on the ability of analysts to choose appropriate parameters.

A final, important word on instream flow quantification methods concerns the results obtained by using these methods. Though results, and patterns of results, vary widely depending on the particular river and instream use, the frequency with which graphs of streamflow level versus resource benefit depict a particular shape, roughly an inverted U, as shown in figure 3.4, is striking. Starting from zero or minimal flows, many instream uses show increasing benefits with increasing flow. As flows continue to increase,

TABLE 3.4
Flow Regime Necessary to Support Designated Resource Values in the Rio Chama

Resource	Flow magnitude and timing
Fish habitat (brown trout)	150–700 cfs October 15–March 31 (400 cfs optimum) 150–300 cfs April 1–August 31 (200 cfs optimum) 75–300 cfs September 1–October 14 (200 cfs optimum)
Macroinvertebrates	185 cfs minimum
Scenic/aesthetic	40 cfs minimum
Whitewater boating	800–1,000 cfs
Scenic boating	500–600 cfs
Fishing	150–300 cfs
Riparian (maintenance flow)	185 cfs April 1–September 30
Riparian (regeneration flow)	5,000 cfs at least one day every 5-10 years, between May 15 and June 15
Bald eagles	150–250 cfs December 1 to March 1

the rate at which benefits increase diminishes. Total benefits eventually plateau and then decrease. Very large flows may even produce negative benefits. This common characteristic supports the claim of water resource managers that flows must be carefully quantified to achieve the greatest possible benefit. It should also be clear that "minimal" flows do not necessarily provide the greatest possible benefit.

Conflicts among Instream Uses

Conflict over the allocation of water between competing uses is at the heart of water law and policy. Conflict between instream and offstream water uses has long made instream flow a controversial topic. It should now be apparent, however, that instream uses are also often in conflict with each other. This occurs not only between hydropower and less disruptive uses such as fish migration, channel maintenance, and recreation, it also occurs even within a particular type of use, as between different species of fish, or between different watercraft and skill levels in recreational boating.

Competing needs make it imperative to define the goals to be achieved by

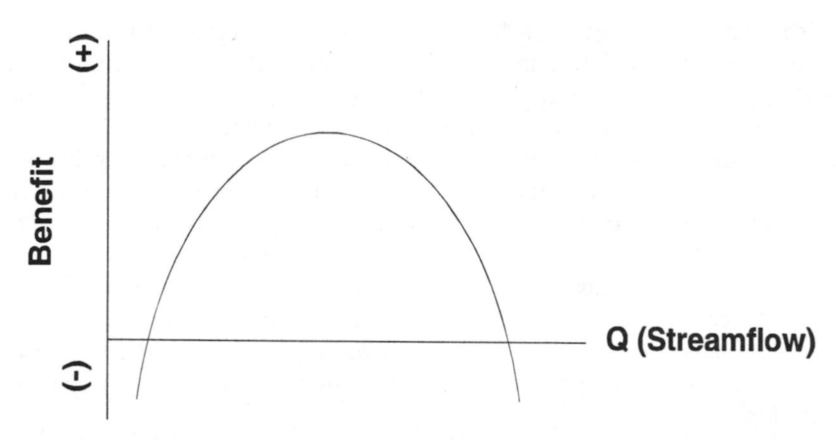

FIGURE 3.4
Common Relation of Benefit to Streamflow Volume

instream flows. To be successful in meeting these goals, instream flows must be carefully tailored to meet the specific needs of each chosen use. It is difficult to emphasize these two points strongly enough. Leaving enough water in streams to meet some instream needs will not automatically meet other instream needs. If some streamflow is good, more is not necessarily better. There is often significant overlap in needs of uses, so that flows conducive to the maintenance of one use are also sufficient to maintain other uses. This is why the needs of fish are so frequently used as a surrogate for other instream uses. But this situation does not always occur. Flows that meet the needs of a particular species of fish may not be suitable for other species of fish, or for boating, swimming, or navigation.

One also must define more than just the quantity of water necessary to maintain an instream use—the timing and location of flows can be crucial. The importance of flow timing was mentioned above, with respect to the propagation of salmon and of riparian vegetation. Flow timing is equally as critical for other instream uses. Location is important because channel size, the flows necessary to accomplish specific objectives, or even the objectives themselves may vary over the length of a river.

Example: The Rio Chama

The study of instream flow needs on the Rio Chama provides a good example.[112] The management plan developed subsequent to the river's designation as a Wild and Scenic River specified the instream uses that were to be studied by the interagency group. Taking these as a starting point, the group then made some refinements before starting the study. There are a total of

14 fish species in the reach of the Rio Chama designated for study, including an extremely productive and popular cold-water game fishery for brown trout, rainbow trout, and kokanee salmon in the upper reaches. The New Mexico Game and Fish Department stocks the rainbows annually but wanted a flow regime that would support a naturally reproducing brown trout population. At the request of the state, the interagency group therefore identified the brown trout as the target management species and decided not to study the needs of other species in detail. The interagency group also decided to divide the needs of recreational boaters into two categories, "scenic" and "whitewater," that would accommodate a mix of skill levels and crafts within each category, and to determine both the low-flow (for maintenance) and high-flow (for regeneration) requirements of riparian vegetation.

The interagency group identified flows necessary to support each of the designated resource values, shown in table 3.4. It is interesting to note the many potential conflicts between these instream uses. For example, flows necessary to support whitewater boating are higher than recommended maximums for fish at any time of the year. Whitewater boating flows are also too high for fishing and for scenic boating.[113] Note also that whereas fisheries can be sustained at some level, they cannot be maintained at "optimum" levels without harming other uses, including bald eagle habitat and whitewater and scenic boating.

So how were the agencies able to determine a single flow regime that would accommodate as many uses as possible? They weren't. Within priorities given by the management plan, the agencies devised two different instream flow scenarios, as depicted in figure 3.5. The "comprehensive environmental" scenario places a high value on resource protection but largely excludes recreational boating flows. This scenario calls for a stable flow of 250 cfs in the winter (October 15–March 31) to support fisheries and the bald eagle; this flow level also supports macroinvertebrates, scenic and aesthetic qualities, and maintenance of riparian habitat. The scenario calls for a minimum flow of 185 cfs the remainder of the year to support riparian areas; this flow level also meets the needs of brown trout and macroinvertebrates, maintains scenic and aesthetic qualities, and supports recreational fishing. The flows do not, however, meet the needs of either scenic or whitewater boaters.

The second scenario, "recreation opportunities," calls for essentially the same flows as the environmental scenario for most of the year, but for increased flows to provide better recreational opportunities when recreational user demands are high in summer, from mid-July through August. Under this scenario flows would be increased from 185 cfs to 250 cfs in mid-June to provide optimum fishing experiences, and to 500 cfs from July 15 to August 31 to accommodate scenic boating experiences. Peak flows of approximately 900 cfs for whitewater boating would be provided on six weekends

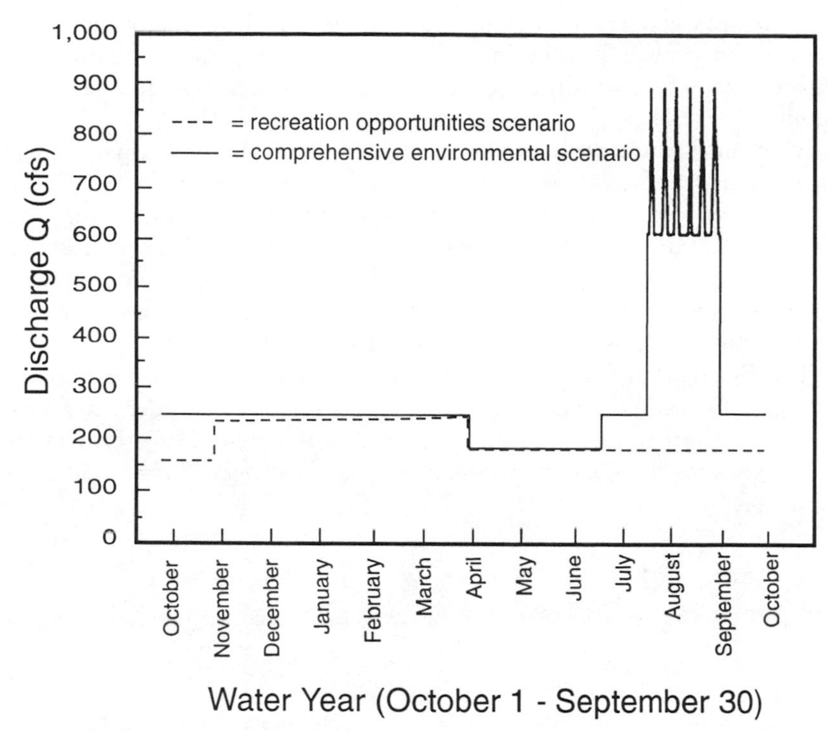

FIGURE 3.5
Alternative Flow Regimes for the Rio Chama Emphasizing Resource Protection and Recreation Opportunities

during this same time period. The interagency group noted that no matter what scenario was chosen, increases in flow should be requested in hourly increments so that impacts to aquatic habitats are minimized, and instream flow hydrographs should approximate the natural hydrograph (i.e., high flows in the spring) to the extent that water is available.

Competing Uses—Concluding Comments

The interagency group was not able to design an instream flow regime that would simultaneously optimize all instream uses, particularly given the constraints of the management plan, which identified the goal of meeting downstream water delivery obligations as the highest priority. After extensive study the group was, however, able to identify the flow-related needs of individual instream uses and to suggest scenarios that would accommodate as many uses as possible. As such, this study was very successful. The study was an example of successful interaction between scientists and policy makers; scientists and technicians provided important information about the choices and trade-offs faced in allocating the river's water, but the actual choices were left to river managers, water users, courts, legislatures, and the public.

Instream flow quantification studies, including the Rio Chama, indicate that a high level of instream benefits can be provided in situations in which water is also used for offstream purposes. Trade-offs between instream and out-of-stream uses exist but do not have to be absolute. In many cases the two types of uses can coexist. The Rio Chama is one of many rivers on which potential water uses are in at least partial conflict, and decisions have to be made about which uses to protect or enhance. Similar situations will be described throughout the book and are so common they can be assumed in virtually all situations where not explicitly addressed. The reader needs to be constantly aware of trade-offs between uses, both between offstream and instream uses, as the conflict is usually cast, and between different instream uses.

CHAPTER FOUR

How Much Water Should Be Left in Streams?

The central instream flow question is: How much flow is enough? All of the instream uses described in chapter 3 have value. Ideally, in a world without scarcity, all water uses would receive, at the appropriate times, sufficient water to maximize benefits. But there is not enough water to supply all instream and offstream uses at levels that would produce maximum benefits from each, so some balancing of needs must occur. Chapters 5 and 6 discuss the amount of water that states actually allow to be protected; the question here is the broader one of how much water should be left in streams relative to the amount of water diverted for offstream use.

General Considerations

Some people believe that because water is scarce, its primary uses should be human consumption and commodity production, that only the amount of water not needed for these purposes—if any—should be left in streams. A competing view is that because water is scarce, we should restrict offstream consumption to only the bare minimum needed to support human populations and thereby preserve the West's comparatively rare aquatic and riparian environments. Most people, to the degree that they think about the issue at all, probably fall somewhere between, recognizing a need to divert water for a variety of purposes but also a need to leave water instream.

Two broad principles often used to guide society in the quest to properly allocate water among competing needs are efficiency and equity. The efficiency principle proposes that water be allocated among competing uses so as to produce the maximum possible benefit to society. "Maximum benefit" does not refer solely to financial returns; financial returns are important, but benefits not typically assigned a dollar value must also be considered. This means that the whole array of use and nonuse values must be considered, including the value of water in producing commodities, providing recreational opportunities, supporting domestic and municipal uses, and maintaining stream channels and water quality, together with the existence and bequest values of preserving fish, wildlife, and a healthy environment.

The goal of producing maximum benefit from a scarce resource has long played a role in western water allocation. The prior appropriation doctrine was created and justified in large part on the principle that the greatest benefit to society would be achieved by allocating the rights to use water, on a first-come, first-served basis, to individuals who stood ready and willing to put the water to a beneficial use. However, this method of allocation will not necessarily produce maximum benefits if there are more proposed uses for water than can be met by the available supply. To produce maximum benefits from a scarce resource, all rights must be transferable to their highest valued use.

On a practical level, it is hard to define the allocation of resources that will produce the maximum social benefits. Some insight into the best allocation can be gained from economics, in which the rule for producing the maximum benefit from a scarce resource is to allocate the resource between uses so that the value produced from the last unit of the resource in each use is equal. For example, as between two competing uses of a limited water supply, say irrigation and instream flow, if the last unit of water used in irrigation produces less value than it would if left in the stream, then that unit of water should be reallocated to instream use. The transfer increases the total returns to society, because the reallocated unit of water now produces greater value. Total returns to society will continue to increase as additional units of water are transferred to the more valuable use, until maximum returns are reached at the point that value produced by the last unit of water in each use is equal.

Implicit in this rule is the concept of diminishing marginal returns. Diminishing marginal returns means that the value produced from use of the first unit of a resource is greater than the value produced from use of the second unit, which in turn is greater than the value produced from use of the third unit, and so on. Diminishing marginal returns have been widely observed in a variety of situations. In the example above, the value produced from the last unit of irrigation increases as additional units of water are transferred from irrigation to instream flow, while the value produced from the last unit of water left instream declines as flows increase. Eventually the value produced from the last unit of water in each use will be equal. Or, less likely but still possible, all units of water will have been transferred from the less valuable use to the more valuable use. Either way, the allocation will result in the maximum total return to society from use of the resource. Though difficult to achieve in practice, there is no doubt that the desire to maximize returns has played a large role in society's efforts to allocate scarce water resources.

Equity is the other broad principle that has guided the allocation of water resources. It is more difficult to articulate a means of achieving equitable allocations than to define rules for maximizing returns. But, in very general terms, society's equity goal is best defined as the effort to be fair.

The desire to achieve fairness, like the goal of producing maximum returns, has played a role throughout the history of western water allocation. The principle of first come, first served seemed as fair as any to miners in the gold camps. Fairness was a goal in the effort to clearly define property rights and to define the circumstances in which those rights could or could not be lost. The creation of miners' committees to establish and adjudicate rules for the use of water, and later use of the courts to establish water policies and protect property rights in water, are other examples of the search for equity in water use.

Considerations of equity have played a highly visible role in the debate over instream flow protection. Some of the more common notions of fairness expressed in these debates include the desire to ensure that all members of society benefit from resource use, that particular individuals are not placed in a position from which to either reap all the benefits or suffer all the costs of water allocation decisions, and that property owners should not be deprived of their property in the absence of just cause and adequate compensation.

Neither the desire to maximize value from use of its resources nor the search for equity can completely explain society's water allocation decisions or the continuing debate over how much water should be left in streams. Both, however, play a vital role in the debate and are useful criteria for evaluating existing conditions and proposed changes to stream resource allocations in the West.

Are Special Measures Needed to Protect Instream Flows?

Several factors must be weighed when considering whether special measures are needed to achieve an efficient and equitable allocation of water between instream and offstream uses. Many of these factors suggest that special mechanisms will be necessary to achieve an efficient and equitable allocation of water, but first we present three arguments that have been made against using special protection measures: (1) instream flows don't have enough value to justify special protection measures; (2) enough water will usually be left in streams without the use of special protection measures; and (3) the operation of market forces will correct misallocations of water between instream and offstream uses.

The first argument is that water left instream, at least in locations with potential for additional offstream diversions, is essentially wasted. Or, as commonly phrased, the "failure" to use water for beneficial offstream purposes leaves water "to waste to the sea." The notion that water left instream is water wasted dates at least to the first days of the prior appropriation doctrine in the mid-1800s, when beneficial use was defined solely in terms of commodity production. The argument is not as widely heard today but is far from dead. It is an argument over values, not facts. The argument is generally based on the personal values of the individual or group making the argument rather than on a more broadly based assessment of social values or on evidence from economic studies. The argument is now more often encountered in relative than in absolute form, in which the value of instream flows is recognized but is viewed as something of a luxury or amenity that should not be allowed to interfere with truly "productive" or "economic" uses. Similarly, some argue that instream flow protection should be limited to only a few exceptional situations. This line of reasoning may seem unpersuasive in light of the description of the instream use values presented in chapter 3, and yet it continues to be heard throughout the West in the dialogue about instream flows.

The second argument, that sufficient water will usually be left in streams even in the absence of specific protection measures, is based on fact rather than on value. The following four factors cause water to be left instream even in the absence of special protection measures.

1. *Geographic factors.* Water in many rivers throughout the West flows at the bottom of deep canyons or through otherwise difficult terrain where it is relatively inaccessible. Other rivers flow through areas that are too far removed from intended sites of use for the water to be economically diverted and transported. Similarly, streams may not contain enough water, particularly in headwater areas, to justify the expense of building diversion and transport facilities. Distance from intended sites of use is an important factor in deciding whether to develop water, especially for smaller streams, because the cost of transporting water in engineered channels is quite high. These geographic factors cause substantial quantities of water to be left instream.

2. *Senior water rights downstream.* Early arrivals in the American West usually settled at relatively low elevations, where access was easier, temperatures were less harsh, and farming was more promising. The senior water rights established by these early settlers were far downstream from the headwaters of western rivers. Later settlers, forced to seek land farther upstream, were legally required to leave enough water in the watercourse to satisfy senior appropriators downstream, thereby assuring instream flow in the upper reaches, even on fully appropriated streams. The existence of senior water rights downstream continues to protect instream flows throughout the West.

3. *Interstate allocations.* Most political boundaries were established without regard to hydrographic features. River systems can extend hundreds, even thousands, of miles and cross numerous political boundaries. Rights to use many of these waters have been allocated between the states by interstate compact, court decisions, congressional statutes, or other mechanisms. Interstate allocations may cause substantial quantities of water to be left instream, because—as with senior water rights downstream—upstream states are required to pass on specified amounts of water to downstream states.

4. *Surplus flows.* In some places there is more water flowing in streams than is needed for offstream use. This situation occurs most often in the more humid portions of the West, such as along the northwest coast and in the eastern portions of the Plains states, or in relatively unpopulated regions, such as much of Alaska or portions of northern and central Idaho. It may also occur virtually anywhere in the West during seasons or years of unusually heavy precipitation, particularly in areas where there are very few storage facilities. When water availability exceeds offstream water demand, water will be left instream.

The third argument against special protection efforts is that existing mechanisms work to ensure that water is reallocated in the correct proportions between instream and offstream uses. In particular, market transactions transfer water from less valuable to more valuable uses. Water rights are transferred to new uses and users through market activities in every state of the West. Temporary transfers of water—effected in what is sometimes called the "water rental" market after the notion that the water right is being temporarily transferred—are also common. Both kinds of transfers have been used to increase instream flows. [1]

The factors cited above do often result in substantial quantities of water being left in streams. However, there are reasons to believe that, notwithstanding the existence of these factors, there will not be enough water left in streams unless specific measures are taken to make that happen. The following seven considerations suggest that special instream flow protection measures are necessary.

1. *Incidental existing protection.* Water left in streams often occurs incidentally rather than as an intended consequence of other conditions and actions. Because conditions change, flows that incidentally exist today—whether owing to geographic conditions, senior water rights downstream, interstate compacts, or surplus conditions—may not exist tomorrow.

Geographic inaccessibility impedes, but does not preclude, water diversion. Indeed, western history is rich with examples of water being diverted from "inaccessible" locations and then transported great distances to sites of use. Though federal funding for such projects is dwindling, municipalities—now the most economically powerful entities in the West—continue to grow. Isolated, distant water bodies are being investigated as new sources of municipal water supply, and additional canals, tunnels, and pipelines will likely be built as the pressure on current supplies increases. In the words of one game and fish official in Wyoming, "One of the lessons we've learned with instream flow filings and possible water development is never to assume a stream is protected by its location alone. Water flows uphill to money. Where there's enough money, there's a feasible project."[2]

Senior water rights downstream—unless established specifically for the purpose of protecting flows—cannot be counted on to protect instream flows, for three reasons.[3] First, water rights may be transferred from one site to another. Sometimes water is even transferred to sites in entirely different watersheds, leading to a complete loss of water in the source stream. Second, deliveries scheduled to meet the needs of the senior water user may not coincide with the needs of the instream use. For example, most stored water in the arid West is delivered to irrigators from July to September, but flows for fish are often most limiting in the winter, when the reservoirs are filling. Third, although water held by senior rights holders may be used at lower elevations, some of it is diverted from points high in the watershed to take advantage of gravity flow when moving water through pipelines and canals. This is particularly true for sites of use located some distance from stream channels.[4]

A major limitation of interstate compacts as protectors of instream flow is that the timing of compact-required water deliveries may not be compatible with particular instream uses. For example, the extensive use of Glen Canyon Dam on the Colorado River and Elephant Butte Dam on the Rio Grande to fulfill obligations to downstream states has significantly altered natural flow regimes. In addition to poor timing, interstate compacts may leave long stretches of river with little flow. Indeed, using the existence of interstate obligations as justification for denying instream flow protection measures is ironic given that interstate apportionments have typically been based on the concept of 100 percent consumptive use. States participating in the compact are allocated a specific amount or proportion of the flow, with the total allocation equaling the total flow of the river. A prime example is the Colorado River, which has been fully allocated to the seven Colorado River basin states and Mexico. Water reaches the Colorado's natural outlet in the Gulf of California only in unusually wet years.[5]

Surplus flows, particularly those due to periods of unusually heavy precipitation, may frequently exist somewhere in the West. However, they may be too sporadic to support instream uses that require more consistent flows. And rivers that consistently contain more water than needed for offstream uses have increasingly become the target of new water development projects. For example, Texas has been working on plans to divert water from rivers flowing in the relatively humid northeastern portion of the state for use in growing municipalities, and the rivers of northern California have long been the target of water developers from the southern half of the state.[6]

2. *Legal obstacles.* Numerous legal obstacles prevent water from being allocated to instream purposes. Instream uses of water were often systematically excluded from the water allocation process and continue to face a variety of legal obstacles. Much of the exclusion occurred during the first one hundred–plus years of allocation under the prior appropriation doctrine, but, as will become apparent in the next chapter, continues to occur today. By far the biggest problem has been that until very recently, and even now in some states, water rights have been granted only for uses that require diversions. Reliance on markets won't lead to the most productive allocation of resources if particular uses and individuals are excluded from participation, or if they face obstacles not encountered by other uses.

Some of the obstacles faced by instream uses are being removed, but most of the West's water has already been allocated to offstream uses. Economic theory predicts that the operation of free markets will lead—in the absence of transaction costs and other market imperfections—to the most efficient allocation of resources, no matter from what initial allocation of property the process begins.[7] However, the theory says nothing about the fairness of the initial allocation. Starting out with all property in the possession of user A means that user B will bear all the costs—in the form of purchasing water rights from user A—necessary to reach the optimal allocation. Granting instream flow advocates the ability to participate in water markets once the initial property rights to the use of water have already been allocated may not be an equitable solution.

3. *Transaction costs.* As noted above, the operation of free markets should theoretically lead to the most efficient (value-maximizing) allocation of resources.[8] But the existence of a truly free market depends on several important conditions, not the least of which is zero transaction costs. Transaction costs are all those costs that make the transaction possible except the purchase price itself. Transaction costs include costs of gathering information, overcoming bureaucratic hurdles, and meeting legal requirements. Water transfers in most locations involve substantial transaction costs, mostly because of the technical difficulty of determining how much of a diversion reenters the stream as return flow, but also because of the administrative or legal processes in place to protect the interests of other water rights holders.[9]

The very real and substantial costs associated with obtaining information and effecting transfers of water from one use to another mean that water will be transferred from use A to use B only when the value of the water in use B exceeds the value of the water in use A *plus* those transaction costs. Because almost all water rights are starting out in the control of offstream water users, and transaction costs effectively raise the price to purchase these water rights, efforts to transfer water from offstream use to instream use through the market will fall short.

4. *Negative externalities.* The presence of externalities makes it unlikely that markets will produce an optimal allocation of water between instream and offstream use. Negative externalities—costs borne by those who are not directly involved in a transaction—occur every time new appropriations or water rights transfers reduce the amount of water available for instream purposes. Water allocations in most of the West have been based on the idea that society will achieve the greatest possible value from use of its resources by relying on the choices made by individuals. Each individual chooses whether to take an action—apply for a new water right, or buy or sell an existing water right—based on an assessment of whether the benefits that would accrue to the individual from taking the action will exceed his or her costs. But efficient allocations—those that would maximize the value to society from use of the resource—will occur only if *all* costs and benefits—not just those affecting the parties to the transaction—are taken into account. Western water law recognizes this principle to some extent by requiring that applications for new water rights and water rights transfers be approved only if they cause no harm to other water rights holders. Yet western water law often fails to consider the very real harms occurring to persons other than water rights holders, an especially damaging situation given that instream water users have been systematically excluded from holding water rights.[10] Negative externalities associated with offstream water use make it likely that too much water will be allocated to offstream water uses.

5. *Positive externalities.* Instream uses may be undersupplied because of positive externalities—benefits received by those who are not directly involved in the transaction. For example, the purchase of an offstream water right and its conversion to instream use produce several benefits beyond those that occur within the protected reach. One of these benefits is that more water will be available for use downstream after the water has passed through the protected reach.[11] Downstream junior water rights holders will be the primary beneficiaries, because senior water rights holders were already entitled to receive their full allocations. But even senior water rights holders will benefit because of the reduced need to put a "call" on the river during periods of water shortage. And, if return flows from the retired use had previously been altered in quality, both senior and junior downstream water rights holders will benefit from receiving water of higher quality once the offstream right is discontinued.

Similarly, the purchase of an offstream water right and its conversion to instream use produce benefits to nonconsumptive users other than the purchaser. For example, a conservation group may purchase a water right to sustain a fishery, but the same water may also be used for swimming, boating, sustaining riparian vegetation and wildlife, and enhancing the aesthetics of the watercourse and landscape throughout the protected reach. Protection of the flow may also enhance existence and bequest values for a large number of people who may never "use" the water at all. And because most instream uses don't consume water in the same way that irrigated crops or city lawns do, the same water—provided that it is not subsequently appropriated for another consumptive use downstream—can provide like kinds of benefits over and over again as it moves downstream. The property of being able to provide benefits through one use without simultaneously reducing the ability to provide benefits through some other use is called "nonrivalness" and is a common characteristic of instream flow water uses.

In spite of all these benefits, the amount of money put forward to buy rights for conversion to instream use will depend only on the benefits to the purchaser. Because of the existence of positive externalities, the amount of money tendered for purchasing rights will be less than the benefits that result from the transaction. This dynamic, applied throughout the West, will result in the purchase of too few offstream rights for conversion to instream purposes relative to the true value of such purchases.

Goods that are easily used by people who don't pay for the provision of the goods are called "nonexcludable" goods.[12] The owner of a protected instream flow might be able to charge a fee to others for use of the water and thereby capture some of the benefits of having provided the flows. For example, it may be possible to make boaters and other recreationists who use the river in the protected reach pay a fee. But in many cases it is not legally, economically, or even physically possible to do this. Collecting fees from even a small proportion of those who benefit from the flows may be difficult in situations in which the flow provider does not own the surrounding land, or the protected reach is very long and has multiple access points, or the primary benefits are received by people who never physically visit the river. It would be virtually impossible to obtain a fee from those who benefit from the aesthetic improvements to the landscape or from the provision of existence and bequest values.

Nonexcludable goods are said to suffer from the "free rider" problem. Free riders are the people who receive benefits without having to pay for them. Free riders don't necessarily evade efforts to make them pay. The more likely situation is that it is too expensive, or physically impossible, to collect payment from every person who benefits from the provision of a good. Even people who would gladly pay for the benefits may not know who, where, or how to pay for them. Who, upon admiring the beauty of a waterfall seen from afar, stops to wonder how it got there and who they should pay for the privilege of viewing it?

The real problem isn't so much that it is difficult to collect payment from all who benefit once the good has been provided, as that the difficulty of collecting payment from all who might benefit prevents the good from being supplied in the first place. Goods such as instream flow that exhibit these nonrivalness and nonexcludability characteristics are what economists refer to as "public goods." Public goods produce more benefits than can easily be captured in private market efforts to supply the appropriate quantity of the good. Market forces alone will not supply enough of the good. Other common examples of public goods include national defense, public health activities, species protection, and space exploration and technology. None of these things are left solely to markets to provide.

Private organizations such as The Nature Conservancy or the Audubon Society sometimes bring together enough like-minded people to protect instream flows through market arrangements. However, this approach is feasible only in isolated situations, for relatively small projects, because of the free-rider problem. Usually it is necessary for a public agency with the ability to collect revenues more widely, through taxation, to intervene if the appropriate amount of the good is to be supplied. This has been the theory behind creation of public agencies such as water conservancy and irrigation districts, which play such important roles in other aspects of water use in the West. Efforts to supply the appropriate amount of instream flows are also likely to require public intervention.

6. *Economic evidence.* Economic studies have indicated that there are numerous circumstances in which water may have greater value if left to flow instream than if used for an out-of-stream consumptive use.[13] This evidence contravenes the arguments made against the need for specific instream flow protection mechanisms, because it indicates that: (1) water used for instream uses not only has value, but may have more value at the margin, than the same water used for an offstream consumptive use; and (2) more water needs to be left instream than has been protected by existing mechanisms, including geographic conditions, senior water rights downstream, interstate allocations, surplus flows, and water markets.

One such study was done on the Bitterroot River in western Montana.[14] Researchers interviewed recreationists along the Bitterroot and related the interview responses to the flows occurring when people were interviewed. Based on this relationship the authors estimated how users' willingness to pay for recreation, over and above their out-of-pocket costs, varied with flow level during the recreation season. Total willingness to pay increased with flow to a point, and then decreased with further increases in flow as fishing, boating, and other activities became less desirable. Researchers added to the recreation value the value to downstream hydropower production (on the Columbia River mainstem and its Clark Fork) of leaving additional water in the stream. In economic terms, the marginal value of these instream flow benefits varied from over $100 per acre-foot at very low flows, to zero at about 1,900 cfs (figure 4.1).[15] This benefit was compared with the marginal benefit of diversions for local agriculture, which was constant at $40 per acre-foot.[16] The marginal benefit of diversions is equivalent to the marginal cost of leaving water in the stream. Figure 4.1 shows that the marginal val-

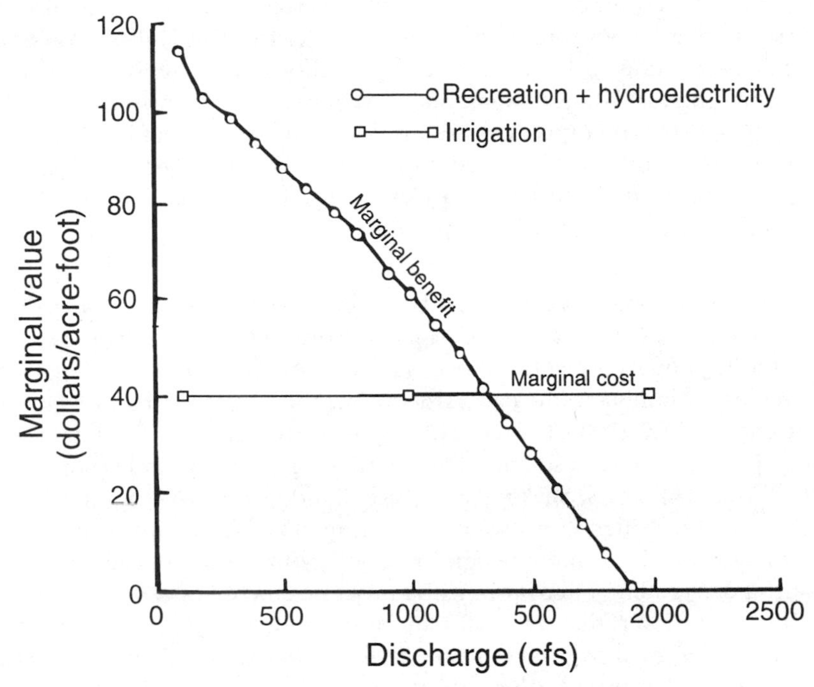

FIGURE 4.1
Marginal Benefit and Marginal Cost of Instream Flow on the Bitterroot River

ues are equalized when flows are at 1,400 cfs. When flows drop below 1,400 cfs, instream flow is a more valuable use of the water than agriculture. In other words, if the willingness of recreationists and electricity consumers to pay for instream flows could be collected without cost, those funds would be more than enough to pay the farmers to reduce diversions for irrigation.

It is important to realize that such studies do not suggest that a complete transfer of water from offstream uses to instream uses is needed. The question is how much value is produced from the last unit of water applied to an offstream use versus how much value would be produced if that unit of water were transferred to instream use. What the studies do suggest is that, *given existing allocations of water between offstream and instream uses,* transferring some water from offstream use to instream use will, in many situations, produce more total value from use of the water. Given the law of diminishing returns, the marginal value of water in offstream uses will rise, and the marginal value of water in instream uses will decline, as water is transferred from offstream use to instream use. Transfers are warranted—if efficiency is the goal—until the marginal values in the two uses are equalized.

Given that 80 percent or more of the water diverted in western states is used for irrigation, a proportionally small shift of water from agriculture could result in a proportionally large increase in the amount of water left instream.

The finding that a unit of water may have more value in instream use than if used for offstream purposes makes intuitive sense given that instream uses have been systematically excluded from water allocation processes in the past, and that water used instream may be used numerous times and still used subsequently for an offstream purpose after passing through the protected reach.

7. *Observed conditions.* People's belief in the need for special measures to protect instream flows may stem more from personal observation than from the arguments presented above. The loss of fish on rivers ranging from the mighty Columbia to the backyard stream, the loss of riparian vegetation throughout the West, the existence of degraded stream channels, diminished water quality, adverse aesthetics, and the drying up of rivers like the Gila, Salt, Powder, Arkansas, Snake, Platte, Rio Grande, and San Joaquin all provide observable evidence of stream dewatering. The lack of instream flows is becoming apparent to more people as they become aware of the issue, and as our understanding of environmental processes increases.

Our conclusion is that water allocations made in the absence of measures specifically designed to protect instream flows will result—from the standpoint of society's current values—in the allocation of too little water to instream purposes. Incidental physical and legal conditions, and the existence of markets, are by themselves insufficient to supply the socially desirable amount of instream flow. This is, of course, not a radical stance; virtually all of the western states and the federal government have implemented at least some measures to keep water in streams, indicating a similar conclusion. The conclusion leaves unanswered, however, the questions of how much public interference, and of what kind, is needed? These questions are discussed in following chapters.

How Much Water Should Be Kept in a Stream?

Having addressed the issue of whether water should in general be left in streams, it is now necessary to look at the question of how much water should be left in any particular stream. As discussed in chapter 3, the benefits of instream use will vary with both the magnitude and the timing of flows, so different flow regimes are associated with different levels of benefit. Protection of only minimal amounts of water, particularly during critical portions of the year, may result in a very low level of instream benefits, whereas more substantial flows could result in a much higher level of benefits. In the words of one commentator:

> True minimums do little in the long term to protect the values that are associated with riverine systems, whether those values be fisheries, riparian habitat, aesthetics, or white-water recreation. To say that one has "preserved for recreational use" an internationally renowned stretch of white water by capturing a flow that represents the low end of recreational utility rings hollow. Optimal flows create celebrated white water; minimum flows do not.[17]

The choice between protecting streamflows at "minimum" or "optimum" levels exists everywhere and is the source of a good deal of confusion. In practice, words like "minimum" and "optimum" have been used to mean different things by different people. Unique circumstances of individual rivers and uses make it impossible to define a single function that accurately depicts the relationship between flow level and benefits, but for the sake of discussing the implications of using terms such as "minimum" and "optimum" to describe instream flows, consider figure 4.2.

Figure 4.2a portrays a "typical" relationship between flow level and total benefits in which the benefits attributable to instream flow rise reach a maximum (at Q_t), and subsequently diminish as streamflow levels further increase.[18] Of course, total benefits are only half of the picture. The other half consists of the costs. The primary component of total cost is the forgone benefit to consumptive water uses of keeping water in streams.[19] A secondary component, adding to total cost at very low and possibly also at very high flow levels, is the permanent or at least long-run damage sustained by fish and other organisms, and possibly by other instream uses.[20]

Net benefit is the difference between the total benefit and total cost of instream flow (figure 4.2b). Figure 4.2b illustrates the possibility that, at very low and very high levels of flow, net instream benefits can be negative. Keep in mind that for actual rivers and uses the shape and position of the curve will vary with the particular resource and circumstances studied. For some uses and rivers even a very small flow may be enough to produce positive net benefit. And in some rivers even very large flows, if infrequent, may yield a positive net benefit if such flows are critical for channel maintenance.

The consequences of choosing to protect minimum rather than optimum flows can be substantial. In figure 4.2b the minimum flow is the flow level at which net benefits first start to become positive (Q_{min}). Protecting instream flows at or near this level will yield a very low level of benefit.[21] Increasing streamflow from Q_{min} to Q_{opt} will yield progressively greater net benefits. At streamflow levels greater than Q_{opt} net benefits decrease but remain positive until reaching Q_{max}, the level at which net benefits subsequently become negative.

Note the very real possibility that the same level of social net benefit could be achieved by two different levels of flow. For example, a benefit level of B_a

(a)

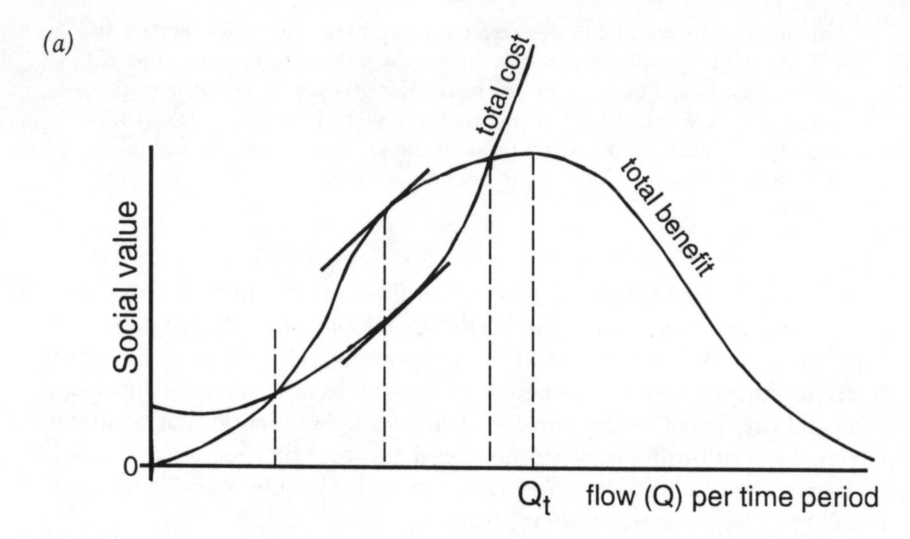

(b)

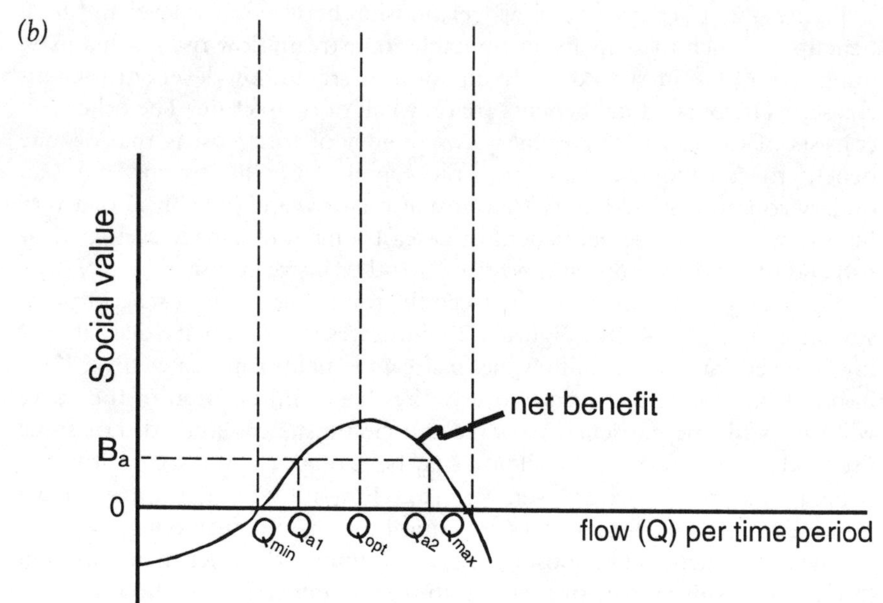

FIGURE 4.2
Hypothetical Benefit and Cost Curves for Instream Flow

could be achieved at a flow level of either Q_{a1} or Q_{a2}. Providing benefits at a streamflow level of Q_{a2} would favor instream uses at the expense of offstream uses, whereas flows at Q_{a1} would do the opposite. Neither situation is optimum.

The timing of flows affects the level of benefits achieved. The need for different flow levels at different times of the day or year is most apparent for biota, from the needs of fish to the needs of riparian vegetation, but also exists for uses such as recreation (activities such as swimming and boating generally take place only at specific times of the day and year), aesthetics (riverine features may be visited only at certain times of the day or year), channel maintenance, navigation, and water quality. Preserving flows year-round at a level needed only during a portion of the year, such as during the period that fish are incubating, is not likely to produce optimum benefits given the very different flows needed at other times of the year. So to achieve maximum benefits the question of how much water should be left in streams needs to be considered in light of when the flows are needed.

It is useful at this point to return to the principle that the most efficient allocation, the allocation that produces maximum benefits, is the one at which the last unit of water allocated to each use produces an equivalent value. This equimarginal principle is evident in figure 4.2, where the equality of value in offstream and instream uses is indicated by the point at which the slopes of the total cost and total benefit curves are equal. At this point, the change in total cost (i.e., in benefit to offstream uses) is equal to the change in total benefit to instream uses. This is also, of course, the point at which the net benefit of instream flow is maximized, signified by Q_{opt}.

CHAPTER FIVE

Instream Flow Protection Issues in the States

The acute drop in western streamflows over the last 150 years was matched by growing awareness of the value of instream flows. Organized interest in water recreation and in the need to protect streams occurred toward the end of the last century, but this early attention was generally limited to few individuals and situations.[1] However, by 1958 the federal Outdoor Recreation Resources Review Commission noted a wider and growing public interest in the preservation of natural areas and rivers and in their recreational potential.[2] The environmental movement of the 1960s and 1970s brought even more recognition to river issues. This lay interest had its scientific counterpart. Led by fisheries and wildlife biologists, instream flow protection became recognized as its own area of research in the 1940s.[3] By the early 1970s instream flow had become a priority item on the research agenda of many water resources professionals, particularly in the Pacific Northwest.[4] The first large conferences on instream flows were held in Logan, Utah, in 1975, and in Boise, Idaho, in 1976, and the subject has since become widely studied and discussed among water resources professionals and the water development community throughout the West.

The states, because of their primary responsibility for the allocation and administration of water resources, have been the venue for much of the debate over river issues as well as the source of many of the West's instream flow protection policies. Each state has had to work within the context of its own geography, climate, and legal, political, and social institutions when addressing the subject, and yet has faced, or will soon face, similar problems and issues. This chapter is devoted to a discussion of these problems and issues, and to a description of the ways in which they have been addressed by the western states. Chapter 6 contains a more detailed discussion of the particular methods used by the states to protect instream flows, such as minimum flows, instream reservations and rights, administrative procedures, and the public trust doctrine.

Who Creates Instream Flow Protection Programs?

All three branches of government have played important roles in water allocation. This section briefly mentions the legislative role and then provides some detail about the less obvious administrative and judicial roles. Citizen initiatives have also been used to create instream flow protection programs on occasion, as when Oregon's Scenic Waterways Program was created by a citizens' ballot initiative in 1970. More often than not just the threat of a citizen initiative has been enough to move legislatures to act. For example, the desire to preempt citizen initiatives that would have created more far-ranging opportunities to appropriate water for instream purposes in Colorado and Wyoming provided much of the motivation for legislatures in those two states to authorize their own, more restrictive, instream flow protection programs.[5]

Legislatures

Legislatures, as elected, representative bodies, seem to be an appropriate venue for debating issues and setting state water policy. Legislatures can also be a difficult place to get instream flow protection measures passed, because of the controversy and strong feelings surrounding instream flow protection issues and the substantial clout that traditional water users often wield in state legislatures. Nonetheless, state legislatures have been the primary source of new instream flow protection measures in eleven western states: Alaska, California, Colorado, Idaho, Kansas, Montana, Nebraska, Oregon, Utah, Washington, and Wyoming. All of these states but California have authorized instream flow rights, minimum flows, and instream reservations as their primary methods of protecting instream flows. The California legislature created a state Wild and Scenic Rivers program in 1973 and authorized the transfer of existing rights to instream purposes in 1991, but it provided the greatest protection to instream flows by directing the state's administrative agency to protect flows through administrative methods described in chapter 6.[6]

State legislatures in Arizona, Nevada, New Mexico, North Dakota, Oklahoma, and South Dakota have not yet acted to provide significant protection for instream flows. The state legislature in Texas has been the source of a limited amount of protection but has not yet fully embraced the concept of creating special measures to protect instream flows. There is a tendency to believe that a state that has not legislatively implemented any instream flow protection programs has no mechanism to protect flows, but, as seen below, instream flows have also been protected through administrative and judicial action.

Administrative Initiative

Administrative initiative has been responsible for the creation of instream flow protection programs in some of the states where legislatures have not yet acted. Administrative initiative has often followed a similar process: (1) private citizens, groups, or public agencies submit an instream water rights application to the state water rights agency; (2) the administrative agency determines that nothing in state law specifically prohibits the use, and so grants the permit; and (3) the process is repeated by other applicants, creating a de facto instream flow protection program. The western states that do not have legislatively created instream flow protection programs are at different stages in this process; some have not yet begun. Examples from Arizona, South Dakota, and Texas illustrate the ways in which administrative agencies have created, or could create, substantive instream flow protection programs.

Arizona. Arizona's program was initiated in 1979, when the Arizona Nature Conservancy submitted two instream flow applications to the Arizona Department of Water Resources (DWR).[7] The Conservancy wanted the water to support wildlife and vegetation in two of its southern Arizona nature preserves. The Conservancy was relying in part on a 1976 court decision that had ruled that *in situ* uses of water were permissible under state law.[8] The decision in that case, which concerned a water impoundment rather than an instream flow, was guided by the fact that the legislature had added "wildlife, including fish" and "recreation" to the state's list of beneficial water uses in 1941 and 1962, respectively. The court believed that the addition of these two uses, which do not require diversions, was evidence that a diversion is not necessary to receive a water right under state law.

DWR applied the same criteria to the Conservancy's instream flow applications as it did to other applications. The agency required the Conservancy to prove the appropriation would be for a beneficial use, would not be in conflict with vested rights, and would not be a menace to the public safety or against the interests and welfare of the public. DWR granted the permits in 1983, following standard public notice and hearing procedures and after concluding that state laws did not expressly forbid instream appropriations.

DWR requires holders of offstream water permits to show proof of having actually applied the water to the designated beneficial use before the right is "certificated" (vested). For the Conservancy's instream permit, DWR decided that such proof would consist of several years' worth of hydrologic monitoring data proving that flows in the requested amount actually existed and were being used instream. DWR issued the Conservancy a certificated water right in 1990 after submission of the required data.[9]

During the course of this process DWR decided that it needed better procedures for handling subsequent instream water rights applications. To this end, DWR created an interagency task force in 1986 to establish guidelines and criteria for evaluating instream flow applications. The task force included both government and nongovernment personnel, including representatives of federal agencies. Task force recommendations were included within the state's "Guide to Filing Applications for Instream Flow Water Rights in Arizona," issued by DWR in 1991.

In 1992 the state legislature considered, but did not enact, codification of the state's instream flow policy. Arizona's instream flow protection program is still wholly a product of the Department of Water Resources. As of April 1994, three instream flow applications had received certificated water rights. Seven more had been issued permits, and 61 applications were awaiting approval by DWR.[10]

South Dakota. The granting of water rights for instream purposes in South Dakota has been so quiet that many people aren't even aware that the state has recognized such water rights. The state administrative agency, the Water Rights Division of the Department of Environment and Natural Resources, and the Water Management Board, which has the responsibility for approving or denying permit applications, have never specifically referred to the rights as "instream rights."[11] Nonetheless, the Water Management Board has granted two permits to the Division of Wildlife, Game, Fish and Parks for what are essentially instream purposes.[12]

The Water Management Board has also recently granted a change-of-use request on a water right held by a private organization that wanted to convert part of its water right to instream use.[13] The Homestake Mining Company had been diverting water at several points in Spearfish Canyon, in the Black Hills, for use in generating electricity. Homestake decided that it could bypass up to 20 cfs of water at one of its diversion points and then subsequently divert the water three or four miles downstream at a second hydropower facility. The increased flows between the two points could be used for fish and to improve the aesthetics of Spearfish Falls. Homestake applied to the Water Management Board to change the use of 20 cfs of its water right to fishery purposes, and the permit was granted by the Board in February 1995. The right may eventually be transferred to the U.S. Forest Service as part of a land swap agreement.

A recent South Dakota water rights application that drew considerably more attention was closely related to instream flows. The U.S. Fish and Wildlife Service was granted a permit by the Water Management Board for a new water right to the natural flow of water from six springs to be used in supporting habitat at the LaCreek National Wildlife Refuge.[14] The application was for the right to use the water to maintain 235 acres of marshes, sloughs, and wet meadows for wildlife habitat rather than for an "instream" purpose per se, but it was related to instream flow because of similar purposes (wildlife) and the fact that the U.S. Fish and Wildlife Service did not intend to divert any of the spring water when putting it to beneficial use. The application was protested by several nearby landowners, who appealed the Water Management Board's decision to grant the permit. The appeal was taken all the way to the South Dakota Supreme Court, which affirmed the Board's decision. The court ruled that no diversion was necessary to secure a water right in South Dakota and that the maintenance of vegetation to support wildlife constituted a beneficial use of water under state law. This decision could provide a foundation for a variety of instream flow claims in the future.[15]

Texas. Texas does not yet recognize instream flow rights. The Texas situation is interesting because there does not appear to be any legal obstacle that would prevent the Texas Natural Resource Conservation Commission (TNRCC, the state's water rights administrative agency) from taking the same path as its counterparts in Arizona and South Dakota and granting a permit for an instream flow right.

Texas statutes list several beneficial uses of water, including domestic, municipal, industrial, irrigation, mining, hydroelectric power, navigation, recreation and pleasure, stock raising, public parks, and game preserves. Should this list of uses not be sufficient to justify a particular instream flow claim, Texas law also allows the use of water for "other beneficial use."[16] No Texas court has ever declared any particular use of water improper.[17] Texas water law does not explicitly authorize the appropriation of water for instream purposes, but it does not explicitly require a diversion, either. Texas law currently defines a water right as "a right acquired under the laws of this state to impound, divert, *or use* state water" (emphasis added).[18]

The absence of statutes or court decisions that would prohibit the granting of an instream flow right has been noticed by state agencies. In 1977 the Texas Parks and Wildlife Department, represented by the state attorney general's office, filed a water rights claim in the general adjudication of water rights in the Medina River subbasin of the San Antonio River for the purpose of supporting fish and wildlife. However, the agency subsequently decided not to pursue the claim, and the claim was withdrawn.[19] Nonetheless, instream flow rights continue to be contemplated at the Parks and Wildlife Department[20] and at the TNRCC.[21] The TNRCC has on occasion issued water rights permits for very similar purposes, including the enhancement of habitats for wildlife, even when permit applicants have been private parties rather than public agencies.[22] But, so far, the agency has not issued any permits for true instream flow rights.

There are undoubtedly a number of strategic and political considerations involved in the decision to apply for and grant an instream flow water rights permit, but agencies in other states have faced the same considerations and decided to proceed. The difference between Texas and the other states may well be a case of bureaucratic inertia.

Other States. None of the other states without statutory programs have gone as far as Arizona in the process of recognizing instream rights through administrative initiative, though some others have made a start. The Nevada state engineer granted a water rights permit for in situ use to the federal Bureau of Land Management, which was upheld by the Nevada Supreme Court in 1988.[23] Like Arizona, Nevada now allows any party, public or private, to appropriate water for instream use. Since 1988, two additional permits for new instream appropriations have been granted, both to the U.S. Fish and Wildlife Service. The Nevada Division of Wildlife has also converted some existing offstream water rights to instream use on ranches the Division has purchased to protect wildlife.[24] A severe lack of available surface water—Nevada is the nation's driest state—may be the main reason more instream water rights have not been granted in the state.[25]

New Mexico may be in the process of becoming the most recent state to recognize instream flow rights through administrative means. The New Mexico Supreme Court ruled in 1972 that a diversion was necessary to establish a water right for the purpose of irrigation,[26] but there are a number of reasons to think that the diversion requirement may not apply to other claims.[27] New Mexico courts have previously recognized recreation and fishing as beneficial uses of water,[28] and an application to change an irrigation right to instream purposes is working its way through the state engineer's office. The application has not been vigorously protested by other water users.[29]

Instream rights may also eventually be established through administrative means in Oklahoma and North Dakota, the only two other states without statutory protection programs. State law makes no explicit provision for instream flow water rights in Oklahoma, but the state attorney general's office has determined that there probably is no legal barrier to such rights.[30] Oklahoma has not yet experienced problems with dewatered rivers to the same degree as the other western states and still has substantial quantities of unallocated storage water,[31] but instream issues are likely to become more important in the future. The Oklahoma Department of Wildlife and Conservation has had some preliminary discussions with the Oklahoma Water Resources Board about recognizing instream water rights claims, but no applications have yet been made.[32]

There definitely are problems with overappropriated rivers in North Dakota, and recreational interests have put some pressure on state agencies to protect instream resources there. The state Game and Fish Department has had a few internal discussions about the possibility of applying for instream rights, but the state engineer's office, which would have to approve the rights, has not yet been very receptive to the idea.[33]

The Courts

The courts have played an important role in instream flow protection. State courts have generally supported efforts to create water rights without diversions and to include a variety of instream water uses within the definition of "beneficial use." Court decisions have encouraged and affirmed the actions of administrative agencies and ratified programs created by state legislatures. Three of these decisions were briefly described above: *McClellan v. Jantzen* in Arizona, *Nevada v. Morros* in Nevada, and *DeKay v. United States Fish and Wildlife Service* in South Dakota. Each of these cases played an instrumental role in advancing the efforts of administrative agencies to grant water rights permits for water uses that didn't require diversions.

A few of the court decisions in other states that have been important in efforts to protect instream flows include: *Colorado Water Conservancy District v. Colorado Water Conservation Board,*[34] in which the Colorado Supreme Court in 1979 affirmed the constitutionality of the state's instream flow rights statute and held that the "right to divert" language in the state constitution was only a rejection of the riparian rights doctrine and not a requirement that a diversion be associated with a water right; *State Department of Parks v. Idaho Department of Water Administration,*[35] in which the Idaho Supreme Court in 1974 upheld the constitutionality of a legislative act that directed the state Department of Parks to appropriate the water of several springs in trust for the people of the state, ruled that no diversion was necessary to create a water right, and held that the water uses listed in the state constitution did not preclude appropriations for other uses, including recreation and aesthetics; and *In Re Application A-16642,*[36] in which the Nebraska Supreme Court in 1990 upheld the legality of the first instream appropriation granted by the Department of Water Resources and affirmed the constitutionality of the state's instream flow appropriation statute.

Courts in at least two states have recognized water rights for instream purposes created through methods other than those specified in instream flow protection statutes. In 1992 in the case of *City of Thornton v. City of Fort Collins,*[37] the Colorado Supreme Court ruled that the City of Fort Collins could appropriate a specified amount of flow of the Poudre River at two control points—the first a combination boat chute and fish ladder and the second a dam that redirected the flow of the river into a historic channel—located on either end of a river reach used extensively for recreation and aesthetics.[38] In Montana, water courts have used similar reasoning in adjudication proceedings to support water rights claims by the state Division of Fish, Wildlife and Parks for instream purposes.[39] In both states the claims have been supported only when diversions of some sort have been present. In Colorado, the court took pains to point out that the appropriation by the City of Fort Collins did not constitute an "instream flow right," which statutes said could only be appropriated and held by the Colorado Water Conservation Board. The right was based instead on the statutory definition of a diversion, which could include any facility that controlled flows.[40] Most water rights for instream purposes in Montana have been granted only in situations in which physical control structures are present, because the state supreme court has been adamant in denying claims that do not depend on such facilities.[41] However, a recent water rights compact between the State of Montana and the National Park Service includes some instream water rights that don't include structures. While the compact has not yet been approved by the courts, the compact has been passed by the state legislature and ratified by Montana's governor, and the parties do not expect any problems from the courts.[42]

In some situations the courts have acted on their own to protect instream flows. The most potent example of such action is application of the public trust doctrine, which is described in more detail in chapter 6. The public trust doctrine is common law, a product of the courts rather than of the legislative or administrative branches of government, and has enormous potential to protect instream flows. Other actions may be much more modest, such as a ruling by the supreme court of Utah that the state engineer must apply criteria protective of instream flows—determining that appropriations will not damage public recreation or the natural stream environment—to water rights change and transfer applications in addition to applications for new appropriations.[43] Whether having a large impact or small, these and other actions by the courts lend some credence to the observation that "courts everywhere now recognize that in-place water uses must be legally protected, and that the prior appropriation system itself fails to provide adequate protection. The real question now is not *whether* but *how* this protection is to be provided."[44]

For What Purposes Should Instream Flows Be Protected?

From the perspective of a society that wants to maximize the value it receives from use of scarce water resources, there should be no question about what uses will be allowed; maximum productivity of the resource demands that water be allocated to whatever uses are most valuable. However, this rule does not necessarily prevail throughout the West, where there have long been restrictions on the purposes for which water can be used. Until the last couple of decades, when courts started to rule that a variety of water uses could be beneficial, the requirement that water must be put to "beneficial use" was widely thought to preclude all but a few standard consumptive uses, such as the use of water for irrigation, mining, municipal, and stock watering purposes. As described in chapter 2, many states have established statutory water-use preferences in which water uses are ranked. Where listed at all, instream uses are often at the bottom of such preference lists,[45] perhaps reflecting the belief that water for instream uses is a luxury that can be afforded only when there are no other uses for the water. Beneficial-use listings and water-use preferences may have reflected earlier conceptions about the most valuable uses, but they have also had the effect of excluding whole classes of valuable uses from legal protection.

Even states that recognize instream uses of water as beneficial often limit the purposes for which flows can be protected. The preservation of fisheries is considered a permissible use in every state that gives legal protection to instream flows. In some states the protection of fisheries has been the only purpose for which instream water rights could be granted, as was true in Utah until 1992 and is still true in Wyoming. Several states recognize wildlife, water quality, and recreation, and a few states explicitly recognize aesthetics as a use that can be protected. Where not listed, instream flow protection proponents often find it difficult, or impossible, to successfully seek legal protection for other water uses. For example, Colorado does not yet recognize recreation as a valid purpose for seeking an instream flow water right, so the Colorado Water Conservation Board will not apply for, or accept donations of, water rights for recreational purposes.

The western states that have explicitly recognized the broadest array of purposes for which instream flows can be protected are Alaska, Idaho, and Washington. These states recognize fish and wildlife, other aquatic life, recreation, scenic and aesthetic, water quality, navigation, and transportation as water uses worthy of protection. In many cases one explicitly specified use can be construed to include other uses. For example, the Arizona Department of Water Resources interprets the statutorily designated beneficial use of "fish and wildlife" to include riparian vegetation and "recreation" to include aesthetics.[46] In Alaska, "recreation and park purposes" has been extended by regulation to include scenic, natural, historic, and cultural values.[47] "Wildlife" in particular is a use that is often interpreted to include a variety of habitat-enhancing measures. Some statutes employ language broad enough to easily encompass a variety of uses. For example, Washington ("all other uses compatible with enjoyment of the public waters of the state") and Texas ("other beneficial uses") define beneficial use so broadly that a variety of uses can qualify. Colorado authorizes the creation of instream water rights for the purpose of protecting "the natural environment to a reasonable degree," and Utah for the purpose of "reasonable preservation or enhancement of the natural stream environment." States defining accepted uses broadly may be the only ones in which instream flows for purposes such as channel maintenance and general environmental protection may be allowed.

Sometimes instream uses that have not been recognized as worthy of protection can be protected indirectly. For example, flows sufficient to sustain fish may also be sufficient to maintain riparian vegetation, improve aesthetics, and protect other wildlife species and water quality. Unfortunately, the needs of fish are not always a sufficient umbrella for other uses, particularly riparian regeneration, channel maintenance, and many forms of recreation. Recreational uses of water, particularly whitewater boating, can require much more water than would be necessary to sustain aquatic species.

There likely are streams in virtually every western state where recreation, or riparian regeneration, or channel maintenance, or aesthetics, or other uses not receiving legal protection are more highly valued uses than some of the offstream water uses that can be legally protected. The narrow construction of purposes for which instream flows can be protected is not conducive to the goal of achieving maximum possible value from our use of rivers and streams.

Who Can Participate in Protection Activities?

The prior appropriation doctrine has emphasized the allocation of water to and by individuals, but in most of the western states the protection of instream flows has been treated quite differently. State government has assumed much of the control over protection activities, and individuals have been highly constrained in their opportunities to protect instream flows.[48] The states, normally fierce defenders of private property rights in water,[49] often seem to work hard to exclude individuals from the most direct methods of participation in the instream flow protection process.[50]

Table 5.1 gives a complete listing, by state, of who is allowed to participate in the protection process. The list shows that only Alaska, Arizona, and Nevada explicitly allow private entities to apply for and own water rights or reservations for instream purposes. California and South Dakota have allowed individuals to change the purpose of existing water rights to instream use and retain ownership of the rights, and in some other states individuals are allowed to donate water rights to the state to be used for instream purposes. But in all other cases these roles have been reserved exclusively for public agencies. The role played by private citizens in many states has essentially been limited to the right to petition one of the authorized public agencies to act on their behalf. Some states are even more restrictive; most instream appropriations or minimum flows recommended by state agencies in Idaho, Kansas, and Utah must also be submitted to the state legislature for approval.

Several arguments have been put forward to support these restrictions on private individuals. Instream uses often benefit the public as a whole rather than particular individuals, so it might make sense that public agencies should have control over instream protection activities. Control of protection activities by public agencies may also enable better coordination of instream flow protection with other state water-use objectives.[51] Public control is also compatible with the idea—also expressed by the public trust

TABLE 5.1
Who Do the States Allow to Participate in the Protection of Instream Flows?

Alaska	Any public or private entity may apply to the Alaska Department of Natural Resources for an instream reservation.
Arizona	"Any person or the state of Arizona or a political subdivision thereof"[a] may apply to the Department of Water Resources for an instream water right. "Any person" has been interpreted to include federal agencies.
California	The state's most potent instream flow protection is a result of administrative activities of the State Water Resources Control Board, which is required to consider the comments of the state Department of Fish and Game when making decisions about appropriation and transfer permits. Since 1991, individuals have been authorized to change the purpose of existing rights to instream purposes. Private individuals and organizations have also taken advantage of the opportunity to initiate public trust proceedings.
Colorado	Only the Colorado Water Conservation Board is allowed to apply for or hold an instream water right. The Board is required to request recommendations from the state Division of Wildlife and Division of Parks and Outdoor Recreation, and, since 1986, from the U.S. Departments of Agriculture and Interior. Individuals can acquire existing rights and donate them to the state for instream purposes.
Idaho	Only the Idaho Water Resources Board is allowed to apply to the Department of Water Resources for an instream water right. State statutes allow "the public" to petition the Board to apply for instream flow rights, but the Board has interpreted this language to mean that it may accept petitions only from state agencies.[b] Applications approved by the Department of Water Resources must be submitted to the state legislature for approval.
Kansas	Minimum streamflows are established by the state legislature and administered by the office of the Chief Engineer of the Division of Water Resources of the Kansas State Board of Agriculture. Several state agencies have been involved in making minimum streamflow recommendations to the state legislature, including the departments of Health and Environment, Wildlife and Parks, Water Resources, and the Kansas Water Office.[c]
Montana	Federal agencies and any political subdivision of the state, including municipalities, conservation districts, and state agencies, may apply to the Board of Natural Resources and Conservation for instream flow reservations. Any public or private entity may lease water for instream purposes.
Nebraska	The state's Natural Resource Districts and the Nebraska Game and Parks Commission may apply to the Department of Water Resources for an instream water right.
Nevada	Any person may apply to the State Engineer for an instream water right. The state supreme court has declared that "any person" must include federal and state agencies.[d]

(*Table Continues*)

TABLE 5.1 (CONTINUED)

New Mexico	Has not specified any method of protecting instream flows.
North Dakota	Has not specified any method of protecting instream flows.
Oklahoma	Has not specified any method of protecting instream flows.
Oregon	Only the Oregon Water Resources Department may hold instream water rights. The Water Resource Department considers requests from the state Fish and Wildlife, Environmental Quality, and Parks and Recreation agencies. Individuals may acquire existing rights and take responsibility for changing the use to instream purposes in an administrative hearing, but then must turn the right over to the Water Resources Department to be held in trust.
South Dakota	No explicit rules have been developed. So far the South Dakota Water Management Board has granted permits for instream uses to South Dakota Game, Fish and Parks, and an existing offstream right held by a private company has been changed to instream use. It is probable that the state will allow federal agencies to hold permits issued for instream purposes.
Texas	The Texas Natural Resource Conservation Commission, with advice from Parks and Wildlife, employs administrative procedures to protect some instream flows. Individuals may protest offstream water right applications that would leave them without enough water for domestic and stock watering purposes.
Utah	The state Division of Wildlife Resources and Division of Parks and Recreation may apply to the State Engineer to change existing rights to instream purposes. The two agencies may purchase rights to be changed to instream use, but only with the approval of the state legislature. Individuals may donate existing water rights to these two agencies, and the agencies must then apply to the State Engineer for the change. No person or entity, public or private, is allowed to make a new appropriation for instream purposes.
Washington	The state Department of Ecology establishes minimum flows either at its own initiative or after request from the Department of Fisheries and Wildlife. Because minimum flows are established through administrative rule-making procedures, public notice and hearings are involved. Individuals may donate rights to the state and specify that they are to be used for instream purposes under the state's trust water rights program, which is administered by the Department of Ecology.
Wyoming	The Wyoming Game and Fish Department identifies priority streams, performs studies, and makes instream flow recommendations to the Wyoming Water Development Commission, which in turn applies to the State Engineer for a water right. Only the state may hold an instream right. Individuals may donate water rights to the state to be used for instream purposes.

[a]Ariz. Rev. Stat. Ann. sec. 45-151.A (1987).

[b]Just (1990).

[c]Rolfs (1993).

[d]Potter (1990).

doctrine—that some things, including the public use of rivers and streams, ought to be beyond the reach of private control and ownership.

Public control over instream flow protection may also prevent speculation in water. Speculative offstream water-use claims are restricted by the need to build water-control structures and prove diligence in putting water to a beneficial use. But no control works are necessary for instream uses, and instream flows are put to "beneficial use" without doing anything at all, so the barriers to speculative instream claims are much lower than for offstream uses. Reliance on public agencies to determine whether flows need to be protected may eliminate the possibility of speculative instream water claims.

The comparative ease of claiming water for instream use might also frustrate efforts to claim water for other uses.[52] Some have expressed the fear that with no need to exert control over river flows or build offstream infrastructure, it would be too easy to appropriate water for instream purposes; offstream water users would suffer a competitive disadvantage. If individuals were allowed to appropriate instream flows, the comparatively low cost of appropriating water for instream purposes might cause too much water to be tied up by those who "enjoy the sound of free-flowing water running past their homes."[53] This seems to be the crux of the argument in favor of restricting instream flow claims to public agencies; many state legislators and existing water users fear the possibility of environmentalists running rampant and claiming all remaining flows for instream use, thereby eliminating the possibility of future offstream uses.

This fear has been a factor behind adoption of restricted-access instream flow protection programs in several states. For example, Colorado's state engineer noted in 1989 that the "water buffaloes" in Colorado had been afraid that "lunatics" would appropriate all of the state's water if individuals were authorized to appropriate water for instream flows, and that this fear helped to motivate passage of the 1973 act that authorized the Colorado Water Conservation Board to appropriate water for instream purposes.[54] Similarly, many members of Wyoming's state legislature believed that allowing only the state to appropriate water for instream purposes would prevent "radicals" from appropriating all available water in the state;[55] Nebraskans expressed fears about "radicals" locking up remaining flows;[56] and water users in Oregon feared that "environmentalists" would buy up all the water, leaving none for agriculture, after 1987 legislation allowed private parties to purchase existing water rights and donate them to the state for instream purposes.[57] One commentator noted that the main reason that only public agencies are allowed to apply for instream flow reservations in Montana is that agricultural interests—which were heavily represented in the state legislature and also generally opposed to instream flow protection measures—believed that the state legislature could more easily control the agencies than it could control private individuals and organizations.[58]

Other considerations suggest that private individuals and groups should be allowed to participate directly in the instream flow protection process.

Foremost among these is evidence that the fears of many existing water users and state legislators are unfounded. For example, as of March 1995, very few water rights had been transferred to instream use in Oregon.[59] In the three states that have explicitly authorized private individuals to make new appropriations for instream purposes—Alaska, Arizona, and Nevada—very few instream water rights have been granted, and almost all of the applications that have been granted have been submitted by public agencies. In Alaska, all but one of the applications granted to date have gone to the Alaska Department of Fish and Game, and the other went to the federal Bureau of Land Management. Most instream applications in Arizona have been submitted by the U.S. Forest Service and Bureau of Land Management; the Arizona Nature Conservancy is the only private organization to have been granted an instream water right in the state. No new appropriations for instream purposes have yet been made by private individuals or organizations in Nevada. Even a quick look at what has happened in these three states indicates that there is little evidence to support the claim that private parties are standing ready to lock up all remaining flowing water if given the opportunity to appropriate water for instream uses.

Substantial barriers exist to prevent the lockup of state water even if private individuals wanted to do so. For example, water rights claims for instream purposes must go through the same administrative application procedures as claims for offstream water uses. Public notice must be given of the proposed claim, other water users and interested parties have the opportunity to file objections, and hearings are held. Perhaps most important, instream flow claims have to meet the same public interest criteria as other water claims. Faced with water claims that would effectively lock up all of a state's remaining water supplies, it is unlikely that an administrative agency or court would find the granting of such claims to be in the public interest. State legislatures could enact statutes that specifically guard against the possibility that remaining water would be locked up by a single use by stating that the state's water supplies should be available for a variety of uses.

There are other barriers to the rapid proliferation of private claims for instream flows. One is cost. Applicants must typically present data on both the amount of water needed for the intended use and the amount of water that is actually available for appropriation. Applicants are often required to submit more data for instream flow claims than for offstream water-use claims.[60] The application process, including data gathering, verification, and defense, if the application is protested, can take several years. It may be necessary to seek legal advice during the application process and to defend the water right from infringement by other water users after the right is granted. Obtaining and then supporting a water right, especially for a type of use or location that may be controversial, is neither simple nor inexpensive.[61]

Another argument for allowing individuals to act directly to protect instream flows is that private individuals and groups will do a better job than public agencies in determining the correct water allocation.[62] People may disagree about the validity of this argument, but it is essentially the same one that has been used to justify our current system of allocating water for off-stream uses. Individuals may be in a better position to evaluate the need for instream flow protection measures. One commentator has argued that "local fishermen, rafting companies, recreational users, and conservation organizations are likely to be far more aware both of impending threats to a particular river and of the benefits of preserving a certain level of streamflow than would a government agency charged with managing all instream appropriations in the state."[63] It would be wise to incorporate this kind of local interest and knowledge in the decision-making process no matter what kind of system is used to allocate water.

The cost of monitoring may suggest a need to incorporate private individuals more directly in the instream flow protection process. Though administrative agencies are now largely responsible for the enforcement of water rights in the West, agencies in most states are still likely to depend on individual water rights holders to let the agencies know when private rights have been infringed. People using the water rights are in the best position to know whether the rights are being damaged by the actions of others. Any attempt by an agency to comprehensively monitor all of a state's water rights would be prohibitively expensive. But if private individuals are not allowed to participate in the administration of protective measures, the state agencies themselves must take on the responsibility for both monitoring and enforcement. This can be a difficult task, especially given that administrative agencies are already finding it difficult to enforce existing water rights.[64] Assigning monitoring responsibility to resource-starved agencies, or to agencies that do not place a very high priority on the enforcement of instream flow protection measures, may significantly lower the probability of enforcement. Enforcement would be enhanced if states authorized private groups and individuals, the actual water users, to help monitor the protective measures.

The possibility that individuals could help administer publicly held rights illustrates the fact that protective measures do not have to be 100 percent public or 100 percent private. There may be many opportunities for state agencies and private individuals to share responsibility for protective activities. One innovative sharing agreement has been worked out in Colorado, where the state legislature authorized the Colorado Water Conservation Board, a state agency, to engage in contractual agreements with private individuals who donate water rights to the state. The state retains control of the right—only the state is allowed to hold instream water rights in Colorado—but individuals and organizations may sign contracts with the state to ensure that the rights they have donated will be used for instream purposes. Should the Conservation Board fail to enforce these rights, the individual or group would then have legal recourse, under the provisions of contract law, to force the agency to take appropriate enforcement measures. The state still achieves many of the benefits of public ownership. The state can ensure, for example, that instream rights are not created for speculative purposes. But individuals are allowed to participate in the administration process and ensure that the rights they have donated to the state will be put to the intended use.

The privileges of individuals who donate water rights to the state are an issue in several states. Once a right is donated, it belongs to the state. The state could conceivably dispose of the right, convert it to some other use, or fail to enforce it. The inability of individuals to specify how the right is used after it is donated to the state may be a major impediment to the more widespread donation of water rights to the states for instream purposes.[65] Arrangements like Colorado's may help to overcome some of the reluctance of water rights holders to donate their rights to the state. Colorado is the only state where donations of water rights to the state to be used for instream purposes have occurred to any significant degree.

Perhaps the most frequently heard argument for allowing private individuals to protect instream flows relates to fairness. The argument asks: If private individuals and organizations are trusted to enhance the public welfare by appropriating water for consumptive offstream uses, why should private individuals not be trusted to enhance the public welfare by appropriating water for nonconsumptive instream uses? And if states allow individuals to appropriate water for offstream uses that produce only private benefits, why should individuals not be allowed to appropriate water for instream uses that produce only private benefits?[66] Further, if it is worth more to a homeowner to hear water flowing past the house than it is to a farmer to irrigate another few acres, then why shouldn't the homeowner be allowed to buy the right for instream purposes? Some states, such as Alaska and Arizona, will grant instream reservations and rights only if the applicant can ensure that the public will have access to the protected reach of river,[67] and Idaho clearly specifies that instream water rights must be in the *public* interest.[68] Thus, the fairness argument suggests that access and opportunity should be the same for both kinds of uses, and that if adequate mechanisms exist to protect the public interest when allowing individuals to make appropriations for offstream purposes, then adequate mechanisms could also exist to protect the public interest in allowing individuals to make appropriations for instream purposes. The argument concludes that there is no more reason to categorically exclude private individuals and organizations from appropriating instream flow to meet their needs than there is reason to exclude private parties from appropriating water to meet private offstream needs.

The public-good characteristics of instream flows must also be taken into consideration during the public versus private debate. As explained earlier, the existence of public-good characteristics makes it unlikely that relying solely on individual initiative to protect instream flows will result in enough flows being protected.[69] Some sort of direct action by the public as a whole, acting through state legislatures and agencies, must also be taken to ensure that enough instream flows are provided. Unfortunately, most states so far seem to assign instream flow protection activities to public agencies for just the opposite reason—to discourage rather than encourage instream flow protection. As described above, this motivation may have been a factor in limiting instream flow protection to public agencies in several states.

The choice of an agency to participate in instream flow protection activities is important. Agencies have different mandates, types of expertise, levels and sources of funding, constituencies, perceptions, and biases, all of which make a difference in the actions that an agency pursues. For example, instream flow protection issues often have greater prominence in environmental resource management agencies—fish, game, parks, recreation, and environmental quality agencies—than in water rights regulatory agencies. The former have mandates to preserve and protect environmental and recreational amenities; are staffed predominantly by fisheries and wildlife biologists, recreational planners, and environmental specialists; and interact frequently with outdoor recreationists and environmentalists. The latter have mandates to administer water rights in an orderly fashion, are staffed primarily by engineers and hydrologists, and spend most of their time interacting with water rights applicants and owners—most of whom use water for agricultural, municipal, and industrial purposes.

Agency differences can affect the manner in which they approach instream flow protection issues. Many states authorize more than one agency to participate in instream flow protection activities, but it is common to find that there is a great deal of variation in the degree to which agencies take advantage of the opportunity. For example, the state of Oregon has authorized three state agencies, Fish and Wildlife, Environmental Quality, and Parks and Recreation, to apply for instream water rights. But only one agency, Fish and Wildlife, has been responsible for submitting all of the approximately 950 instream flow applications currently pending with the Water Resources Department.[70] Public agencies may even be in direct conflict with each other over instream flow protection policies. For example, Nebraska authorizes both the state Game and Parks Commission and the state's Natural Resources Districts to apply for instream flow water rights. The Central Platte Natural Resource District applied for and received an instream flow right on the Platte River to protect endangered species. The Game and Parks Commission determined that those flows were inadequate to protect the species and subsequently applied for an instream right with higher flows. Though supported by some downstream Natural Resource Districts, the Commission's application has been contested by the Central Platte Natural Resource District.[71]

Probably the most common agency conflicts occur between the agencies authorized to apply for instream protection measures and those given the authority to decide whether to approve the applications. One agency has been given the mandate to serve as an advocate for fish, game, parks, recreation, or the environment, and the other has been given a mandate to find an appropriate balance between competing uses.[72] There often is a great deal of cooperation between agencies, and one frequently finds that the balance shifts from conflict to cooperation the longer agencies work together on an instream flow protection program. But differing instream flow protection mandates can be a source of great tension.

To this point we have been discussing only one half of the control issue: the unwillingness of the states to allow private individuals to participate more directly in the instream flow protection process. The other half concerns the efforts of the states to restrict control over water policies by the federal government. The continuing tension between the two levels of government dates from the earliest periods of our history. The federal government has substantial interests in the West, being the West's largest landowner and administrator of numerous programs that affect water use, as detailed in chapters 8–10. The states often take exception to the "intrusion" of the federal government into matters affecting "state" waters, and the national interests to which the federal government responds sometimes conflict with the interests of the individual states. The states frequently perceive the federal government to be a real or potential competitor for control over local water resources, and therefore a threat to state sovereignty. Instream flow protection activities have often been the focus of battles between the federal and state governments for control of the West's water resources.

Some of the states have taken steps to accommodate federal agencies in their efforts to protect instream flows. For example, Alaska, Arizona, and Nevada—the same states that explicitly allow individuals to apply for instream rights and reservations—allow federal agencies to apply for instream rights and reservations under state law. Montana allows federal agencies to apply for instream reservations, and the Colorado state legislature directed the Colorado Water Conservation Board to request instream flow protection recommendations from the federal departments of Agriculture and Interior. Federal agencies have often taken notice, and advantage, of these opportunities. For example, federal agencies have been much more active in pursuing instream flow protection through state mechanisms in Alaska, Arizona, Montana, and Nevada than in most of the other western states.

Other states have given federal agencies little opportunity to protect instream flows via state programs. This increases the prospect that federal agencies will attempt to bypass state law and achieve their goals through federal law.[73] Ironically, this causes even more state consternation over the perceived intrusiveness of the federal government into state water policy.

The desire of states to retain control over water resources is frequently behind efforts to oppose, or to at least treat with great caution, the whole notion of instream flow protection. Anyone that works with instream flow protection issues quickly senses the tension between state governments and private citizens, and between state governments and the federal government, and must work within the context of these dynamics.

Minimum versus Optimum Instream Flow

Despite the greater benefits associated with "optimal" flows, the term *minimum* appears frequently in instream flow protection statutes and policies.[74] Repeated references to preserving minimum instream flows may simply represent the goal of preventing waste, in the same way that water rights have always been limited—at least in theory, if not typically in practice—to the amount of water that is beneficially used without waste. In this case, it would mean that only the amount of water necessary to achieve a given level of benefits would be kept instream.

However, it seems more likely that the frequent appearance of the term *minimum* in instream flow protection statutes and policies—where the term seems much more prominent than in policies that apply to other water uses—is a result of political compromise. "Minimum" language in the instream flow protection statutes of states such as Colorado, Idaho, Nebraska, and Wyoming seems primarily to reflect the limited acceptance of these programs at the time the statutes were passed rather than some desire to ensure that water is not used in quantities beyond those necessary to achieve a chosen level of benefit.

In practice, widespread confusion about the meanings of the terms used to describe flow levels has meant that flows have been protected at a variety of levels, notwithstanding the terms used in statutes and policies. For example, Washington's Water Resources Act of 1971 called for the establishment of "base flows"—often referred to as "minimum" flows—in perennial streams, but recommendations have generally been made for flow levels that would provide "optimum" conditions for fish.[75] On occasion, the state's water administration agency has limited protected flows to the volume of a river's calculated median daily flow—which is a completely arbitrary level with respect to the value of instream benefits that will be produced—but the agency has also generally acted to ensure that at least 90 percent of the optimum habitat of the species of concern will be protected by the established flow level.

The availability of enough water to meet the stated need is a criterion used by most of the western states to evaluate new water-use applications, whether for instream or offstream use. This criterion is very much in evidence in Arizona, where the amount of water that may be protected for instream uses is often limited to the minimum amount of water verified to actually be flowing in a stream. However, Arizona law does not explicitly address the subject of whether minimum or optimum flows are allowed, so the Department of Water Resources has left to water rights applicants the decision as to how much water to request.[76] In Oregon, administratively set minimum flows previously reflected physical low-flow characteristics of the stream under consideration, but, more recently, water availability has not been a factor when reviewing applications for instream water rights.[77] Oregon now allows minimum flows to be set at levels higher than actual water availability. If the minimum flow exceeds water availability, the specified flow is a goal rather than a mandate. Each of the three state agencies that has authority to apply for instream water rights in Oregon has its own methodology for quantifying the necessary flows, which is likely to lead to different levels of protection no matter what quantification terms appear in protection statutes.

Even in states where the original intent of "minimum" language may have been to limit the amount of water kept instream, it has not always worked that way in practice. There are examples in virtually all states in which flows have been protected at levels closer to optimal than to minimal. For example, the policy in Colorado, where the guiding statute allows protection of only the "minimum stream flow" necessary to preserve the natural environment to a reasonable degree, is to protect only those flows necessary to maintain *existing* resources. Flows at levels that would *enhance* resource levels are denied. But in practice, flows identified by the Division of Wildlife as necessary for fish fall somewhere between what one might consider "bare survival" and optimum.[78] Bare survival flows would allow survival of a small population in the short run. Optimum flows would include occasional very large habitat-modifying flows. Flows identified by the Department of Wildlife are greater than those required to sustain minimal populations in the short run, and so exceed bare survival requirements. But identified flows are insufficient to protect habitat in the long run and thus fall short of optimal. The state depends upon natural hydrologic variability, particularly the existence of occasional high flows, to maintain fish habitat. In the long run, flows at the level protected by water rights, if not augmented by these occasional high flows, would allow the fishery to decline. In some situations more water can be protected. For example, the Colorado Water Conservation Board approved the Division of Wildlife's flow recommendation to protect the substantial—perhaps optimal—flows necessary to maintain an existing high-quality, gold-medal fishery in the Blue River.[79]

In Wyoming, the Game and Fish Department is required to determine the minimum amount of water necessary to maintain *or improve* fisheries, and the amount of water necessary to improve fisheries can be substantial.[80] Even flow levels that the agency has recommended to the state engineer to maintain existing fisheries—and which have for the most part been accepted by the state engineer—are closer to what many would consider optimum rather than minimal.[81] In Nebraska, where the word *minimum* features prominently in the state's instream flow statute, the state supreme court has ruled that the quantity appropriated for instream purposes does not need to be the absolute minimum required to allow the use, but rather the amount needed to maintain the existing level of use.[82] And in Idaho, where the term *minimum* is equally emphasized, the Department of Water Resources accepted The Nature Conservancy's application for the *entire* flow of Minnie Miller Springs in the Snake River Canyon, reasoning that the springs could be considered a small part of a much larger complex.[83]

Despite deviations from strict adherence to minimum principles, there is no doubt that efforts to protect instream flows in many states are constrained by policies containing minimum language. There is also no doubt that only a truly minimal level of flow has been preserved in many of the West's "protected" streams. "Minimum" policies may be most limiting in states in which only the needs of fish are considered, since minimal flows for fish are not likely to support many other instream uses. Such policies may prohibit a greater level of benefit even in situations in which more water is available. For example, rules established by the Colorado Water Conservation Board prohibit the agency from protecting flows at levels higher than necessary to sustain existing uses, even if a private party is willing to donate the rights necessary to achieve a higher level of benefit. Arbitrarily limiting the amount of benefit that can be achieved from a given use, without reference to the value of the water in competing uses, serves no logical purpose from the perspective of a society that desires to maximize returns from its use of a scarce resource.

Instream flow policies would produce less confusion and be more effective if they contained clear quantification goals.[84] The general goal of most instream flow protection programs is to preserve a particular resource, use, or value. A more specific goal would also identify the level of the resource, use, or value that is to be protected. For example, one commentator has proposed the following set of fishery protection levels, ranging from most protective to least protective, that may be appropriate protection goals:

1. enhancement above pristine conditions, which may be realistic only on river reaches downstream from storage reservoirs;
2. nondegradation with restoration, which in most situations will be the highest possible goal;
3. nondegradation, in which an already reduced level of the resource is accepted and maintained;
4. no net loss, in which certain local, seasonal, or categorical losses will be accepted if offset in some other place or time;
5. set percentage of loss from existing conditions, which is a common choice but may serve only to postpone hard decisions;
6. no loss of genetic diversity, in which sufficient numbers of organisms are maintained to prevent inbreeding; and

7. population survival, which, in the extreme, may protect just one male and one female. This is an extremely risky strategy, since it is likely to lead to species loss within a few generations. In practice, options 6 and 7 may be the same.[85]

Levels of protection may be measured directly, in units of the actual resource of concern—such as the number of fish or recreation days that are to be provided—or in units of flow or habitat, which are indirect measures of the resource. For example, to protect a fish population of a given size, analysts may first determine how much habitat is needed to maintain that population, and then determine how much water is necessary to provide that amount of habitat. Flow level becomes a surrogate for habitat, which in turn is a surrogate for fish population. But since the allocation and the administration of water resources are accomplished through reference to units of flow, most instream flow protection measures, even if the goal of the protection was originally phrased in units of the resource of interest, are eventually translated into units of flow. However, it is important to realize both that resource levels are affected by factors other than flow level, and that analysts are not always able to establish definitive relationships between resource levels, habitat, and flows. This means that protection of the identified flow will not necessarily produce the desired level of the resource.

Barriers to the Further Protection of Instream Flows

Instream flows have been protected in a variety of circumstances using an assortment of methods. However, the degree of protection has often been limited, and there are a great many barriers to further protection.

Lack of water is the first obstacle to the further protection of instream flows. Water in most of the West was scarce to begin with, and the instream flow protection measures described above were implemented only after the rights to use most of the available water had already been allocated to individuals as private property. Instream flow protection programs have been implemented slowly, sporadically, and incrementally through most of the West, and not at all in some states, and water rights administration systems in the West are still heavily slanted toward the offstream use of water. If more water is to be kept in streams, limits will need to be put on new diversions, and at least some water will need to be reallocated from offstream use to instream use.

Lack of information may be the single largest factor preventing the further protection of instream flows. Water users, agency personnel, and the general public all lack information that would likely lead to a broader implementation of instream flow protection measures. Offstream water users often believe certain "facts" about instream flow protection that are actually fallacies. One common fallacy—usually assumed rather than stated directly—is that water allocated to instream uses somehow disappears, that it is no longer available for offstream use. The fact is that except for losses due to evaporation and transpiration as the water flows downstream, instream water uses are nonconsumptive. Water that flows through a protected reach is subsequently available for other uses, including offstream use. Another common misperception is that instream uses are additive; for example, that if a minimum of 10 cfs is needed for fish, 15 cfs for swimming, and 25 cfs for boating, that instream uses require flows of 50 cfs. In fact, the amount of water used for one instream use is simultaneously available for other uses. In the example, and assuming that all instream needs occur at the same time, the first 10 cfs would not only satisfy the fish requirement but would provide most of what is needed for swimming and a substantial amount of what is required for boating. A flow of 25 cfs may well satisfy all three uses—though it is possible that 25 cfs would be too high for fish and/or swimming—but a flow of 50 cfs would be much more water than necessary to satisfy all three instream needs.

Water users often fear what they don't know, and many water users don't yet have much personal experience with instream flow protection measures. Water users often think, or fear, that they will be hurt by instream flow protection measures in situations in which there actually would be no harm, or might even be some benefit. State Game and Fish officials in Wyoming discovered that holding public information meetings a week prior to scheduled instream water rights hearings helped substantially to ease the fears of existing water users, reduce the number of protests filed against instream water rights applications, and increase support for the instream water rights program.[86]

The widespread talk of transferring water to municipalities and other "new" uses like instream flows has led to a siege mentality and a high degree of defensiveness among many existing water users, who understandably are very protective of their property rights. Water users often seem to fear that new water users or the state will simply take their water rights, though of all the protection measures discussed in the following chapter only one—the public trust doctrine—has the potential to effect an involuntary, uncompensated transfer. And, as many have argued, it is likely that public trust methods are employed primarily where other effective options have not been made available. As discussed at further length in chapter 7, it is possible that some senior water rights holders could lose some opportunities for transferring their water rights because of instream flow protections, but it is also possible for a reallocation of water to instream uses to take place on an entirely voluntary basis that will have very little impact—or, depending on the circumstances, even a positive impact—on existing water users.

Many state agency personnel are unaware of instream flow protection issues and opportunities. Knowledge of instream flow issues varies enormously among individuals, professions, agencies, and the states. In some agencies instream flow protection issues have a very high profile and there are several people who have experience in working with at least one facet of the subject. In others the level of awareness is quite low. Even those who are familiar with instream flow issues in their own agency or state are often unfamiliar with the ways in which other agencies and states have dealt with similar issues. Each of the states must find solutions that satisfy their own unique set of circumstances, but from a broad perspective it appears that all of the states are facing, or will soon face, variations of the same issues and problems. This makes it surprising to discover how little most people know about the methods and solutions that have been tried in other states. Each of the states—sometimes each agency—seems to be trying to solve problems with little awareness of what has already been tried or discussed in other places rather than taking advantage of the combined wisdom and experiences of those who have already been working on the subject.

Lack of information among the general public is even more widespread and is probably a larger obstacle to the protection of instream flows. Many people still don't know what "instream flow" is, why it's threatened, or how it can be protected. The general public doesn't know much about water policy and water law. Only a relatively small group of people, among whom existing water users and their attorneys seem conspicuous, are conversant with the often arcane details of state water law. It is easy for members of this group to argue that the topic is too complex for the general public to understand. Water law and policy have their complexities, to be sure, but it is not that hard to understand the fundamental principles. As information and awareness about both water law and instream flow protection spread, it is likely that instream flow protection efforts will become both more widespread and more effective. In the meantime, the lack of information and awareness is a barrier to the further protection of instream flows.

Opposition is also due to a genuine dislike of instream flow protection. People have different values, and some people don't place much value on the protection of instream flows, at least when compared to the value of using water for offstream purposes. Much of the opposition of agricultural interests—who account for the bulk of offstream water use in the West—is due to fears that their way of life is gradually being squeezed out by the growing concerns, and numbers, of urban and suburban residents. Instream flow protection is often cast as an interest of "city folk" rather than of "farming folk." The political power of instream flow protection opponents is often substantial. For example, agriculturalists are represented in state legislatures throughout the West in numbers far larger than their proportion of the population or share of the economy would suggest.

Instream flow protection measures continue to be blocked by a variety of legal obstacles. Limitations on the participation of private citizens, organizations, and federal agencies; the purposes for which flows can be protected; and the amount of water that can be protected is widespread and has been described in detail above. It is hard to overestimate the impact that these restrictions have had on the amount of water that has been protected for instream use.

Numerous other conditions, limitations, and qualifications have been applied to the protection of instream flows in individual states. For example, instream reservations are reviewed and can be revoked in Alaska and Montana. Instream water rights have been conditioned for later review in Idaho and Wyoming. Wyoming allows other water users to recover all costs of litigation from the holder of an instream flow right if the water users can prove in court that an instream right has caused them harm. Several states require applicants for instream water rights and reservations to submit far more data in support of their applications than is required for most offstream water rights. Instream rights are also subjected to many more levels of review than offstream rights. For example, Colorado's Division of Wildlife must submit applications to the staff of the Colorado Water Conservation Board, which then makes a recommendation to the Board. If approved by the Board, the application is submitted to the state attorney general's office, which then takes the application before one of the state's water courts. Offstream water rights applicants can go directly before a water court.

It is possible that all of these measures can be justified on some grounds. For example, water users and state officials are much more familiar with the use and measurement of water for growing corn or satisfying municipal needs than with the use and quantification of flows for instream uses, and thus may not feel comfortable with instream water uses in the absence of these special measures. But it is not necessary to debate the wisdom of applying these measures so much as to point out that the same conditions have not been applied to offstream water uses. The prevalence of requirements above and beyond those applied to offstream water uses creates a systematic bias in our water allocation systems that works against the allocation of water to instream uses.

CHAPTER SIX

Methods the States Use to Protect Instream Flows

The loss of instream flows and an evolving awareness of their importance have engendered numerous protection efforts by the states. The first such efforts were isolated and sporadic, and concentrated primarily in the Pacific Northwest. Oregon was probably the first state to act when in 1915 it took measures to protect flows at waterfalls along the scenic Columbia River Gorge. Oregon also defined its public interest in water to include purposes of public recreation and commercial and game fishing,[1] and in 1929 took steps to protect flows in the Rogue River.[2] In the 1920s Idaho moved to include aesthetics, health, and recreation in its list of beneficial uses,[3] and it enacted measures to protect water levels in several scenic lakes in the northern part of the state. Oregon enacted a minimum-flow program in 1955 to protect salmon during spawning seasons,[4] and a successful ballot initiative in 1970 created Oregon's Scenic Waterways program, which indirectly protects instream flows.

Pressure to consider environmental values in public policies increased dramatically during the environmental movement of the 1960s and 1970s. Most western states initially resisted adding streamflow protection options to their traditional water allocation laws and procedures, but by the 1970s they were starting to take instream flow protection more seriously.[5] Montana created instream flow water rights on 12 blue-ribbon trout streams in 1964, declared that water resources were to be protected and conserved for fish, wildlife, and public recreational purposes in 1967, and created a process for reserving flows for instream purposes in 1973.[6] Idaho acted to protect the flows of several springs in the Snake River Gorge in 1971. Washington created a minimum-flow program in 1967 and strengthened the program in 1971. California created a state Wild and Scenic Rivers program in 1972, the same year that Colorado established a procedure for creating instream water rights. Other protective measures were adopted in Idaho in 1978, Alaska and Kansas in 1980, Arizona and California in 1983, Nebraska in 1984, Utah and Wyoming in 1986, and Nevada in 1988.

Even this quick listing of events demonstrates that the methods used by the states to protect instream flows are many and varied. Some of the more common methods used to protect instream flows include the setting of minimum flows and instream reservations, special administrative measures, creation of instream water rights, water rights transfers, protected rivers programs, and the public trust doctrine. We now turn to a description of these and other methods implemented by the states to protect instream flows.

Minimum Streamflows, Base Flows, and Instream Reservations

One method of protecting streamflows is to set aside a specific quantity of flow that will not subsequently be available for appropriation. This method has been applied in different forms. "Minimum streamflows," sometimes referred to as "base flows," or even occasionally—and confusingly—as "instream flows," are defined flow levels below which new appropriations will not be allowed. Similarly, "instream reservations" reserve water for use instream, thereby preventing the water from being appropriated for other uses.

Minimum flows and instream reservations have been integrated into existing systems of appropriative rights through use of priority dates. New minimum flows and reservations are junior to existing appropriations, and so are effective only with regard to future appropriations and to changes in existing, senior appropriations. Examples from Washington, which uses minimum streamflows, and Montana, which uses instream reservations, further illustrate the use of these methods.

Washington's 1971 Water Resources Act authorizes the state's Department of Ecology to establish "base flows" in all of the state's perennial streams, excepting only those streams where there is an overriding public interest to the contrary.[7] The Department of Ecology establishes minimum flows when requested to do so by the Department of Fish and Wildlife or may do so at its own initiative. The flows are set through administrative rule-making procedures that include comprehensive public notice and hearing provisions. The primary goal of the minimum streamflow policy is to protect the state's anadromous fisheries, and minimum streamflow levels are based on the needs of fish.

The Department of Ecology administers minimum streamflows by closing streams to further appropriations when water availability falls below the defined minimum-flow level. On streams in which water supplies are sufficient to allow new appropriations, new permits are conditioned to require that diversions or storage cease when streamflow falls below the established minimum-flow level. Administration through reference to priority date makes minimum streamflows look much like appropriations, and in fact Washington amended its state water code in 1979 to clarify that minimum flows established by administrative rule are appropriations.

The Montana Water Use Act of 1973 authorized the state, any political subdivision of the state, or the federal government to reserve up to 50 percent of the average annual flow of a river for instream use.[8] Public entities wishing to reserve a particular amount of a river's flow must apply to the state's Board of Natural Resources and Conservation for the reservation, and demonstrate that the reservation is needed and in the public interest. The Board may then, after appropriate notice and hearings, grant the reservation. Once reserved, flows are protected from appropriation for other purposes.

The Montana Water Use Act actually allows streamflows to be reserved for any future purpose, including irrigation, municipal, storage, and other beneficial uses. The reservation statute was motivated by fears that the coal industry would soon appropriate all of the state's previously unappropriated water.[9] Reservations thus became a method of protecting the state's unappropriated waters by "pre-allocating" the water to other uses. The state reviews reservations every 10 years to ensure that they are still needed. At that time reservations can be extended, modified, or revoked. However, the Board is authorized to modify reservations made for instream purposes every five years. The state's first reservations were made in the Yellowstone River Basin, for future irrigation, storage, and instream uses. Instream reservations were assigned a higher priority than reservations for irrigation and storage in the upper basin, whereas irrigation was given the highest priority in the lower basin.[10] Reservations have also been made in the Missouri and Upper Clark Fork river basins.

Minimum streamflows and reservations have also been used in other states. Oregon was the first state to establish minimum flows, but in 1987 it converted all existing minimum flows into instream rights.[11] Kansas continues to use minimum streamflows as its primary method of protecting instream flows, and as of June 1995, had established minimums on 23 streams, mostly in the eastern part of the state. The state has no plans to establish minimum streamflows on additional rivers, but continues to actively administer existing ones.[12] California does not have a minimum streamflow program but achieves a similar result in some circumstances through application of a Fish and Game Code provision that requires dam owners to allow enough flows to pass the structures to "keep in good condition" any fish that exist below the dam.[13] Alaska uses instream reservations to protect flows, though the distinction between Alaska's reservations and actual water rights is minimal; reservations, unlike water rights, can be reviewed periodically.

Minimum streamflows and instream reservations can be an effective method of protecting instream flows. The methods set clear flow-based limits on a river-by-river basis. Neither method interferes with the allocation and administration of water rights through application of prior appropriation principles. Use of this method can also provide opportunities for public participation, because minimum streamflows and reservations are established through administrative rule-making processes, which are open to the public. For the same reason, use of the methods can be compatible with systematic efforts to solve water resource problems through watershed or regional planning efforts.

The use of minimum streamflows and reservations to protect instream flows also has significant drawbacks. One is that, owing to their relatively junior priority dates, they have no effect on senior appropriations, and so do not provide a means of putting water back into streams that have been dewatered. Minimum streamflows and reservations that are not backed by instream water rights lack the same legal protection as water rights, and so may be more vulnerable to changes in administration and policy, or to federal water development programs that recognize property rights but not other state river designations and programs.[14] Reservations in particular have the additional weakness that they are subject to periodic review and can be modified or revoked. The process of establishing minimum flows and reservations can be time consuming and expensive,[15] especially given the controversy surrounding any measure that reduces opportunities for future offstream appropriations.[16]

Administrative Protection

> Perhaps the most effective use of existing law for stream preservation could be made not by granting appropriations for recreational purposes but by denying appropriations that destroy them.
>
> —Frank Trelease[17]

All of the western states have criteria to evaluate new appropriations or changes to existing appropriations.[18] All except Colorado and Oklahoma require that new appropriations be in the public interest, and many of the states apply public interest criteria to water rights transfer and change applications. The application of instream flow criteria to appropriations and transfers as part of public interest standards can be an effective method of protecting instream flows.

The use of public interest criteria to protect instream flows is perhaps the easiest method available to the states. In states that already have such criteria, it doesn't require changes to existing law. Most western public interest statutes date to the late 1800s and early 1900s,[19] and now either include instream flow considerations explicitly or could easily be interpreted to include them.[20] Public interest criteria are a source of flexibility within the prior appropriation doctrine. Court decisions during the early years of the criteria essentially confirmed that economic development was in the public interest, and that projects that would impede maximum economic development were against the public interest.[21] By those standards, public interest criteria could have been—and may have been—used to deny appropriations for instream uses. But public interest criteria have increasingly been used to achieve the opposite result, to protect instream flows from development.[22]

California is the state best known for using administrative criteria to protect instream flows. California first authorized its administrative agency, now the State Water Resources Control Board (SWRCB), to consider the public interest when acting on applications to appropriate water in 1917. The SWRCB was directed to "allow the appropriation . . . under such terms and conditions as in its judgment will best develop, conserve, and utilize in the public interest the water sought to be appropriated."[23] California takes these public interest criteria seriously. A state court of appeal declared in 1965 that the public interest is "the primary statutory standard guiding the Water Rights Board in acting upon applications to appropriate water."[24] The SWRCB even has substantial authority to modify the terms and conditions of existing permits if necessary to achieve state goals.[25]

Instream flows are protected by other features of California's water law. Since 1965, section 1257 of the California Water Code has required applicants for new appropriations to set forth all data and information reasonably available concerning the extent to which fish and wildlife would be affected by a proposed appropriation, and to describe any measures that will be taken to protect fish and wildlife. The SWRCB relies heavily on the Department of Fish and Game to review and comment on new applications. The Department of Fish and Game recommends flow conditions, release requirements, and other conditions necessary to protect instream values,[26] and the Board routinely uses these comments to condition or deny permits.[27] The SWRCB is required to weigh the relative benefit of a new appropriation with respect to other uses, including preservation and enhancement of recreation, fish, wildlife, and water quality.[28] The SWRCB has shown considerable interest in protecting recreational and environmental resources since at least the early 1970s,[29] and administrative protection has become the state's most effective tool for protecting instream flows.[30]

California's Fish and Game Code is a source of additional protection. Section 5937 of the code states that "the owner of any dam shall allow sufficient water at all time to pass through a fishway, or in the absence of a fishway, allow sufficient water to pass over, around or through the dam, to keep in good condition any fish that may be planted or exist below the dam." The California Department of Fish and Game initially had only limited success in enforcing this requirement, but a successful lawsuit by the conservation group California Trout in 1990[31] has given section 5937 greater prominence.[32]

Administrative measures are also employed in other states to protect instream flows. For example, a rivers planning bill passed by the Idaho legislature in 1988 states that minimum flows in the state's rivers are to be fostered and encouraged.[33] The Utah state engineer has been explicitly directed since 1971 to determine whether a proposed use of water "will unreasonably affect public recreation or the natural stream environment," and to reject or condition applications as necessary to protect environmental values.[34] This feature of the code was first invoked in 1983, and has been used several times since.[35] These same standards have been applied to water rights change and transfer applications in Utah since 1989.

Similarly, Texas water code requires the Texas Natural Resource Conservation Commission (TNRCC) to consider the environmental consequences of proposed appropriations, including effects on water quality and instream flows, during the permitting process.[36] The TNRCC actively reviews new permit applications for environmental impacts.[37] The TNRCC is required to assess permit applications to store, take, or divert water in excess of 5,000 acre-feet per year for the effects the issuance of a permit might have on fish or wildlife habitats.[38] In practice, the TNRCC routinely performs the same analysis for quantities less than 5,000 acre-feet per year, exempting only domestic and livestock uses.[39] Both provisions may be applied to additional watercourses if there are proceedings to change the terms of a permit, as recently occurred in Texas' Lower Colorado River basin.[40]

Other administrative measures may protect instream flows indirectly. For example, anyone wanting to alter the bed or banks of a stream in Utah or Idaho must first obtain a permit from the state engineer (Utah) or Department of Water Resources (Idaho). These regulations do not specifically address questions of instream flow but are designed at least in part to protect fisheries, wildlife, and the natural stream environment. The regulations may protect instream flows because placement of virtually any diversion structure will require alteration of the bed or banks of a stream and therefore require a permit.[41]

Oklahoma, South Dakota, and Texas may indirectly protect instream flows through their protection of domestic and stock-watering uses. The three states do not require permits to use water for these purposes, but the administrative agencies in Oklahoma and South Dakota routinely require at least minimal flows in streams to protect domestic and stock-watering uses that exist downstream,[42] and people using water for domestic or stock-watering purposes in Texas are allowed to protest applications for new upstream appropriations that would leave them without water.[43]

Administrative protection, though effective and relatively easy to implement, has some drawbacks. One is that administrative review can't be used to modify existing conditions unless agencies are explicitly given the authority to review past appropriations, as in California. Administrative review could have resulted in a good balance between instream and offstream use if consistently applied since the first appropriations were made, but the use of administrative review to protect instream flows is relatively recent. Second, administrative review continues to be applied only sporadically in many states, especially where public interest criteria haven't been defined explicitly. Further, relying on administrative agencies to protect instream flows may be a problem given that regulatory agencies are often more sympathetic to—or at least knowledgeable about—offstream water uses than instream water uses.[44] California may be the only state in which administrative measures have been consistently applied to evaluate new appropriations for their impact on instream values. And even in California the administrative agency is required only to consider, not necessarily to act upon, recommendations made to protect instream flows.[45] In the absence of consistent review and ability to review past appropriations, administrative review will result in protection of only the smallest flow that has ever been protected; flows protected by one administration may be partially or entirely appropriated during the tenure of a subsequent, less protective administration.[46]

Administrative protection is relatively inefficient, as it may require that multiple streamflow determinations be made for the same reach of river. The value of the flow for instream needs is reevaluated relative to an offstream use each time that a new use is proposed.[47] Reviews may be less than thorough given that resources available to administrative agencies for permit review are typically scarce. Lack of a direct filing to protect a specified quantity of flow also increases uncertainty, in that the process may not provide adequate notice to other potential water users about the availability of water in a particular watercourse.[48]

Instream-Flow Water Rights

A legal right to use water is a property right and has much the same legal protection as other forms of property. Protecting instream flows through the creation of instream-flow water rights provides one of the strongest legal forms of protection.

A water right issued for instream use is a right to maintain a specified level of flow at specified times through a specified reach of river. The specified level and timing of the flow depend on the needs of the particular use, the availability of water, and policies used by the states to regulate appropriations of water.[49] Like other water rights, an instream right has a priority date that protects the right from water depletions by junior users but not from the exercise of senior rights. The creation of an instream right is subject to the same criteria as creation of an offstream right, including the availability of enough water to satisfy the right, the prevention of harm to other water rights holders, and public interest standards. Unlike offstream water rights, for which a specific diversion *point* is defined, an instream right is defined for a particular *reach* of the river.

The following twelve western states have recognized water rights created under state law for instream use: Arizona, California, Colorado, Idaho, Montana, Nebraska, Nevada, Oregon, South Dakota, Utah, Washington, and Wyoming.[50] Some of these states have recognized instream water rights only in very limited situations. For example, Montana has recognized only 12 rights, known as "Murphy" rights after the sponsor of the 1969 legislation that created them, used to protect a dozen of the state's blue-ribbon trout streams. Washington recognizes water rights for instream purposes only through its "trust water rights" program, but trust water rights have yet to be created.[51] California and Utah do not grant rights for new appropriations made for instream purposes but have authorized the transfer of offstream rights to instream use. And Nebraska and South Dakota have recognized so few instream rights that, as of June 1995, there were not yet a half dozen rights between them. No instream water rights created under state law have yet been recognized in Alaska, Kansas, Oklahoma, New Mexico, North Dakota, or Texas, though Alaska's instream reservations are very similar to water rights.

A major advantage of using water rights to protect instream flows is that it puts instream uses on the same legal basis as offstream uses. Equity suggests that if water is to be allocated through the creation of water rights, and water rights are created to protect offstream uses, then water rights should also be created to protect instream uses. Water rights, as property rights, receive strong legal protection and may be more permanent than other methods of protecting instream flows.[52] State water rights are also recognized by the federal government, unlike some other methods that have been used by the states to protect instream flows.

The primary disadvantage of protecting instream flows by creating water rights is that, like other methods discussed above, the water rights would be of fairly recent priority. Very little water may be left instream to protect by the time rights are created. By themselves, new water rights do nothing to put water back into dewatered streams. The value of instream rights may therefore be limited to protecting the status quo by providing legal standing to protest new rights or transfers of existing rights that would be harmful to the instream use.[53] Getting states to recognize instream water rights has also cost instream flow proponents substantial quantities of time, money, and political capital. Nonetheless, instream water rights may be a valuable tool for protecting existing flows.

Water Rights Transfers

Perhaps the greatest advantage of recognizing water rights for instream uses is that rights can facilitate the transfer of water from offstream uses to instream uses. Water rights transfers have a long history, though it is only recently that transfers have been used to move water from offstream uses to instream flows.

A water right typically specifies the amount of water that will be used, the purpose of the use, the timing of the use, the point from which the water will be diverted (if applicable), and the site of use. Once issued, the right can be used in perpetuity under the specified conditions.[54] However, it is possible to change the conditions on the right so that, for example, a water user could change the site at which the water is used, or the point from which it is diverted. A water right may also be sold, leased, traded, or donated. The only caveat is that these "changes" and "transfers"—often, as here, considered and referred to jointly as "transfers"—must be harmless to other existing water rights holders and be approved by the appropriate state administrative agency or court, just the same as applications for new appropriations.[55]

Water rights transfers are relatively common. Water is routinely shifted to new sites of use, diversion points are moved up or down the river, and water rights are sold or leased. Most transfers have occurred between people using water for essentially the same purposes, as between irrigators. But water rights transfers are also a topic of great current interest because they provide what is often the only feasible means of moving water from old uses to new or expanded uses now that much of the West's water has already been allocated. The transfer of water from agriculture to municipalities has become quite common, but water has also been transferred to environmental use, including instream flows.

The relationship between instream flows and water rights transfers can be quite complex, as discussed at greater length in chapter 7. Chapter 7 contains information about water rights transfers as a potential source of harm to instream flows, additional information about the effect of instream flow rights on water rights transfers, and information about the effect of transfers to instream use on other water users. Here we consider three different issues related to instream flows and transfers: (1) protection of instream flows through transfer of water rights from offstream to instream use; (2) protection of instream flows through transfer of a senior water right to a new location further downstream; and (3) the issue of whether transfers from instream to offstream use should be allowed.

From Offstream Use to Instream Use

The most obvious method of reallocating water from offstream use to instream use is to transfer water rights from one use to the other. For example, a state wildlife agency might purchase a water right from an irrigator and use the water to maintain an instream fishery. Or the holder of a water right used for industrial purposes might donate part of the right to a conservation organization to be used in maintaining riparian vegetation in a nature preserve. The ability to transfer water rights has enormous implications for instream flow protection, because the transfer of water rights from offstream use to instream use is one of the few methods available for increasing streamflow in rivers that have previously been dewatered. Transfers may occur through market transactions or donation, but either way they are voluntary, respecting private property rights. Neither the seller-donator nor buyer-recipient is forced to participate. The transfer of rights through voluntary means accommodates shifting values while protecting vested rights and is therefore an attractive option for reallocating water from offstream use to instream use.

Notwithstanding this immense advantage, the transfer of water rights from offstream to instream use is not always allowed. For example, such transfers cannot occur in states that don't recognize instream water rights. This may lead to situations in which there are willing sellers or donators, and willing buyers or recipients, but no transaction because the state won't approve the transfer if the water is to be used for instream purposes. Some states put other restrictions on transfers that effectively preclude their use as a method for protecting instream flows. For example, Nebraska recognizes instream flow rights but allows transfers only if the water will be used for the same purpose as in the original appropriation. This catch-22 effectively eliminates the possibility of transferring an offstream water right to an instream use.[56] Agricultural water rights in South Dakota may be transferred only to municipal use.[57] Arizona allows water rights to be transferred to other uses, but only the state and political subdivisions of the state may make the transfer if the water is to be used for fish, wildlife, or recreational purposes. Arizona law also gives irrigation districts the right to veto any proposed water rights transfers that occur in the watershed where the districts are located. As of June 1995, no transfers from offstream to instream use had occurred in Arizona.[58]

Other states have made explicit provision for the transfer of water from offstream to instream use. For example, California and Utah recognize instream flow rights but allow them to be created *only* through transfers. Colorado's legislature in 1986 authorized the Colorado Water Conservation

Board to acquire instream flow rights through transfer, 13 years after having first authorized new appropriations for instream purposes. And in Oregon the same 1987 legislation that converted minimum streamflows to instream rights also explicitly authorized the purchase or lease of existing water rights for conversion to instream use. Explicit authorization of transfers is not sufficient to ensure that they take place,[59] but it does create the opportunity.

In some states the authority to transfer water rights from offstream to instream use is unclear. For example, South Dakota has on a couple of occasions approved the transfer of existing water rights to instream purposes but has never explicitly recognized instream rights. Kansas does not recognize instream rights but could achieve a result akin to a water rights transfer through other means; the state has been authorized since 1988 to purchase and hold in "custodial care" existing surface water or groundwater rights in order to restore streamflow in over-appropriated areas. However, the state has never funded or used this program.[60] Montana allows the transfer of water reservations from one use to another but so far has no provision for purchases or donations of water rights for instream use. A bill to authorize such donations and purchases was defeated by the Montana legislature in 1991.[61]

Water rights transfers do not have to be permanent. Water rights can be leased or temporarily transferred in any number of other ways that could protect instream flows. For example, Montana passed a law in 1994 that allows any person or government agency to lease water from an existing right holder for the purpose of enhancing instream flows. A water lease is much like the lease of land or other real property. The lessee obtains use of the water for a specified period of time for an agreed-upon price, but the lessor retains ownership of the actual water right and resumes use of the water, or leases the water again, after the original lease expires.

Another form of temporary transfer is the "dry-year option," increasingly used to enhance municipal supplies but may also be used to protect instream flows. Under the terms of a dry-year option the purchaser pays the right owner a fee to maintain the option of leasing water during years in which water supplies fall below a specified amount. In years in which ample water is available, no lease occurs, but the right owner receives the agreed-upon option fee. In years in which flows fall below the specified amount, the lease goes into effect and the purchaser pays the right owner an additional amount for the water that is transferred into the new use.

"Water banking" is another method of temporarily transferring water, as illustrated by the case of the upper Snake River water bank in Idaho. This water bank has been operating on an informal basis since at least the 1930s, and its existence and operations have been formally recognized since 1978.[62] Farmers holding storage rights in excess of their anticipated needs for any given year make the excess water available to the bank, which then sells the water to users who desire additional supplies. The storage rights themselves are not sold, and rentals are valid for only one year. Most of the excess water has historically been sold to a downstream hydropower producer, but on at least one occasion it has been used for wildlife purposes. The winter of 1988–89 was so cold that ice formed over large areas of Henry's Fork of the Snake River, blocking the access of trumpeter swans to their aquatic food supplies. The Nature Conservancy and Trumpeter Swan Society each rented 3,200 acre-feet of water from the bank, and the Snake River Water District No. 1 donated an additional 10,000 acre-feet. The water was then released to help break up the ice.[63] Water banks have been used in other areas, most notably in California during the extended drought of the late 1980s and early 1990s.

Water leases, dry-year options, water banking, and other temporary transfers have the advantage that they transfer water to the new use only in years in which the water is truly needed, otherwise leaving the water in its previous use. From the perspective of offstream water users, these temporary arrangements may prove to be one of the least disruptive and threatening methods of protecting instream flows. The obvious disadvantage of temporary transfers is that they do not provide permanent protection for instream flows, because short-term sources of water may not always be available for instream use when needed. And even if water continues to be made available after the original lease or rental has expired, new leases and options must continually be negotiated, adding to the long-term cost of protecting flows.

Transfer of Senior Water Rights Downstream

Just as the existence of senior water rights downstream can provide incidental protection for instream flows, the transfer of senior water rights to a downstream location can be used as an intentional method of protecting instream flows. An individual or organization who wishes to protect streamflows can purchase a senior right in an upstream location and transfer it to a downstream location. Because the water is then diverted farther down in the watershed, the intervening reach between the sites of the previous and new diversions will have additional flows. Only the site of consumptive use, not total consumptive use, is affected. This method requires no change in existing laws, works fully within—and because of—prior appropriation principles, and can be used even in states that don't recognize water rights for instream purposes.

The method is not without its limitations, however. It is imperative that the water associated with the right continues to be used in the downstream location, because the right will be lost if the water is not used for a legally recognized purpose. The opportunity to use the method therefore depends on the existence of suitable locations for diversion downstream, the availability or construction of necessary infrastructure downstream (including diversion works, canals, pipelines, and access roads, among others), the timing of diversions and river flow, and economics. There may be few situations with senior rights upstream, unmet use sites downstream, and willing buyers and sellers. On many rivers the best—or only—points of diversion are already taken.

The most promising way to use this method may be for the instream advocate to merely facilitate, perhaps financially, the swap of priorities between an upstream senior water user and an existing downstream junior user. But this situation too has its drawbacks, for it leaves the instream advocate—who does not own any rights—without legal standing to protest future appropriations or transfers that would cause injury to the newly created instream flow. Because the instream protection is indirect—the newly transferred downstream right is for an offstream, not instream, use—it may not even be possible to argue that an injury has occurred; only the use for which the right is issued is legally protected from injury.[64]

Another way the method could be used to protect instream flows is the purchase of offstream water rights by downstream hydropower producers. A hydropower facility does not face problems with land use or suitable diversion points, and generating electricity can often produce enough revenue to make the transfer economically feasible without financial incentives from instream flow advocates. The use of streamflow to generate electricity may in fact provide considerably more value than the use of water for offstream purposes, so that a transfer not only would generate incidental instream benefits—instream protection would be a positive externality of the transaction—but would also increase the total financial return attributable to the use of water.[65] The transfer of water rights to hydropower producers as a means of increasing streamflow may seem ironic to river advocates who often find themselves in conflict with hydropower facilities,[66] but it may nonetheless prove to be an effective means of protecting instream flows.

Transfers from Instream Use to Offstream Use

Most discussions about instream flows and transfers focus on whether transfers from offstream use to instream use should be allowed. An equally valid question, however, is whether transfers from instream use to offstream use should be allowed.

To date, many of those arguing in favor of the ability to transfer instream rights to offstream use have been opponents of instream flow protection. The rationale seems to be that allowing instream flow rights to be converted to offstream rights will limit the permanence of instream protection and help to limit the total amount of water that is protected for instream use. A more considered reason for allowing the transfer of water rights from instream to offstream use is the same one that is advanced for transfers in the other direction: Water rights should be free to move in whatever direction is necessary to accommodate shifting values and needs. Current allocations, water supply conditions, and values have led to calls for protecting additional instream flows, but if conditions were to change in the future, the goal of maximizing total return from use of the resource might indicate that some of the protected flows should be diverted for offstream use.[67]

The question of whether instream rights should be convertible to offstream rights frequently gets tangled up in other issues and can become quite complex. The answer may, for example, depend on the purposes for which the water has been left instream, or on whether the instream right is held by private individuals or collectively by the public through a government agency. It may also depend on how much water is being discussed. For example, it may be desirable to deny transfers of water to offstream purposes if only the minimum streamflow necessary to preserve the instream use has been protected, but to allow the transfer of water in excess of the minimum.

Some people oppose the idea of allowing instream rights to be converted to offstream rights because they fear that speculators will use instream purposes as a ruse to claim and hold water at little cost in order to sell it later for a large profit. Fear of water speculation has been omnipresent in the West and is partly responsible for the requirement that people build diversion works—essentially that they spend money—before receiving a water right. Fear of speculation is largely responsible for the requirement that water be forfeited if not put to continuous beneficial use.

The speculation issue might be solved by specifying that if the water were no longer needed for an instream use the right would be forfeited and the water made available for new appropriation. Unfortunately, this would also remove the incentive for those holding legitimate instream rights to pay attention to market signals and transfer the rights when the water became more valuable in another use.[68] The difficulty of solving this speculation issue is one rationale that has been advanced for allowing only public agencies to hold instream flow rights. Perhaps a public agency would be more responsive to societal, rather than individual, expressions of value and would have the resources and judgment necessary to determine if or when the water would be more valuable in some other use.

Restricting ownership to a public agency has its own disadvantages, many of which were described in chapter 5. One act to be avoided is the elimination of an instream right by a public agency in response to pressures stemming from what would later prove to have been only a temporary shift in public sentiment. One commentator has suggested that the following criteria be met before all or part of an instream right is transferred to offstream use:

1. the applicant demonstrates that there are no other sources of supply available for the proposed use, including transfers from other consumptive uses, water conservation, and the use of reclaimed water;
2. there are no simple physical solutions, such as moving the point of diversion to a point below the instream use, that would allow the new appropriation to occur without damaging the instream right;

3. it is determined that the new use is more socially valuable than the existing instream use;
4. other existing rights, and not just instream rights, are also evaluated for their reasonableness and value relative to the proposed use; and
5. the reallocation is in the public interest.[69]

A public agency holding the instream right would be a logical entity to apply these criteria, but most of the criteria might just as easily be applied to a privately-held right during a transfer application proceeding before a regulatory agency or court.

Water Transfers: Summary

Water rights transfers are an important component of water resources policy. Transfers introduce flexibility into the water allocation process and provide a means of accommodating changing needs while protecting existing rights. Transfers may be used to protect instream flows but, as will be described in chapter 7, may in some cases damage instream flow values. It is impossible to issue blanket statements about the effect of water rights transfers on instream flows, or the effect of instream flows on water rights transfers. But it is clear that policymakers in all of the western states will need to be aware of the complex relationship between instream flows and water rights transfers when establishing water resources policy.

The Public Trust Doctrine

An ancient philosophy sometimes applied to natural resource management maintains that certain resources have too much public value to be left in private control and thus should be held by the state for everyone's benefit. This concept was applied in the Roman Empire to the ownership of great navigable rivers and harbors, which were held by the state to protect the public's participation in navigation, commerce, and fishing.[70] The precept was subsequently applied by the English courts to lands beneath tidal waters and lands subject to tidal action. The principle that the government should hold certain critical resources in trust for the people—now known as the public trust doctrine—is also a part of American water law.[71]

The public trust doctrine in this country is primarily a matter of state law,[72] and the degree to which it has been recognized varies considerably among the states. The doctrine has been recognized most frequently in California, where it was first applied to lands bordering San Francisco Bay in the 1850s.[73] The California courts have since expanded the scope of the doctrine to include additional lands, waters, and water uses. A significant advance in the doctrine occurred in the 1971 case of *Marks v. Whitney,* in which the supreme court of California declared that the state "is not burdened with an outmoded classification favoring one mode of utilization over another" when administering the trust, and that uses subject to the trust could include not only the classic trio of navigation, commerce, and fishing, but also recreation, scientific study, open space, scenery, and habitat for birds, fish, and wildlife.[74] Later cases extended the scope of the trust to navigable waters not affected by tides, including inland lakes.[75]

The National Audubon Case

The link between the public trust doctrine and the protection of rivers came to the forefront following the landmark case of *National Audubon Society v. Superior Court of Alpine County,* decided by the supreme court of California in 1983.[76] This case involved the allocation of waters flowing from the east side of the Sierra Nevadas into Mono Lake. Mono Lake has no outlet, and its saline waters support large numbers of brine shrimp and brine flies, which in turn support extensive breeding and migratory bird populations. The lake also contains numerous natural features of interest to scientists and visitors. In 1940 the City of Los Angeles applied for a permit to appropriate virtually all of the water in four of the lake's five tributary streams. Though protested by a number of people who argued that the diversions would prove catastrophic to uses currently being made of the lake, the Division of Water Resources granted the permit on the grounds that California law gave the agency no basis for denying the permit.[77] Los Angeles then began diverting about half the flow of the streams, approximately 50,000 acre-feet per year, and transporting the water over 300 miles to Los Angeles via an extension of the Owens Valley Aqueduct. The water was used to generate electricity en route.

Los Angeles constructed an additional diversion tunnel and began taking its full appropriation of approximately 100,000 acre-feet per year in 1970. As predicted by the original permit objectors, this action dewatered the four tributary streams and substantially reduced the volume of Mono Lake. Many of the lake's natural features were left stranded as the lake's surface elevation dropped, including an island where breeding bird colonies had formerly been protected from predators by the surrounding water. The remaining lake water also became more saline after the freshwater sources were eliminated, which caused the numbers of brine shrimp and brine flies in the lake to decline. This in turn threatened the lake's bird populations.

These damages led three environmental groups—the National Audubon Society, Mono Lake Committee, and Friends of the Earth—to bring suit in 1979, arguing that the Los Angeles diversions should be constrained because of neglected public trust values. The state supreme court handed down a unanimous decision in favor of the environmental groups in 1983, ruling that the State Water Resources Control Board (SWRCB, successor to the Division of Water Resources in the administration of water rights) has an affirmative duty to consider public trust values not only in permitting initial allocations, but in continuing to review past allocations. The court then ordered the SWRCB to determine whether the diversions by Los Angeles should be cut back to maintain public trust values in Mono Lake.

The *National Audubon* decision has the potential to profoundly affect water allocation decisions. The court expanded the scope of the doctrine to cover some nonnavigable waters, holding that "the public trust doctrine, as recognized and developed in California decisions, protects navigable waters from harm caused by diversion of nonnavigable tributaries." The California court also found that:

> The public trust is more than an affirmation of state power to use public property for public purposes. It is an affirmation of the duty of the state to protect the people's common heritage of streams, lakes, marshlands and tidelands, surrendering the right of protection only in rare cases when the abandonment of that right is consistent with the purposes of the trust. . . . The state has an affirmative duty to take the public trust into account in the planning and allocation of water resources, and to protect public trust uses whenever feasible.[78]

The court left no doubt that it intended for the doctrine to be applied even to past allocation decisions:

> Once the state has approved an appropriation, the public trust imposes a duty of continuing supervision over the taking and use of the appropriated water. In exercising its sovereign power to allocate water resources in the public interest, the state is not confined by past allocation decisions which may be incorrect in light of current knowledge or inconsistent with current needs.

The SWRCB decision on Mono Lake,[79] issued in 1994 pursuant to the court's directive and incorporating the holdings from another case decided in the interim,[80] required Los Angeles to cease diverting water from the four streams until the surface elevation of the lake recovered to an elevation of 6,391 feet above sea level. The Board ruled that Los Angeles could resume diverting water from the tributary streams after that level was reached, but only at the rate of 32,000 acre-feet per year. In effect, consideration of its public trust responsibilities led the SWRCB to cut the amount of Los Angeles' appropriative right, dating to 1940, by two-thirds.

Implications for Instream Flow Protection

It is possible that the public trust doctrine will be extended to cover a variety of waters and uses relevant to instream flow protection in other parts of the West. The doctrine has great appeal for many who believe that states should play an affirmative role in the protection of natural resources. Several commentators have noted that the public trust doctrine "is a dynamic concept that courts can adapt to meet new public needs."[81] But in other states it is uncertain to what extent the doctrine has been recognized. On the one hand, although courts in most of the states have made reference to "the public trust," it is not always clear that they are referring to the same concepts described here.[82] On the other hand, the Utah Supreme Court has incorporated principles of the public trust doctrine as described here in some recent decisions but has strictly avoided using the term *public trust*.[83] This makes it difficult to determine exactly where, how, and to what degree the public trust doctrine has been recognized.

Nevertheless, the public trust doctrine has clearly begun to make inroads in states outside of California. For example, the North Dakota Supreme Court held in 1976 that resource allocation decisions must take public trust interests into account, and required the state's Water Conservation Commission to consider the potential effect of a water allocation on the public interest before issuing a permit.[84] In 1984 the Montana Supreme Court cited public trust principles to sustain the right of the people to use streams for recreational purposes.[85] Article VIII, Section 3, of the Alaska Constitution, which states, "Wherever occurring in the natural state, fish, wildlife, and waters are reserved to the people for common use," is said to be the primary source of the public trust doctrine in Alaska.[86] The Alaska Supreme Court made this view explicit in a 1988 case by recognizing that the constitution had imposed upon the state "a public trust duty with regard to the management of fish, wildlife and waters."[87]

The public trust doctrine has had an interesting history in Idaho. The Idaho Supreme Court recognized public trust requirements in a 1983 case involving the disposition of lakeside lands, stating that the public trust covers "property, navigation, fish and wildlife habitat, aquatic life, recreation, aesthetic beauty or water quality."[88] As in California, the Idaho court asserted that "the public trust doctrine takes precedence even over vested rights."[89] Two years later the Idaho Supreme Court directed the state's Department of Water Resources to apply public trust criteria to every new water rights application prior to approval.[90] And in early 1996 the Idaho Land Board cited the public trust doctrine in rejecting a proposed dam on the Snake River.[91] But in 1996 Idaho's governor signed into law a bill that limits the scope of the public trust doctrine.[92] Proponents of the bill argued that it was needed to remove much of the discretion from the judiciary in deciding where the public trust doctrine should be applied. Among other things, the bill prohibits the judiciary from applying the doctrine to "the appropriation or use of water, or the granting, transfer, administration, or adjudication of water or water rights . . . or under any other procedure or law applicable to water rights in the state; [or to] the protection or exercise of private property rights within the state."[93]

A recent case in Colorado appeared to embrace the public trust doctrine but was quickly amended to eliminate all public trust language. Colorado is one of only two western states that does not yet apply even minimal public interest criteria to the allocation of water. But in a 1995 case over whether the Colorado Water Conservation Board could decide to not fully enforce one of its instream flow rights, the state supreme court ruled that "the Conservation Board, unlike other appropriators of water, is imbued with a public trust and, because its authority is circumscribed by statute, it therefore must be held to a different standard than other appropriators."[94] The supreme court further stated that "the Conservation Board has a fiduciary duty to protect the public in the administration of its water rights decreed to preserve the natural environment."[95] Though falling well short of requiring the application of public trust principles to all water allocation decisions, the decision caused an uproar among certain elements of the water development community. On petition for rehearing, the state supreme court backed down from its "public trust" language, replacing it with the apparently less threatening, and more ambiguous, term "unique statutory fiduciary duty." As of this writing one can say little more than that there appears to be some effort by the state supreme court to insert public values in some way into the process of appropriating and administering instream flows, but this effort should not be construed to have brought the public trust doctrine to Colorado.

None of these decisions has been as far reaching as the *National Audubon* decision, and it remains to be seen whether the western states will recognize the public trust doctrine as completely as, over a century ago, the western states followed California in adopting the prior appropriation doctrine. Courts may choose to apply the public trust doctrine only to lands beneath tidal navigable waters, as the doctrine was originally formulated, or, citing California court precedents, to lands beneath all navigable waters, to the water itself, and even to the water flowing in nonnavigable tributaries of navigable waters. Similarly, courts may choose to protect only the trust's classic uses of navigation, commerce, and fishing or may expand the list of protected uses to include a variety of contemporary activities.

It is not entirely clear what actions are required once the trust is invoked. There is a variety of possibilities.[96] The trust may be interpreted to require only that applications for new appropriations be scrutinized to ensure that they are consistent with public uses of water protected by the trust, as described above in Idaho and Montana, or as already occurs in the western states that employ a public interest review. Or the trust may be interpreted to require ongoing evaluations of the public interest in the use of water, and even retroactive reallocations of water if uses are found to be incompatible with an expanded view of the public trust, as happened in the *National Audubon* case. The number of possibilities is limited only by the perceptions and reasoning of the state courts.

Should the Public Trust Doctrine Be Invoked?

There continues to be active debate about the merits and demerits of the public trust doctrine. To some, the public trust doctrine seems at odds with the prior appropriation doctrine, as the former emphasizes public rights and the latter private property rights.[97] To others, the public trust doctrine brings a measure of sanity to what they perceive as a water allocation system that fails to recognize the public value of water.[98] Probably no water issue has been as controversial, or the source of as much acrimony, as that of whether public trust criteria should be retroactively applied to existing appropriations. Many proponents favor the doctrine precisely because the trust can be used to correct what they see as earlier mistakes in the allocation of water. They argue that individual uses of water have always been subject to conditions necessary to protect the public trust, and that water uses can and should be changed in light of changed circumstances or new information. This argument was determinative in the *National Audubon* case.

Opponents of the trust view the retroactive application of the doctrine as unfair, as an unwelcome change in the rules long used to allocate resources, as a source of disruption to long-held expectations, and as a violation of constitutionally protected property rights. Though theoretically always present as a condition of water use by individuals, few water rights holders have been aware of potential trust responsibilities. The public trust doctrine was largely unknown at the time that many of the West's appropriations were first made, and even now few water rights holders know much about the doctrine. Opponents of retroactive application of the public trust doctrine argue that to evaluate water uses that have been occurring for decades or longer using criteria made explicit only recently is not only unfair but also detrimental to the last 150 years of effort to make water rights more secure. They argue that if new uses of water subsequently become more valuable, then individuals acting on their own or collectively should take steps to acquire water through voluntary means, such as the purchase of existing rights from willing sellers.

Retroactive application of the public trust doctrine could impose huge costs on individual water rights holders.[99] For example, the cost of finding water and power supplies to replace those formerly supplied to Los Angeles by diversions from the Mono Basin is expected to total at least several tens of millions of dollars.[100] Other property holders may experience a loss of security, since they have no way of knowing if or when public trust provisions will be applied to reallocate their holdings. This insecurity may lead them to severely restrict new water-dependent investments, thereby diminishing the total value produced from the use of water resources.

The Public Trust As a Balancing Mechanism

Experience to date indicates that courts will use the doctrine—if it is invoked at all—to balance competing public and private needs rather than to make a clear choice between the two. Even the *National Audubon* court recognized this need for balance, noting that "the prosperity and habitability of much of this state requires the diversion of great quantities of water from its streams for purposes unconnected to any navigation, commerce, fishing, recreation, or ecological use relating to the source stream."[101] The court did not intend to deny all appropriations that would affect public trust values. Rather, it sought to ensure that such values were considered during the process of evaluating proposed appropriations.[102] What the court proposed, in effect, was not a dominance of one doctrine over the other, but a merging of the public trust and prior appropriation doctrines into a single, cohesive system of allocating water that would recognize both public and private values.[103]

It is not possible to know how all of these issues will be resolved until more courts apply the doctrine. For now, the public trust doctrine, for better or worse, looms more as a potentially powerful tool for affecting the allocation of water than as reality, especially outside of California. Given the intensity of the controversy surrounding retroactive application of the public trust doctrine, it is interesting to speculate as to how much support retroactive application of the trust would garner if there were adequate opportunities for river protection advocates to achieve their goals through other means. Because many states provide only limited opportunities for citizens to participate in the protection of instream flows, the public trust doctrine may be a last resort for advocates of river protection. Retroactive application of the public trust criteria can be a radical solution to the problem of inadequate protection of instream flows, but in some states it may be virtually the only method available for forcing changes in the ways that water resources are used and allocated.

State Rivers Programs

Several western states have programs designating certain rivers as having special characteristics worthy of preservation. Often these designations are patterned on the federal Wild and Scenic Rivers Program initiated in 1968, but characteristics of the programs vary widely.[104] The variety of the programs, and of their relationship to instream flow, can be seen in a quick survey of the protected rivers programs in Oregon, Alaska, California, Idaho, Oklahoma, Washington, and Montana.

The voters of Oregon established a Scenic Waterways Program in 1970 following a successful initiative campaign.[105] The Oregon statute declares that the highest and best uses of designated rivers are recreation, fisheries, and wildlife and requires that the free-flowing character of the rivers be maintained in quantities necessary for these purposes. This requirement is implemented by quantifying the amount of water necessary to accomplish the designated purposes and then prohibiting or conditioning offstream water rights applications that would reduce flows below the specified amount. Essentially, this emulates the method used by the state on other rivers to set minimum streamflows (prior to 1987) or instream water rights (after 1987). As of June 1995, the Scenic Waterway Program included 1,100 miles of streams. No new river designations had been made since 1987.

Alaska is the only other western state to specify a direct method of protecting instream flows in state-designated rivers. In 1988 the state legislature directed that instream flows on six "recreational" rivers located in the south central portion of the state, where most of the population lives,[106] be reserved to achieve the specified purposes. As of April 1995, the collection of data to support instream flow reservations on the six rivers had not been completed and the instream flows had not yet been reserved.[107]

A common but less direct method of protecting instream flows is to prohibit the construction of new dams or other facilities that would threaten instream flows in the designated rivers. This method has been used in state rivers programs in California, Idaho, Oklahoma, and Washington. The California Wild and Scenic Rivers Act prohibits the construction of dams and other impoundment or diversion facilities on designated rivers and bars state agencies from assisting any entity in the planning or construction of such facilities. "Assistance" precluded by the act includes the granting of licenses required before construction can proceed. Though potentially providing substantial protection for instream flows, these protective features were later weakened considerably. A 1982 amendment to the Act eliminated the restrictions on development above designated reaches, leaving open the possibility of upstream diversions. Together with a very narrow interpretation of the Act by the State Water Resources Control Board, the amendment left the state's designated rivers with little more instream flow protection than was already available through other means.[108]

The 1988 Idaho Protected Rivers Act authorizes the designation of certain rivers as either "natural" or "recreational." Designation of a river as recreational does not automatically result in the implementation of any instream flow protection measures, but a natural designation prohibits the construction of dams, impoundments, hydroelectric projects, and diversion works on the river. As of March 1993, there were 581 miles of rivers in the system, roughly double the total mileage protected by instream water rights held by the Idaho Water Resources Board but still considerably less than 1 percent of Idaho's 93,000 miles of rivers and streams.[109]

Oklahoma's Scenic Rivers Act prohibits the construction of large dams on designated rivers without legislative approval.[110] As of March 1995, only the Illinois River and some of its tributaries had been designated under the Act. The Illinois River, used extensively for recreation, runs through the well-watered eastern part of the state, where there is little pressure for diversion.[111]

In practice, the protection may therefore apply only to large hydropower facilities. The Act does not state explicitly whether water is reserved for instream purposes, but the state's Water Resources Board is likely to investigate the issue in the near future.[112] Oklahoma has not previously recognized water rights for instream purposes on any of the state's rivers or streams.

Washington's Scenic Rivers Program, initiated in 1977, also protects designated rivers from the construction of new dams, impoundments, and diversions.[113] The goal of the program is to keep designated rivers in a natural condition. Only two rivers have been designated to date: the Skykomish system, totaling 67 river miles and designated in 1977, and the Little Spokane, totaling 7.5 river miles and designated in 1986. Seventeen other rivers have been listed as candidates for designation, though there are no ongoing efforts to have any of them designated in the near future.

Some state rivers programs may provide even less direct protection for instream flows. For example, the Recreational Waterway Program created in 1972 by the Montana Division of Fish, Wildlife, and Parks through administrative rule-making procedures serves primarily to identify the state's most valuable streams and rank them according to certain criteria.[114] Some of the rivers identified through the program may subsequently receive protection through other means—two of the rivers identified by the program were later added to the federal Wild and Scenic Rivers system, and another river received an instream flow reservation under state law—but the program itself contains no mechanism for protecting instream flows.

Private Agreements and Modified Reservoir Operations

A variety of other means can be used to protect instream flows, including the modification of reservoir operations. Dams and reservoirs play an enormous role in regulating streamflow on rivers throughout the West. Most rivers of any consequence have been dammed at least once somewhere along their length, whether on the mainstem, tributaries, or, for the most part, both. Reservoirs store water for release when needed by downstream water users. These water users are virtually always protected by water rights, but it is often possible for private individuals or groups, or even public agencies, to make arrangements with the water right holders to release water from storage in quantities and at times that will enhance the value of the water for instream use.

In some cases this can be accomplished quite easily. Many reservoir owners and managers are simply unaware of either the instream uses that could be made of the water released from their dams or the particular flow needs of those instream uses. Once reservoir operators are made aware of these needs, they are sometimes able to modify release schedules in ways that enhance instream uses at very little cost to themselves. For example, the state of Montana has successfully negotiated informal agreements with a number of federal and state dam operators for voluntary releases that will enhance instream flows.[115]

In other situations, tailoring reservoir releases more closely to the needs of instream water uses imposes significant costs on reservoir owners or off-stream water users. To successfully negotiate such an agreement, instream water users may need to offer compensation. Such transactions may be quite simple, as in the payment made by The Nature Conservancy and the Trumpeter Swan Society to lease water from the Snake River water bank to benefit trumpeter swans on the Henry's Fork River.

A more complex arrangement exists on North St. Vrain Creek in Colorado, where a consortium of interests got together in an effort to provide flows necessary to improve a fishery above the town of Lyons.[116] The creek's water is fully allocated to agricultural and municipal water users, which store water upstream in Ralph Price Reservoir during winter and spring. Much of the water released from the reservoir is diverted two miles downstream at Longmont Reservoir, which was built primarily to enable the diversion and has little storage capacity. Streamflow in the seven-mile reach below Longmont Reservoir, extending to Lyons, is adequate in the summer, because water is released from the upstream reservoirs to meet agricultural needs below Lyons. But no more than 2.5 cfs is left in the creek during the winter months, when most of the flow is diverted out of the creek at Longmont Reservoir for use in generating electricity and supplying the municipal needs of the City of Longmont.

The effort to enhance winter flows in the creek below Longmont Reservoir began with formation of the St. Vrain Corridor Committee in 1991. The Committee is composed of representatives of the City of Lyons, City of Longmont, Boulder County, the St. Vrain and Lefthand Water Conservancy District, and the St. Vrain Anglers, a local chapter of Trout Unlimited. Supporting roles in the Committee were played by representatives of the Colorado Water Conservation Board and the Colorado Division of Wildlife. The committee's short-term objective was to provide winter flows of 8 cfs for a period of five years, the minimum amount of time necessary to adequately assess the influence of the enhanced flows on the fishery. The long-term objective, provided that the flows were successful in improving the fishery, was to find a permanent source of water for the fishery.

The Committee has purchased from 600 to 1,000 acre-feet of water per year, starting with the winter of 1992–93, to enhance instream flows in the river. Money for the purchases, approximately $8,000–14,000 per year, has come primarily from public funds made available to the City of Lyons and Boulder County for economic development and the provision of open space. Members of the Committee have been looking for additional sources of funding to supply water on a permanent basis if the experiment proves a success. Most of the water for the instream flows has been purchased from the

City of Longmont, which owns Ralph Price Reservoir. On at least one occasion water was provided through a paper transaction in which the Committee purchased water from the Colorado–Big Thompson Project, a major transmountain diversion project built by the Bureau of Reclamation in the 1930s, and traded it for water stored in Ralph Price Reservoir. The Committee has not always been successful in maintaining a minimum flow of 8 cfs, but it has been able to maintain flows of at least 6 cfs, which has been enough to keep the river from completely freezing over. As of the summer of 1995, the Committee was just beginning to make a preliminary assessment of the effectiveness of the flows in improving the fishery.

Private arrangements to protect instream flows may be even more complex, as was the case with The Nature Conservancy's successful effort to increase flows through its Phantom Canyon Preserve on the North Fork of the Poudre River in northern Colorado.[117] Halligan Reservoir, located just upstream from the preserve, is owned by the North Poudre Irrigation Company, which stores water in the reservoir during winter and spring mainly for release during the summer irrigation season. This pattern of reservoir releases put stress on the preserve's rainbow and brown trout fisheries during the fall and winter. The Conservancy negotiated with the irrigation company to provide a more favorable pattern of releases, including a drawn-out emptying of the reservoir at the end of the irrigation season, to provide greater flows in the fall. The irrigation company also agreed to release a small survival flow during the winter and to increase releases more gradually at the start of the irrigation season to avoid flushing brown trout fry that had hatched in the fall.

The new pattern of releases was not nearly as complex as the arrangements that provided it. The Conservancy first considered buying shares in the irrigation company, but discovered that ownership of the shares would not entitle the Conservancy to release of flows in the fall and winter. The Conservancy then agreed to lease, but not take delivery on, an amount of shares in the irrigation company that would be equivalent to the increased amount of water lost in transit as a result of the drawn-out releases from the reservoir. The amount of water released in the winter is small enough that it usually doesn't prevent the reservoir from refilling in the spring. But to pay for the water in years in which the winter release does prevent the reservoir from filling, the Conservancy leased an equivalent number of shares from the Eastman Kodak Company, which owned shares in the irrigation company. Kodak leased the shares to the Conservancy without cost, because of credits given to Kodak by the City of Greeley in recognition of the fact that the winter releases from Halligan Reservoir were captured downstream in Greeley's Seamen Reservoir. The Conservancy actually had to agree to purchase options to lease shares to cover shortages resulting from the bypass flows on a 2-to-1 basis, and not take delivery on those shares, in order to maintain a uniform yield per share throughout the irrigation company in the event that the reservoir did not refill in the spring. The Conservancy accomplished this by purchasing an option to lease one block of irrigation company shares from the City of Fort Collins and an option to lease a number of Colorado–Big Thompson units from the City of Loveland for trade to the City of Fort Collins for the balance of the North Poudre shares required under the agreement.

One of the advantages of arrangements such as those used on the North St. Vrain and North Fork of the Poudre is that they do not require the involvement of the state.[118] Nor do they depend on the recognition of instream flow rights. All parties who participate in the agreements do so on a voluntary basis, and no changes in public policy are required. These kinds of arrangements may not be possible in all situations, because they depend on the existence of cooperative and favorably located facility owners, and may provide flows only temporarily. Flows provided through these agreements may also be vulnerable to diversion by other water users who are not parties to the agreement, because the flows are not protected by direct-flow water rights. Nonetheless, the success of these arrangements is proof that in any given situation there may be a variety of opportunities available for preserving river flows if people are both creative and determined in their efforts.

Enhancing Instream Flows through Conservation

Conservation is one of the most promising sources of water to meet new and expanding needs. Using less water to meet current consumptive needs frees up water for other users. Water saved through elimination of waste or the use of more efficient methods and technology has been mentioned as a possible source of instream flows in all of the western states.

A water right under the prior appropriation doctrine has always been limited to the amount of water put to beneficial use. Water that is wasted, or used unreasonably, is not considered to have been put to beneficial use. Theoretically, rights to the amount of water that is determined to have been used in a wasteful manner revert to the state, and the water becomes available to junior or new appropriators.[119] This raises the potential to put water back into streams by arranging for water saved through conservation practices to be applied to instream use.

Conservation has been linked directly to instream flow protection in Oregon and Washington. Oregon passed a law in 1987 that allows a water user to submit a water conservation proposal to the Oregon Water Resources Commission.[120] If the proposal is approved and the conservation measures are implemented, the law authorizes the water user to keep 75 percent of the saved water for additional use, sale, or lease. The remaining 25 percent is dedicated to instream use.[121] Both new uses are protected by water rights; the water user's new right has the same priority date as the original water right, and the instream water right, which is held by the state, has a priority date one minute later than the original water right.

Washington's "trust water rights" program,[122] enacted in 1991, also provides a means of enhancing instream flows using water saved through conservation. The trust water rights program was developed by the state's Department of Ecology in response to the legislature's call for a means "to facilitate the voluntary transfer of water and water rights, including conserved water, to provide for presently unmet and emerging needs."[123] The program authorizes the temporary or permanent transfer of water or water rights to the state for a wide array of purposes, including instream flows. The law states that "any agency of the state may purchase, lease, or accept a gift of an existing water right to be dedicated to instream use in the form of a trust water right. . . . A voluntary transferee may designate up to 100% of a trust water right for instream use."[124] The link between conservation and instream flows is that conserved water ordinarily subject to relinquishment under the state's waste and beneficial use laws can instead be managed through the trust water rights program.[125] The program was designed in part to encourage conservation and enhance instream flows by making the transfer of a water right to the state trust water rights program a condition of federal or state aid given to water users to implement conservation measures. In return for financial assistance in implementing conservation measures, the saved water, or a negotiated part thereof, is donated to the state. As in Oregon, water rights transferred to the state retain the same priority date as the water right from which they originated, so that trust water rights can be much more effective than a new instream appropriation.

Though these two programs seem very promising, neither the Oregon nor the Washington program has yet been used to transfer any water from offstream use to instream use.[126] Part of the reason may lie in the difficulty of administering the programs. Given that much of the water "wasted" in existing uses—for example, water lost from leaky pipes or unlined canals during transport, or water applied to fields in excess of what can be taken up by the irrigated crop—eventually ends up back in the stream as return flow, it is difficult to determine how much water is actually saved and available for new use after implementing conservation measures. Administrators have to ensure that water "savings" credited to the conservation measures and allocated to new uses are not in fact already being used by others. The amount of time and money necessary to evaluate proposed conservation projects in the effort to ensure that other water rights are not damaged may make the process feasible only for very large conservation projects.[127]

There are other reasons that water users may be hesitant to sign up for the programs. Neither water users nor administrators yet have any practical experience with them, so the process is not likely to work very smoothly at first. Also, water users may fear that the new rights to saved water are vulnerable to legal challenge from junior appropriators who could argue that the original water right owner never had the right to use water inefficiently in the first place, and that the saved water should therefore be made available to junior appropriators rather than to the original water right owner or the state.[128] Water users tend to think of their water right in terms of the amount that is diverted rather than the amount that is consumptively used and may resist the notion of having to give up any of the conserved water just because they implement measures to use the water more efficiently. Historically lax enforcement of beneficial-use waste provisions lends some credence to the belief that if water users resist giving up the conserved water, the state will eventually let them keep it all.

There is still hope that the programs will transfer water from offstream to instream use. Oregon's law has been amended in an effort to facilitate the process,[129] and at least one private, nonprofit organization, the Oregon Water Trust, is currently emphasizing the benefits of the program on the theory that if they can get one water user through the program, others will follow.[130] The same theory may work in Washington, where there are currently at least a couple of trust water rights in the Methow Valley slowly making their way through the system.[131] The programs may become more popular if these initial efforts are successful.

Instream Flow Protection—The Use of Multiple Methods

An important point to consider when contemplating use of these instream flow protection methods is that use of one method does not necessarily preclude use of others. For example, California places primary emphasis on the protection of instream flows through administrative procedures but also authorizes the transfer of offstream water rights to instream purposes, has implemented a protected rivers program, and has witnessed extensive and important applications of the public trust doctrine. Most of the western states have already employed more than one method to protect streamflows. Virtually all states can use administrative procedures in at least some situations; most states recognize water rights for instream purposes; all but one of the states with protected rivers programs use other methods of protecting streamflow; private arrangements can occur anywhere; and all of the western states are potentially subject to application of the public trust doctrine. The methods described here can be viewed as tools for accomplishing the goal of leaving water in streams. Viable instream flow protection efforts may well involve the use of several of these tools.

CHAPTER SEVEN

Effect of Instream Flow Protection on Other Water Uses

Instream flows may be protected by, among other means, water rights, minimum flows, and reservations. All have the potential to affect other water uses. Conversely, other water uses, and particularly the transfer of existing offstream rights to new locations or uses, have the potential to affect instream flows. This chapter discusses the interaction of instream flow protection measures with other water uses. All references in this chapter are to instream "rights" for simplicity, but many of the same points apply as well to minimum flows and reservations.

Effect of Creating an Instream Flow Right

The effect of an instream water right on other water uses can be summarized quite easily. Fundamentally, an instream water right prevents subsequent appropriators from diverting water within or above the protected reach in amounts that would reduce the level of flow in the protected reach at a specified time below the amount protected during that time. Uses of water initiated prior to the creation of an instream water right are unaffected, because they are fully protected by their senior status no matter where they are located. Even some junior water rights holders are unaffected, because an instream water right "consumes" very little water, leaving the water for further appropriation downstream. And even those who wish to initiate new water uses upstream from or within the protected reach may be able to do so if there is more water flowing through the protected reach at the specified time than is necessary to satisfy the instream right.

Ignoring for now the possibility of future water rights transfers, the impact of an instream flow water right on other water users is summarized in table 7.1. It is important to emphasize that the effects discussed here, and summarized in the table, are those that occur in situations in which the possibility of transferring water rights does not exist. Instream rights can affect, and be affected by, the transfer of other water rights to new locations on a stream. The relationship between instream rights and transfers is discussed later in this chapter.

Table 7.1 indicates that the only water rights that may be adversely affected by an instream right are junior water rights located upstream from the instream right. Whether an upstream junior is adversely affected will depend on the circumstances of the particular situation. For example, the presence of a senior offstream right below the instream right, if for an amount of water equal to or greater than the amount specified in the instream right, would make the instream right irrelevant to the upstream junior's situation. The upstream junior would be forced to pass enough water by his or her diversion to satisfy the downstream senior right whether or not there was an instream right located between the two offstream rights. Even if the instream right is the only senior right downstream there may be enough water in the stream to satisfy both the instream right and the upstream junior right. In those situations in which the instream right causes less water to be available for subsequent diversions upstream, the situation of the upstream junior is the same as if the downstream right were for offstream rather than instream purposes—the situation of the upstream junior is a result of the application of standard prior appropriation principles that protect senior appropriations from junior appropriations, not of the fact that the downstream water is being used for instream rather than offstream purposes.

These factors suggest that (1) the existence of an instream right is not likely to have any adverse impact on most other water users; and (2) to the extent that other water users are affected, the effect will depend on location, priority, and other circumstances unique to the river at issue—the same as is

TABLE 7.1

Effects on Other Water Users of a New Instream Flow Appropriation, Assuming No Potential for Subsequent Water Rights Transfers

Priority and position on the stream relative to instream right	Effect of instream appropriation
Senior	
upstream	no effect
downstream	no effect
Junior	
upstream	additional senior right downstream
downstream	no effect

true of offstream water rights. Because only water users who initiate their uses subsequent to, and upstream from, the instream water right may be adversely affected, an instream right has the greatest potential to affect other water users the earlier the priority date of the instream right and the farther downstream it is located. An instream water right located in the headwaters will have no impact on other water users if there are no such users above the protected reach.[1]

The impact of instream water rights on the total consumptive use of a river depends on the relative locations of the instream right and the points of diversion. An instream right prevents a specified amount of water from being diverted *at a location within or above the protected reach* but does not prevent the water from being diverted once it has flowed through the protected reach. An instream right can result in the absolute loss of water for consumptive purposes only where the instream right is located so far downstream that diversions below the instream reach are not possible, as when an instream right is located just above a river's outlet to the sea.

The placement of an instream right immediately above a river's outlet to the sea doesn't happen often, but there are situations in which an instream right could defer consumption to a point downstream where it is effectively lost to the local community.[2] An example would be an instream right located immediately above some physical obstacle, such as a canyon, which would effectively prevent the water from being used for an offstream purpose until reaching a point several miles downstream. A more common situation would be an instream right located immediately above a political boundary that prevents the local community from using the water once it has passed through the protected reach. Fear of just this possibility has aroused opposition to instream rights in many locations. For example, instream rights created to protect endangered species in Colorado's Colorado and Yampa rivers have aroused opposition from those who fear that the rights will prevent Colorado from taking full advantage of its interstate water allocation to the waters of those rivers. The Colorado and Yampa flow from Colorado into Utah, and the instream rights on both rivers occur on reaches directly upstream from the Utah-Colorado border. Another example comes from the Snake River, where irrigators in Idaho consider water as "lost" if it is allowed to cross over the border into Washington.[3] From a regional, national, or international standpoint the water may still be available for consumptive use after being protected instream, but from a local or state standpoint the water may be forever lost if not diverted before passing over the political boundary.

Even this situation could be turned to a state's advantage, however, if the water bypassed to another state could be credited to the upstream state's interstate compact obligations, or if instream rights could be used strategically to protect water delivery obligations to downstream states.[4] For example, New Mexico's Interstate Stream Commission has been authorized to purchase and retire offstream water rights in the Pecos River to help meet the state's water delivery obligations to Texas under the Pecos River Compact.[5] Though not really creating "instream rights," this program creates a similar effect because the purchased rights are not subject to forfeiture and reappropriation due to nonuse.[6]

It is important to remember that just because water is left instream does not mean that the water is left "unused." Leaving a portion of the river's flow instream may produce more total value from use of the river than would diverting the water for use offstream, because nonconsumptive uses are sometimes more valuable than consumptive uses. Any additional water that flows into Utah due to instream water rights on the Colorado and Yampa, or into Washington due to instream rights on the Snake, may not be available for offstream use in Colorado and Idaho but is used instream in Colorado and Idaho.

Effect of Transfer to Instream Use

It is axiomatic that both parties to a voluntary transfer benefit, but what happens to other water users when a water right is transferred to instream use? First, let us compare a transfer to the creation of a new instream right. Recall from table 7.1 that the only class of water user that might suffer adverse effects from the creation of a new instream right are junior (i.e., future) water users located upstream from the protected reach. The purchase of an existing offstream right for transfer to instream use eliminates the possibility of that harm. The upstream junior has to let pass the same quantity of water to the downstream senior as before; it makes no difference whether the downstream senior uses water instream or offstream. There is still a senior water user downstream after the transfer, but not an *additional* water user downstream.

A most attractive feature of transfers to instream use is that users not party to the transfer may actually *benefit.* Because the instream use is nonconsumptive, retirement of the offstream water use will make more water available downstream from the protected reach. This is particularly advantageous to downstream juniors, who may not previously have had enough water to fill their existing rights, or who may even be able to initiate a new water use. The transfer can also benefit downstream seniors, because the additional supplies available downstream reduce the chance that the senior will need to place calls on the river. In addition, all downstream users, both senior and junior, may benefit from improvements to downstream water quality once return flows from the former offstream use—most of which are of lower quality than the original water diversion—are eliminated.

For situations in which there is no possibility of subsequent water rights transfers, the effects on other water users of transferring an offstream right to an instream right are summarized in table 7.2. These results apply only to situations in which there is *no* opportunity for subsequent transfer. As de-

TABLE 7.2
Effects on Other Water Users of Replacing an Existing Offstream Right with an Instream Right, Assuming No Possibility of Subsequent Water Rights Transfer Upstream

Priority and position of water right relative to instream right	Effect of replacing an offstream right with an instream right[a]
Senior	
upstream	no effect
downstream	(+) better water quality (+) reduced need to place calls on river
Junior	
upstream	no effect
downstream	(+) more water available (+) better water quality

[a](+) = beneficial effect, (–) = adverse effect.

scribed in the next section, the transfer of an offstream water right to instream use can affect the ability of other offstream water right holders to subsequently transfer their rights from a position below the instream right to a position above the instream right. But there are many reasons that subsequent transfer opportunities may not exist and in which table 7.2 would be relevant. For example, the instream right may be either at the very top or very bottom of the watershed, in which case either there are no rights below the protected reach to transfer, or there is nowhere above the protected reach to transfer the rights to. There may not be any sites from which to physically divert the water higher in the watershed, or it may not be economically feasible to move the diversion point higher in the watershed. Water rights transfers are also constrained by a number of other legal and institutional barriers that have nothing to do with instream flow protection.[7]

More about Transfers and Instream Flow

The relationship between water rights transfers and instream flows can be more complex than described so far. Most of the complexity stems from differences in the impacts that offstream and instream water uses have on other water rights, and from guidelines used in evaluating transfers.

Water rights transfers in all of the western states must be approved by a designated administrative agency or court (see chapter 6 for details). These agencies may consider a variety of criteria before approving a transfer, but the predominant one is that the proposed transfer may not cause injury to other appropriators. Under the prior appropriation doctrine every water user is entitled to the stream conditions in existence at the time their water use commences. Changes or transfers that would alter conditions so as to harm other water rights holders are not allowed.[8] This "no injury" rule is applied without regard to seniority, so even the most senior water user, while legally protected in his or her use so long as the water continues to be used according to the terms of the original water right, is not allowed to change or transfer the right if the alteration would cause harm to any other water right holder.

A water user who believes that his or her water right would be harmed by a proposed transfer may lodge a protest with the appropriate regulatory agency or court. If the parties are not able to settle the dispute among themselves, the agency or court must then issue a determination on the matter, usually after a hearing. Some states allow parties other than other water right holders to protest a transfer, so long as the protesting party presents evidence indicating that he or she has interests at risk from the proposed transfer.

The impact of a transfer on offstream water rights may differ from the impact of a transfer on instream water rights. To illustrate, first consider a river segment on which only offstream water rights exist. Even if all water rights are being satisfied, as we will assume here, the sum of all water right amounts on the river may still exceed the total amount of water available from the river. For example, ten water rights may total 100 cfs though the river contains flows of only 60 cfs. This discrepancy is accounted for in part by the difference between the amount of water diverted from the river and the amount that is actually consumed.[9] Water right amounts are usually specified in terms of the volume of water that is diverted rather than the volume that is consumptively used. The amount of water consumed is the amount that evaporates or transpires, or that is otherwise permanently removed from the stream. Few water uses are 100 percent consumptive, so some of the diverted water usually returns to the stream.[10] The water may return through a pipe or canal, such as effluent from a municipal waste treatment plant, or through the ground, as frequently occurs when water is diverted for irrigation. No matter how it is returned, the "return flow" can be withdrawn again downstream. The same water may be diverted several times on its way downstream as a portion of each diversion returns to the stream.[11]

To continue the example, assume that the river is fully allocated, the most senior water right on the river is used for irrigation, and that at least some of the other nine water rights holders are located downstream from the senior user. Further assume that the senior water right is for a diversion of 10 cfs, of which an average of 6 cfs is consumed by crops, and the remaining 4 cfs percolates back into the stream. If the owner of the senior water right sold the right to an industrial user who diverted the entire 10 cfs and transported it to a neighboring watershed, from which there would be no return flow to the source stream, the amount of water available in the river downstream from the then-retired senior water user would be 4 cfs less than what it was before the sale. Because this river was already fully allocated, at least one of the junior water users downstream would be injured by the transfer.

The potential for this type of harm has led to the development of a general rule that transfers will be limited to historical *consumptive* use.[12] In the example presented here, the owner of the irrigation water right would be allowed to transfer the right to the industrial user, but the right would entitle the industrial user to divert only 6 cfs rather than 10 cfs. Only the proportion of the water that was never available for consumption by other users in the first place—the amount that was permanently removed from the river by the senior irrigator—is transferred. The rest of the water, which formerly was returned to the river as return flow, is instead never diverted from the river at all, and hence remains available for use downstream.

An out-of-basin transfer is the most extreme case. In theory, the full diversion amount, rather than just the consumptive portion, could be transferred to a new use or new location in the same watershed without causing injury to other offstream water rights holders, *so long as the amount of water that is consumptively used does not increase.* To illustrate, assume that each of the 10 water rights results in an actual consumption of 6 cfs. The senior water user could transfer his or her entire 10 cfs water diversion upstream, to the top of the river segment, without causing injury to other water users so long as 4 cfs still returned to the river as return flow. With the return flow, 54 cfs—6 cfs for each of the 9 other water rights holders—would be available to satisfy the consumptive needs of the other water rights holders downstream.[13]

In practice, however, things rarely work so easily. In particular, it can be very difficult to determine how much of the water diverted for a new use or location will be consumptively used and how much will return to the river. Other complicating factors are that (1) water flows are seldom constant throughout an extended river segment, (2) return flows may reenter the river below the site of the next diversions, and (3) both water flows and diversion rates vary uniquely over time. In short, it can be very difficult to calculate the impact of a proposed transfer on other water users. To simplify the evaluation and administration of water rights transfers, it has been common to limit all transfers, not just those in which water is transported out of the basin, to the amount of water historically consumed.[14] By limiting transfers to the quantity of water known to already have been permanently removed from the river, administrators improve the odds of protecting other water users from injury resulting from transfers.

The historical consumptive-use rule was established to protect offstream water users, not the natural stream environment. Offstream water users are unaffected by a change in the location of a diversion when the historical consumptive-use rule is applied. However, from an instream flow perspective it matters very much where water is withdrawn from a river. For example, consider a situation in which all of the water rights on a river fully allocated to offstream water uses are clustered at either the top or the bottom of a river segment. If they are all clustered at the top, the river will be dry for the remaining length of the segment. But if they are clustered at the bottom, that same length of river will contain its full flow until reaching the diversion sites down below. All of the offstream water rights are satisfied in both cases, but the impact on instream flows varies dramatically.

Likewise, the transfer of an offstream water right from a downstream diversion point to an upstream diversion point will not adversely affect other offstream water rights holders, particularly if the transfer is limited to the amount of historical consumptive use. But the transfer will result in decreased flows in the reach between the new and former points of diversion. If an instream-flow water right is located between those two points, there is the potential for the instream right to be damaged. The possibility that a transfer could injure an instream right in the same situation in which an offstream right would be left unscathed is critical to an understanding of the full relationship between instream flows and transfers, and it needs to be explained in more detail.

First consider the example of a river on which only offstream water rights exist, as illustrated in figure 7.1. The river, which has an average flow of 15 cfs, has been allocated among three water rights: right A, located farthest upstream, is a 10-cfs water right with an 1896 priority date and return flow of 4 cfs; right B, located farthest downstream, is a 10-cfs right with a priority date of 1920 and no return flow; and right C, located between the other two, is a 5-cfs water right with a priority date of 1940 and no return flow.

In years with average streamflow of 15 cfs, there is not enough water to satisfy all of the rights—not an uncommon situation in the West. The most senior water user, A, would receive the full measure of her right, 10 cfs. With the 4-cfs return flow, the river then would have 9 cfs. User C might try to divert and consume all or part of his 5-cfs right, but this would leave as little as 4 cfs in the river. Once B realized that she was not receiving the full measure of her more senior 10-cfs right, she would put a call on the river,

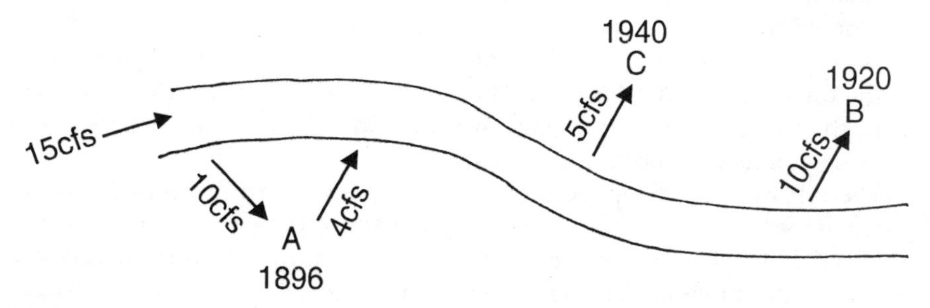

FIGURE 7.1
A Fully Appropriated Stream

forcing user C to completely shut off his diversion and allow all 9 cfs to flow down the river. The call would still not enable B to receive the full measure of her 10-cfs right—there simply would not be enough water available in the river. So in a year with average streamflow, A would divert 10 cfs and consumptively use 6 cfs, C would not receive any water, and B would divert and consume 9 cfs.

Now consider what would happen if B transferred her site of diversion to a point between users A and C to take advantage of greater opportunities for gravity flow. In the average year A would, as before, divert 10 cfs and consumptively use 6 cfs, leaving 9 cfs in the river, and B would then divert the entire remaining 9 cfs, leaving nothing for C. All three water users would continue to receive the same amount of water they received before the transfer occurred, so neither A nor C would have cause to protest the transfer. The only difference for C is that before the transfer he was forced by B's call on the river to allow the entire 9 cfs to flow past his diversion gates, whereas after the transfer B simply took the 9 cfs before it ever reached C's diversion structure.

The outcome would be quite different if right C were an instream rather than an offstream right. As in the first scenario, in an average year A's diversion of 10 cfs would leave 9 cfs to flow through the protected reach and user B would receive and consume the 9 cfs once it reached her headgate near the bottom of the river segment. But replacing the offstream right with an instream right for the same quantity of water would result in the fulfillment of an additional water right—C's—when compared to the situation in which all of the water rights were for offstream purposes. Because an instream right is entirely nonconsumptive, C could use the entire volume of water flowing in the river to fulfill his right without causing any harm to the more senior downstream user. Note also that user B would be able to receive 9 cfs without having to place a call on the river. Placing a call on the river would produce no extra water, because a call would not affect the more senior water user A, and instream rights—which involve no physical control structures—cannot physically produce any more water.

However, the transfer of B's site of diversion to a point above user C would cause injury to C if C were an instream right. If the transfer occurred, user B would divert and use the entire 9 cfs remaining in the stream after user A's diversion and return flows, leaving nothing for user C. Without the transfer, user C had more than enough water flowing through the protected reach to fulfill his water right. With the transfer, there would be no water flowing through the protected reach in the average year.

The possibility and degree of damage to an instream right resulting from the transfer of an offstream right from a position below to a position above the instream right will depend on the amount of water transferred, the amount of water specified by the instream right, and the amount of water flowing in the river. Transfer of the offstream right to a position above the instream right will definitely reduce the amount of water flowing through the protected reach but may not diminish flows to a level less than that specified by the instream right. In the example above, a water flow of 21 cfs or greater is sufficient to satisfy users A and B and still leave enough to satisfy the instream water right C, even if B's point of diversion were above the protected reach. But if at least 21 cfs cannot always be assured—if flows might fall below that amount in dryer years—injury would still be possible. Limiting transfers to the amount of water historically consumed reduces, but does not eliminate, the potential for injury.

The damage done to instream rights through transfer of other rights gives rise to several problems and issues. Among these is a possible conflict between the goal of enhancing flexibility in water use and the goal of protecting instream flows.[15] States have sought to meet new needs by facilitating the free movement of water between uses and sites of use, because it is often not possible, or feasible, to meet new water needs through the development of new water supplies. Much of this effort has been focused on the elimination of barriers to the transfer of water rights. Unfortunately, measures taken to protect instream flows may constitute one of these barriers. Instream flow rights, because of the no injury rule, can be a barrier to the free transfer of water rights up or off a stream—and indeed were intended to be. The creation of legal standing to protest transfers that would damage instream values is a major benefit of an instream flow right. One could argue that the creation of instream flow rights helps to set the boundaries within which transfers will be allowed, and that these boundaries are entirely appropriate given the desire to protect society's environmental and recreational amenities. Instream rights might also be considered a useful tool for offsetting the negative externalities arising from transfers, by providing a means of interjecting instream flow values into the transfer process.[16] But it must be recognized that these instream flow protection measures are not without cost. In some cases instream flow protection will be in conflict with continuing efforts to reduce barriers to the free transfer of water to new uses and sites of use. Even if transfers are not absolutely prevented, the need to evaluate possible damages of a transfer on instream rights will raise the cost of effecting a transfer, and thereby reduce the number of transfers that occur. The conflict between instream flow protection and flexibility in water use is sure to become more intense as society continues to place additional emphasis on both goals.[17]

The negative effect of transfers on instream rights also creates a problem for water rights administrators.[18] The historical consumptive-use rule and its variations were created as a method of simplifying implementation and enforcement of the no injury rule governing water rights changes and transfers. But the consumptive-use rule wasn't designed with instream, nonconsumptive rights in mind and cannot protect such rights from injury during the transfer process. For streams with instream rights this leaves administrators with a choice between rigorous enforcement of the no injury rule—

which would protect instream rights but make transfers more complex, and thus harder and more expensive to evaluate—and continued use of the historical consumptive-use method of evaluating transfers—which is easier to administer but wouldn't protect instream rights from injury.

The effort to protect instream flows not protected by water rights may also require extra effort of administrators. The impact of transfers on instream flows that are not protected by water rights may or may not be considered during a transfer hearing, depending on the state. Where considered, the usual method has been to allow a designated state agency or official to contest a proposed transfer on environmental grounds, or for the state administrative agency to use the impact of a transfer on instream flow as one factor to be considered when making a determination as to whether to approve a transfer. In the past, administrators seldom expended much energy in this effort. The overriding concern at the state level has been to protect other water rights holders, not third parties, from the impacts of water rights transfers.[19] Whether this will change in the future is unknown. Though more emphasis is now being placed on environmental values, it remains to be seen whether instream flows lacking actual water rights will be protected from changes and transfers of senior water rights in the future.[20]

Finally, it should be noted that the lost transfer opportunities caused by the existence of an instream flow right upstream may reduce the value of the offstream right. Before an instream right is created, other water right holders have the opportunity to transfer their water rights upstream. The ability to move a water right adds value to it because the market value of a right reflects not only the value of current use but also that of future opportunities. A water right that may be moved has more value than an otherwise identical water right that may not be moved. Of course, factors such as water availability, location of existing and possible future uses, future water demand, and the alternatives available to other water rights holders all influence transfer opportunities and may make transfers unattractive. Nonetheless, it is fair to say that in many situations creation of an instream water right has the potential, if the no injury rule is enforced, to decrease the value of water rights that are located downstream from the protected reach.

The temporal priority and location of rights relative to the instream right and the question of whether the instream right is a new appropriation or replaces an offstream use are among the most important determinants of the impact of instream rights on other water users. When transfers are involved, the location of a right is especially important. Rights located upstream from the instream right are unaffected by the instream right, while those located downstream bear the full brunt of any limitations. The threat posed to transfers by the presence of instream rights is highest when there are multiple opportunities to transfer a consumptive right to a point above the protected reach, such as when the instream right is located midway between the highest and lowest sites of use on the river.[21] Assuming that there are physical, legal, and economic opportunities available for moving offstream water

TABLE 7.3
Effect of Instream Water Rights on Other Water Rights in Situations in Which There Is a Possibility of Transferring Water Rights Upstream

Priority and position of water right relative to instream right	New instream appropriation	Replacement of offstream right by instream right
Senior		
upstream	no effect	no effect
downstream	(–) potential to lose upstream transfer opportunity	(+) better water quality (+) need to place fewer calls on river (–) potential to lose upstream transfer opportunity
Junior		
upstream	(–) additional senior right downstream	no effect
downstream	(–) potential to lose upstream transfer opportunity	(+) more water available (+) better water quality (–) potential to lose upstream transfer opportunity

[a](+) = beneficial effect, (–) = adverse effect.

rights from below the instream right to above it, the impact of instream water rights on other water rights holders is summarized in table 7.3.

Table 7.3 indicates that the situation of upstream water rights holders, whether senior or junior to the instream right, is the same as depicted in table 7.2. Upstream water rights holders haven't lost any transfer opportunities because they are still able to transfer their rights farther upstream, or to any location downstream, without causing injury to the instream right (as explained above, transferring the water right downstream would actually benefit the instream right). Because downstream water rights holders continue to receive certain benefits from the presence of an instream right, the net impact on downstream water rights holders will depend on the relative value placed on the different benefits and costs for each water right holder. Those for whom transfer opportunities were relatively unimportant will suffer little cost but continue to receive all the benefits of having an instream right located above them. Those for whom upstream transfer opportunities were very important may find that the cost of losing this opportunity outweighs the benefits of increased water, better water quality, and the need to place fewer calls on the river. As in the situation where transfer opportunities are not available, the transfer of an offstream right to instream purposes produces greater benefits and fewer costs than does the creation of a new instream water right.

CHAPTER EIGHT

Federal Authorities and Approaches for Protecting Instream Flows

In the latter part of the 19th century the federal government acquiesced to the states' assertions that the states would have primary responsibility for the allocation and administration of water resources. However, the federal government retained some of its water management authorities and, especially in the last few decades, has used these authorities to take on an increasing number of water management responsibilities. The federal government operates extensive water development programs, manages water resources in conjunction with its management of the federal lands, regulates environmental quality, and actively promotes the protection of fish, wildlife, and wild and scenic rivers. These activities have a profound impact on the status of western rivers.

The following sections describe federal proprietary rights and land management activities that affect instream flow protection in the West. The chapter begins with the sources of federal authority in the management of western water.

Constitutional Foundations

There is no question that the federal government has the authority to make laws and set policies affecting water. These authorities are derived from the Constitution, which specifies the relative authorities of the federal and state governments. The commerce, property, and supremacy clauses have had the most impact on water policy.

The commerce clause, Article I, Section 8, of the Constitution, gives the federal government the power "to regulate commerce with foreign nations, and among the several states," and "to make all Laws which shall be necessary and proper for carrying into Execution the foregoing Powers." The clause has been linked to water management primarily through the use of water for navigation.

Federal involvement in navigation was initially limited to rather narrow purposes, such as the prevention of activities causing obstruction of navigable watercourses. A leading example was the Rivers and Harbors Act of 1899, which authorized the army to regulate discharges into navigable waterways. Originally, rivers subject to federal authority under the commerce clause were those that were "navigable in fact"—rivers that were "used or . . . susceptible of being used in their ordinary condition as highways of commerce, over which trade and travel are or may be conducted in the customary modes of trade and travel over water."[1]

The federal courts subsequently expanded the definition of "navigable" to accommodate more ambitious water development programs requested by the states. The definition was first expanded to include nonnavigable portions of navigable rivers, and then rivers that could be made navigable through "reasonable improvement." Eventually the courts determined that rivers did not actually have to be navigable for the Congress to assert its commerce powers. By the time Congress passed the Clean Water Act in 1972, the courts had allowed the federal government to assert jurisdiction over all the waters of the United States, including rivers, lakes, streams, estuaries, and even wetlands.[2]

The property clause, Article IV, Section 3, states that "The Congress shall have power to dispose of and make all needful Rules and Regulations respecting the Territory or other Property belonging to the United States." Because the federal government owns and manages so much land in the West, including the headwaters of many of the West's rivers and streams, the property clause has been used to justify a wide range of water management activities on the federal lands, including instream flow protection.

The war power, general welfare power, and power to make treaties and approve interstate compacts have also been invoked to justify federal water resource activities.[3] The strength of these provisions, and of the commerce and property clauses, stems from Article VI, the supremacy clause: "The Constitution and the Laws of the United States which shall be made in Pursuance thereof; and all Treaties made, or which shall be made, under the Authority of the United States, shall be the supreme law of the Land; and the Judges in every state shall be bound thereby, any thing in the Constitution or Laws of any State to the Contrary notwithstanding." There is a long history of federal deference to the states in water allocation matters; the supremacy clause makes clear that this deference is a matter of choice rather than of obligation. There really is no question that Congress has the authority to make laws and policies affecting the use of western water if the Congress chooses to do so. Controversy surrounds only the extent to which Congress has actually exercised these authorities in specific situations.

The Federal Reserved Water Rights Doctrine

The federal government needs water to accomplish many of its land management, regulatory, and programmatic goals and has strong, constitutionally authorized powers to secure and manage water resources to meet these needs. However, Congress has rarely been explicit about either the federal government's water needs or its desired methods of meeting them.[4] Federal agencies, and the courts, have therefore been left to determine how much water is needed for specific federal purposes, and how this water is to be obtained. One method created by the courts is the "federal reserved water rights" doctrine. Federal reserved water rights are one of the most powerful—and contentious—tools used by the federal government in its efforts to utilize or protect water resources.

The federal reserved water rights doctrine was created to address the unspecified status of water resources on federal reserved lands. The government first began to set aside, or "reserve," large tracts of the public domain for specific purposes in the late 1800s. Early reservations established army forts and Indian homelands. Subsequent reservations included national forests, parks, wildlife refuges, historic sites, monuments, wilderness areas, among others. Water issues were seldom addressed in the legislation, proclamations, and treaties establishing these reservations.

Federal reserve rights are usually said to have been created by the court decision in *Winters v. United States,* heard by the U.S. Supreme Court in 1908.[5] The *Winters* court held that when the federal government established the Fort Belknap Indian Reservation for members of the Gros Ventre and Assiniboine tribes in Montana in 1888, the government had at the same time *impliedly* reserved, from waters then unappropriated, an amount of water sufficient to meet the purposes of the reservation. To rule otherwise, said the court, would put the court at odds with the intent of the treaty. Like most treaties signed with the Indians during that time, the intent of the treaty was to allow the Indians to become a pastoral people. The reservation was on arid land, and agriculture—the occupation intended for the Indians as an attempt to "civilize" them—would not be possible without water.

The decision in *Winters* turned on an interpretation of the particular 1888 treaty between the federal government and the Gros Ventre and Assiniboine tribes. For many years after the decision it was widely assumed that these new "federal reserved water rights"—sometimes referred to as Winters rights—were simply a quirk of Indian law that applied only to Indian reservations. This assumption was shattered over half a century later, in 1963, when the U.S. Supreme Court handed down its decision in the landmark case of *Arizona v. California*.[6] The *Arizona* court, without seemingly giving the matter much consideration,[7] explicitly ruled that federal water rights of the type discussed in *Winters* accrued to *all* federal land reservations, not just to Indian reservations. Among the lands potentially eligible for federal water rights in the *Arizona* case were Lake Mead National Recreation Area, Havasu Lake National Wildlife Refuge, and the Gila National Forest.

The ruling in *Arizona* stunned many in the water resources community.[8] Federal reserved water rights were viewed as a threat not only to future development of water resources under state law, but also to existing water uses, for the priority date of reserved rights would be the date that the land reservations were first established, not the date at which the reserved rights were subsequently claimed. Federal reserved water rights would thus be senior to all water rights initiated after the date at which the land reservation was made. Some considered federal reserved water rights an attempt to bypass state sovereignty over water resources and a serious disruption to the orderly administration of water use under state law.

Reserved water rights were reaffirmed and further defined by the U.S. Supreme Court in *Cappaert v. United States*, decided in 1976.[9] In *Cappaert* the court ruled that the water forming a subterranean pool in the Devil's Hole National Monument in Nevada should be protected from the negative effects of nearby groundwater pumping. The groundwater withdrawals were initiated subsequent to the designation of the lands as a national monument, and legislation creating the monument explicitly mentioned the existence of a race of desert pupfish in the subterranean pool as a reason the lands were being withdrawn from the public domain. The *Cappaert* ruling was significant not only for its protection of an *in situ* water use—of particular relevance to issues of instream flow protection—but also for its elucidation of principles composing the federal reserved water rights doctrine. The court stated that: (1) congressional intent to reserve water should be inferred whenever water is necessary to accomplish the purposes for which a reservation has been created; (2) the amount of water reserved is limited to the minimum amount of previously unappropriated water necessary to fulfill the purpose of the reservation; (3) federal reserved water rights are created at the time that lands are reserved, and are thus senior to rights based on uses initiated subsequent to that time; and (4) federal reserved water rights arise under federal law, and do not need to be perfected under state law.[10] Finally, the court reaffirmed its *Arizona* ruling that the federal reserved water rights doctrine applies to both Indian and other federal reservations, and that the doctrine encompasses the waters of both navigable and nonnavigable streams.

Federal reserved water rights are very powerful. Reserved rights are often among the most senior rights in a watershed, because many federal land reservations were made quite early, in the late 1800s or early 1900s. Federal reserved water rights are based on the purposes for which they were reserved rather than actual use and cannot be lost through nonuse. And, perhaps most important from the perspective of federal land management agencies, reserved rights are based on federal, rather than state, law, and are presumably not subject to diminishment by the states.

Limits of Federal Reserved Water Rights

The extensive involvement of the federal government in water development programs caused the states to become concerned about sovereignty issues even prior to the *Arizona* and *Cappaert* decisions. Congress passed the McCarran Amendment to the 1902 Reclamation Act in 1952 largely in response to these concerns.[11] The McCarran Amendment waived the immunity of the federal government from being joined in suits designed to establish and quantify water rights. The effect of the Amendment was to make possible the adjudication of water rights held by the federal government together with all other water rights in a single setting, namely the state courts. The McCarran Amendment specifies, however, that the federal government's immunity from suit will be waived only for "general" adjudications, proceedings in which all water rights in a water source—not just the rights of the federal government—are adjudicated. The state courts are required to apply federal law when adjudicating federal reserved water rights, and decisions of the state courts ultimately may be reviewed by the United States Supreme Court. Nevertheless, the McCarran Amendment limited the impact of reserved rights and represented a substantial victory for the states.[12]

Opponents of the reserved water rights doctrine were heartened by the U.S. Supreme Court's 1978 ruling in *United States v. New Mexico.*[13] The decision upheld New Mexico state court decisions denying federal reserved water rights claims for instream purposes in the Gila National Forest.[14] Justice Rehnquist, writing for the majority, stated that because the doctrine of federal reserved water rights is "built on implication and is an exception to Congress' explicit deference to state water law in other areas,"[15] the doctrine must be interpreted narrowly to avoid conflicts with the administration of water rights by the states. The court made a distinction between the "primary" and "secondary" purposes of federal land reservations, holding that federal reserved water rights exist only for the former. The court ruled that the instream purposes asserted by the Forest Service for the Gila National Forest, including flows for recreation, aesthetics, and wildlife preservation, were all secondary to the primary purposes of the national forests, as specified in the 1897 Organic Act.[16] Water for secondary purposes, said the court, could be obtained only under state law, on the same basis and through the same procedures established for other water users. The distinction between primary and secondary purposes of federal land reservations, and the availability of reserved rights for only the former, have had an enormous impact on the degree to which federal reserved water rights have been granted for instream purposes.

National Parks

The use of federal reserved water rights to protect instream flows has varied substantially among the different federal agencies and lands. The rights have been most readily recognized on lands managed by the National Park Service.

The National Park Service manages an array of lands that are among the most prominent and popular of all federal reservations. Some of the best-known units of the national park system include the national parks, national monuments, national recreation areas, and national preserves.[17] The national park system covers over 80 million acres of land, most of it—74 million acres—in the 18 western states, including 55 million acres in Alaska.[18]

The Park Service is on much stronger footing to claim federal reserved water rights for instream purposes than many of the other federal agencies because of the purposes of the national parks. The National Park Service Organic Act of 1916 states:

> The fundamental purpose of the said parks, monuments, and reservations . . . is to conserve the scenery and the natural and historic objects and the wild life therein and to provide for the enjoyment of the same in such manner and by such means as will leave them unimpaired for the enjoyment of future generations.[19]

The preservation and public enjoyment mandates strongly imply a need to keep water in streams. Rivers and streams are integral components of the natural environment and are popular recreation sites. Park Service policy is that the agency will "perpetuate surface and ground waters as integral components of park aquatic and terrestrial ecosystems," and that park waters "will be withdrawn for consumptive use only where such withdrawal is absolutely necessary for the use and management of the park and when studies show that it will not significantly alter natural processes and ecosystems."[20]

The preservation and public enjoyment purposes are likely to be interpreted as primary purposes of the national park system.[21] The purposes appear prominently in legislation defining congressional intent, such as the 1916 Organic Act. Though water rights for the national parks were not at issue in the *New Mexico* case, the *New Mexico* Court stated that "any doubt as to the relatively narrow purposes for which national forests were to be reserved is removed by comparing the broader language Congress used to authorize the establishment of national parks."[22] This part of the decision implies that federal reserved water rights for instream purposes in the parks will receive more favorable treatment by the courts.

No claims for federal instream-flow reserved rights in the national parks have yet reached the U.S. Supreme Court. But other courts have heard such cases and issued decisions favorable to the parks. For example, the Colorado Supreme Court ruled in 1982 that the Park Service was entitled to federal reserved water rights for certain water sources in Rocky Mountain National Park.[23] Similarly, the Colorado District Court for Water Division No. 1 stated in 1993, "It appears that Congress in setting aside Rocky Mountain National Park intended to reserve all of the unappropriated water in the park for park purposes. Only by so doing can the underlying purposes of the creation of the park be achieved."[24] In both cases the Colorado courts followed the example set by the U.S. Supreme Court by contrasting the broader purposes of the national parks to the more restrictive purposes of the national forests. The district court stated that the situation with regard to the national park was "entirely different" from that existing for the national forests.[25]

State courts often refer to the National Park Service Organic Act of 1916 as a general statement of park philosophy and purpose. But the courts primarily look to legislation creating individual units of the national parks system to determine the primary purposes of those units. The goal of the courts has been to determine whether Congress intended, explicitly or implicitly, to reserve streamflows when including particular pieces of land in the national parks system. The statute creating Rocky Mountain National Park stated that regulations of the Park Service should be "primarily aimed at the freest use of the said park for recreation purposes by the public and for the preservation of the natural conditions and scenic beauties thereof."[26] This language does not single out the flow of rivers and streams for particular emphasis, but it is a strong statement of congressional intent to preserve all aspects of the environment in their natural state. The statement underscores the special regard in which the national parks are held.

Park Service policy is to generally rely on appropriative rights obtained under state law to meet park needs. However, Park Service policies also state that the Park Service may assert federal reserved water rights "for water quantities determined to be the minimum amounts needed to protect the primary purposes" of a given park.[27] In practice, because the junior priority dates of appropriative rights are not always sufficient to safeguard park resources, the Park Service often relies on federal reserved water rights rather than appropriative rights to meet its needs.[28] The Park Service has, however, rarely taken *instream* federal reserved water rights claims to court.[29] Park Service personnel believe that it is rarely necessary to claim water rights, given the relative absence of threats to park rivers and streams.[30] Many parks are located at the top of watersheds, or in remote settings where there is little opportunity or desire for upstream development.[31] There is usually little potential for conflict with water users downstream, because most park water uses are nonconsumptive, meaning that the water continues to be available to downstream water users after it flows through the parks.[32] Perhaps most important, strong land-use controls preclude most water development within the parks. Parks are to be left "unimpaired" for the enjoyment of future generations, and, with but a few carefully circumscribed exceptions, development activity beyond that required to accommodate visitors is not allowed.[33]

Fortuitous geographic settings and strong land-use controls often obviate the need for the Park Service to initiate additional streamflow protection activities.[34] However, when joined in general or other adjudications, the Park Service is required to assert whatever water rights claims it has. The availability of federal reserved water rights also has value in that the threat of a federal reserved water rights claim is sufficient to ward off some potential threats to park rivers and gives the Park Service a stronger bargaining position in negotiations with the states and other water users.

Other Units of the Parks System

Probably the second most popular designation within the national parks system is the national monument. The president of the United States is authorized by the Antiquities Act of 1906 to withdraw lands from the public domain and designate them as national monuments.[35] Specifically, lands containing "historic landmarks, historic and prehistoric structures, and other objects of historic or scientific interest" may be withdrawn and designated as monuments. The Antiquities Act focused more on the preservation of archaeological and historical sites than on preservation of the environment, but natural features were often designated as objects of scientific interest under the Act. In 1916 the monuments were brought within the more environmentally oriented rubric of the National Park Service Organic Act, which applies to "parks, monuments, and reservations" managed by the Park Service.

There has not been much litigation over water rights in the national monuments. In the few such cases that have been tried, the courts seem to have decided the status of water rights based largely on the language of individual proclamations creating specific monuments, or on the Antiquities Act, rather than considering monument purposes within the broader context of the National Park Service Organic Act.

This approach has produced very different results in different monuments. For example, the Devil's Hole component of the Death Valley National Monument provided the setting for the important *Cappaert* decision described previously, in which the U.S. Supreme Court ruled that the water forming a subterranean pool in the monument should be protected from the negative effects of groundwater pumping occurring outside the monument. The proclamation creating the monument specifically mentioned the fish in the pool—said to be of "outstanding scientific importance"—several times.[36]

Dinosaur National Monument fared less well in a case decided by the Colorado Supreme Court in 1982. In *United States v. City and County of Denver*—the same case that awarded federal reserved water rights to Rocky Mountain National Park—the court ruled that federal reserved water rights could not be awarded for recreational boating in the monument.[37] The court relied on its interpretation of the Antiquities Act rather than the National Park Service Organic Act to determine that recreation was but a secondary purpose of the monument. On remand to a lower court, federal reserved water rights to protect fish in the monument were also denied, on the grounds that preservation of biological, aesthetic, scenic, and recreational features were all secondary purposes.[38] This result surprised many observers, and was considered a major setback to the Park Service's efforts to protect natural resources.[39] Nonetheless, the Park Service continues to assert a policy of claiming federal reserved water rights whenever necessary to protect the purposes of its lands, including the national monuments.

A federal reserved water right has also been recognized for a National Recreation Area (NRA), the Bighorn Canyon NRA in Montana. But NRAs are often administratively designated rather than withdrawn by Congress from the public domain, and may not always be eligible for federal *reserved water* rights.[40] The Park Service is studying the specific enabling legislation for several NRAs to determine the primary purposes and the nature of the actions that might be taken if those purposes are threatened.[41]

National Wildlife Refuges and BLM Lands

There are almost 500 national wildlife refuges nationwide, half of them in the 18 western states. Refuge acreage totals almost 91 million acres, the bulk of that—76.4 million acres—in Alaska.[42] The refuges were established to protect wildlife, primarily migratory birds. There is no organic act for the national wildlife refuges, but management of the refuges is influenced by the Migratory Bird Treaty Act of 1918, the Fish and Wildlife Act of 1956, and the National Wildlife Refuge Administration Act of 1966, among others. The wildlife refuges are managed by the Interior Department's U.S. Fish and Wildlife Service.

There has not been much instream flow protection activity on the national wildlife refuges. Few refuges require instream flows to meet the purposes for which they were established. Three-quarters of the refuges were created to protect migratory waterfowl, which are dependent on water found in marshes, ponds, and wetlands rather than in flowing streams.[43]

Water used to form ponds and marshes may come from flowing streams, but on refuges it has often been diverted from natural watercourses and transported to new sites of use or impounded by dikes.

Many refuges are probably not eligible for federal reserved water rights because they are on land acquired, rather than reserved, by the federal government. Federal reserved water rights for instream purposes have been considered or claimed for some refuges, but, at least until recently, not on a large scale. Ongoing general adjudications on Idaho's Snake River and Oregon's Klamath River may result in more extensive federal reserved water rights for the refuges.[44] Variation in the degree to which instream-flow water rights have been asserted in the different U.S. Fish and Wildlife regions is discussed in more detail in a later section of this chapter.

The Interior Department's Bureau of Land Management (BLM) manages over 268 million acres of land, virtually all of it located in the 12 westernmost states.[45] Half of all BLM land is located in just two states, Alaska (90 million acres) and Nevada (50 million acres, comprising 68 percent of the entire state). The BLM manages 8–22 million acres in each of the other 10 westernmost states. The BLM lands are essentially the federal lands left over after claims by homesteaders, the states, miners, and railroad companies were settled and after land reservations were designated as national parks, forests, wildlife refuges, and other purposes.

The BLM asserts federal reserved water rights only for lands that Congress has withdrawn for specific federal purposes,[46] *and* in which the assertion of federal reserved water rights would be more effective than the assertion of rights under state law.[47] Because most BLM lands were never "reserved," they are generally ineligible for federal reserved water rights. Most of the reserved BLM lands that do exist were created by the Stock-Raising and Homestead Act of 1916. The goal of this act was to prevent monopolization of scarce water supplies, and the means for doing so was to authorize the president to withdraw from the public domain the lands surrounding certain critical water sources. In 1926 the president issued an executive order withdrawing land within one-quarter mile of any spring or water hole on BLM lands, so long as that land was then vacant or otherwise unappropriated by any other party. This order was known as Public Water Reserve No. 107, and the reserved water holes are often referred to as PWR 107s.[48] Though potentially providing some instream value, most water holes are used for domestic and livestock purposes and produce very little flowing water.

National Conservation Areas (NCAs) are the only other significant reservations on BLM lands (aside from wilderness areas and wild and scenic rivers, discussed in more detail in chapter 10). Congressionally designated NCAs are eligible for federal reserved water rights only if Congress specifically states a need for water to accomplish intended purposes of the areas. Of the nine BLM National Conservation Areas, Congress had explicitly noted the water needs of only one, the San Pedro Riparian National Conservation Area in southern Arizona.[49] A federal reserved water rights claim for the San Pedro would have a very junior priority date of 1988, the year the Conservation Area was established.

U.S. Forest Service

Rivers flowing through national forests have been at the heart of many controversies concerning instream flow protection. The history of activity and conflict related to federal reserved water right claims for instream flows by the Forest Service is more extensive than for any other federal land management agency.

The national forests cover over 163 million acres of land in the 18 western states.[50] The vast majority of these lands were withdrawn from the public domain and designated as national forests during the late 1800s and early 1900s. Most national forest land is located in high mountain areas, which explains the connection between the national forests and western water issues; national forests are the source of most of the West's water supply. In 1970 the Public Land Law Review Commission estimated that 61 percent of the natural runoff in the eleven conterminous western states originated on federal lands, with runoff from the national forests accounting for 88 percent of that amount.[51]

Though streams within the national forests had long been used for a variety of instream purposes, as of the early 1960s the Forest Service did not have an instream flow policy and had not ever claimed water rights for instream purposes. It was then that the U.S. Supreme Court delivered its surprise *Arizona v. California* decision.[52] The Forest Service had been a party to the court proceedings because of its holdings of forest land within the lower basin of the Colorado River watershed;[53] it had not been a particularly active participant in the case. It came as something of a surprise when the Supreme Court ruled that federal reserved water rights might be attached to non-Indian federal reservations, specifically including the national forests.[54]

Based on *Arizona v. California,* the Forest Service decided to pursue federal reserved rights, including rights for instream uses such as fish, wildlife, recreation, and aesthetics, in locations where the agency was party to an adjudication. Several of the western states had initiated general adjudications for the purpose of settling water rights. The first of those cases to go to court was an adjudication taking place on the Rio Mimbres, in New Mexico.

The Forest Service claimed federal reserved water rights in the Gila National Forest based on the implied reservation of water that took place when Congress passed the Creative Act of 1891 and the Organic Administration Act of 1897.[55] The Forest Service argued that instream flows were compatible with the purposes of the acts, as confirmed by the broad purposes listed in the 1960 Multiple-Use Sustained-Yield Act.[56] The special master appointed by the New Mexico district court to take evidence in the case found that water in the forest was in fact being used for the purposes claimed by the United States, and that these uses fell within the scope of the reserved rights doctrine. The district court, however, rejected these findings, as did the New Mexico Supreme Court on appeal. The New Mexico Supreme Court held that the Forest Service could not claim federal reserved water rights for instream purposes or for stock watering.

This ruling was upheld by the U.S. Supreme Court in its 1978 *United States v. New Mexico* decision.[57] The court found that a close examination of the 1897 Organic Act—"No national forest shall be established, except to improve and protect the forest within the boundaries, or for the purpose of securing favorable conditions of water flows, and to furnish a continuous supply of timber for the use and necessities of citizens of the United States"—revealed that "Congress only intended national forests to be established for two purposes. Forests would be created only 'to improve and protect the forest within the boundaries,' or, *in other words,* 'for the purpose of securing favorable conditions of water flows, and to furnish a continuous supply of timber" (emphasis in the original).[58] Neither of these purposes encompassed water used for stock watering, fish, wildlife, recreation, or aesthetics.

This interpretation was by no means obvious and was opposed by four members of the nine-member court, including Justice Powell, who wrote:

> I do not agree, however, that the forests which Congress intended to "improve and protect" are the still, silent, lifeless places envisioned by the Court. In my view, the forests consist of the birds, animals, and fish—the wildlife—that inhabit them, as well as the trees, flowers, shrubs, and grasses. I therefore would hold that the United States is entitled to so much water as is necessary to sustain the wildlife of the forests, as well as the plants.[59]

Justice Powell went on to state that:

> A natural reading would attribute to Congress an intent to authorize the establishment of national forests for three purposes, not the two discerned by the Court . . . (1) improving and protecting the forest, (2) securing favorable conditions of water flows, and (3) furnishing a continuous supply of timber. . . . The Court believes that its "reading of the Act is confirmed by its legislative history." The matter is not so clear to me. From early times in English law, the forest has included the creatures that live there. . . . It is inconceivable that Congress envisioned the forests it sought to preserve as including only inanimate components such as the timber and flora. Insofar as the Court holds otherwise, the 55th Congress is maligned and the Nation is the poorer, and I dissent.[60]

Nonetheless, the decision of the majority prevailed, and the outcome of the Rio Mimbres adjudication was a major setback for the Forest Service in its effort to accomplish what it understood to be its resource management responsibility.[61]

The Rio Mimbres decision also took the Forest Service very much by surprise, especially given the strong suggestion by the court in *Arizona v. California* 15 years earlier that federal reserved water rights would be available for water uses in the national forests.[62] It was not long, however, before the Forest Service tried again to assert its claims for federal reserved water rights.[63] This time the action moved farther north, where the state of Colorado was pursuing general water rights adjudications in several of its water divisions.[64] As in the Rio Mimbres adjudication, the Forest Service was made a party to the proceedings under the McCarren Amendment because of its land holdings within the water divisions.

The first of these adjudications to come to trial was in the Colorado water court for water divisions 4, 5, and 6 in the northwestern portion of the state, where the federal government had first been asked to quantify its claims for water in 1967. Once again the Forest Service claimed water for a variety of purposes, including some taking place within stream channels. This time, however, the United States tried to avoid some of the obstacles it had run into in the *New Mexico* case. First, it claimed water under the primary purposes of the Organic Act—securing favorable conditions of water flows and a continuous timber supply—for uses such as fire fighting, fire protection, and flood, soil, and erosion control. Second, it claimed additional water under the Multiple-Use Sustained-Yield Act, with a priority date of 1960, for recreational and wildlife purposes.[65] The primary objectors to these claims were the City and County of Denver, which argued that federal reserved water rights didn't exist at all, as a matter of law.[66]

The water court ruled, in 1978, that the federal government did possess federal reserved water rights to the use of water in the national forests, but that these rights were much more limited than the United States had claimed. On appeal, the Colorado Supreme Court ruled in the case of *United States v. City and County of Denver* that the Forest Service could not obtain federal reserved water rights—at least as claimed in this case[67]—for instream flow purposes under the primary purposes of the 1897 Organic Act.[68] With reasoning closely paralleling the U.S. Supreme Court's ruling in the New Mexico decision, the Colorado Supreme Court stated that "the congressional intent behind this act was to further economic development of the West by enhancing the quantity of water available to appropriators, not to enlarge the consumption of water by protecting instream flows."[69] The court also rejected claims based on the Multiple-Use Sustained-Yield Act. The court held that the 1960 Act did not expand the primary purposes of the 1897 Organic Act, meaning that the purposes of the 1960 Act were secondary and thus could not be used to assert additional federal reserved water rights claims in the national forests.[70]

Once again the Forest Service had been unsuccessful in its attempts to secure federal reserved water rights for instream purposes. It was also precluded by legal principle from further attempts to assert instream flow claims under the reserved water rights doctrine for national forests in Colorado Water Divisions 4, 5, and 6. Forest Service claims in other water divisions, however, were still pending, so the Forest Service was not yet through. Its efforts did, however, take a new turn, based on emerging developments taking place in the field of fluvial geomorphology.

The field of fluvial geomorphology grew through the research and interaction of hydrologists and geologists interested in the relationships among streamflow, geology, river sediment loads, and the structure and shape of river channels.[71] Given the longstanding interest of the Forest Service in issues affecting stream channel sedimentation, it was not surprising that some of the agency's employees were also involved in the field. Forest Service hydrologists soon developed a methodology for determining the amount of streamflow needed to maintain stream channels in a condition suitable for "securing favorable conditions of water flows."[72] It was hoped that this "channel maintenance" approach would support a reserved water right claim based on one of the primary purposes of the Organic Act, the securing of "favorable conditions of water flows."

The Forest Service first tried out its new theory in a general adjudication taking place in Wyoming, in the watershed of the Big Horn River and its tributaries.[73] Among other claims made by the United States in that adjudication were rights to the natural flows of several watercourses in the Bighorn and Shoshone national forests.[74] For particular watercourses identified by the Forest Service the United States claimed a substantial portion of the snowmelt runoff in the spring, plus a quantified baseflow discharge during all other portions of the year. These claims, which amounted to 78 percent of the total average runoff from the basin, were made under authority of both the 1897 Organic Act and the 1960 Multiple-Use Sustained-Yield Act.[75] The claims were vehemently protested by the State of Wyoming, which argued that:

> This claim . . . is an affront to Wyoming and the Court. The purpose of the [Organic] Act, as described above . . . , demands the delivery of water conserved by the forest to downstream users. To secret away waters deep within the lonesome boundaries of the national forests to suit the active imaginations of the Forest Service would utterly frustrate this fundamental forest purpose.[76]

Most observers believed the results of the proceedings constituted another loss for the Forest Service. The State of Wyoming demonstrated during pretrial depositions that the methodology used by the Forest Service was still in its formative stages and would be difficult to support in court.[77] Wyoming agreed to recognize some of the national forest claims in a negotiated settlement of all non-Indian federal claims,[78] but the United States agreed to subordinate its rights to those of both existing and future water development projects. This subordination almost completely eliminated the value of the rights.[79] The agreement did, however, recognize instream rights in some form and left intact the United States' claims to the natural levels of certain springs and seeps claimed under both the Organic Act and the Multiple-Use Sustained-Yield Act.[80]

The Forest Service made some adjustments to its methodology before taking it back to the Colorado courts. The stage was then set for a full hearing on the United States' federal reserved water rights claims for instream purposes in the national forests based on Organic Act purposes and the science of fluvial geomorphology.[81] The State of Colorado chose to first pursue the adjudication in Water Division 1, encompassing the South Platte River and its tributaries. Water Division 1 is Colorado's largest and most heavily populated water division, including over 2.3 million people in such population centers as Denver, Boulder, Longmont, Loveland, Fort Collins, and Greeley. The watershed also encompasses what the court referred to as a "vibrant agricultural economy," with between 1.3 and 1.8 million acres of irrigated land, most of it located on the plains in the eastern half of the state.[82] The South Platte River, and most of its tributaries, have their headwaters within the Arapaho, Pike, Roosevelt, and San Isabel national forests. The forests are the site of numerous storage reservoirs and diversion points for water used for municipal and agricultural purposes throughout the watershed. The choice of this setting virtually guaranteed that the trial would have a high profile, complete with a large number of objectors, lots of press coverage, and the establishment of a significant precedent for future water rights adjudications.

The Forest Service engaged in a major, multiyear effort to quantify its channel maintenance claims. Once at trial, the claims were vigorously resisted by the objectors. The actual trial commenced early in 1991, lasted until the end of the year, and was marked by extensive legal maneuvering and the discussion of highly technical and scientific theories and methodologies. Arrays of nationally recognized experts were brought in to testify for each side, and the trial in many ways became an intensive and extended seminar on the principles of fluvial geomorphology and associated sciences, complete with field trips.[83] The effort was costly to both sides. The result, however, was familiar. Once again, the Forest Service was surprised to discover that the vast majority of its claims were denied. The court granted the Forest Service federal reserved water rights for administrative and fire-fighting purposes but denied all of its claims for instream flows.

The reasons given for the results were also familiar. The court found that the claims made by the Forest Service were not necessary to fulfill the primary purposes of the national forests and in fact worked against fulfillment of the primary purposes as those purposes were understood by the court. The Colorado court found that the water-related purpose of the national forests was to enhance the ability of downstream water users to use the water for irrigation and domestic purposes. The court noted the great importance of reservoir and diversion sites in the national forests to downstream water users[84] and reasoned that it was "inconceivable that it was intended [by Congress] that water rights were to be reserved to an extent that they would interfere with efficient use of the 'favorable water flows' for irrigation and domestic purposes."[85] Though the court recognized that instream claims made by the Forest Service would not lead to an increase in water consumption, the timing of flows—the channel maintenance claims called for flows in the range between mean annual flow and bankfull flow occurring during the spring runoff and much lower flows through the rest of the year[86]—would conflict with opportunities for storage of water to meet late-season water needs:

> Municipalities need water all year long, and agriculturists generally have better supplies of water in the spring but are particularly in need of irrigation water later in the growing season. Storage of water in the upper part of the watershed promotes these equable flows. Such equable flows were sought by those whose ideas are reflected in the creation of the national forests, and are exactly what they meant when they referred to "favorable water flows."[87]

The court also found other problems with the Forest Service claims. Though it recognized that the Forest Service had been faced with an enormous challenge in attempting to quantify its claims, the court noted that the challenge had not been adequately met—and perhaps could not have been.[88] Questions about the methodology included the debatable application of certain equations and methods to circumstances for which they were not originally intended, the inconsistency of results derived using different methods at a single quantification point, and the fact that in some situations the rights claimed by the Forest Service were misstimed and therefore would not actually lead to the flows necessary to achieve intended goals.[89] The court found that many of the streams chosen for quantification already contained diversions and yet didn't exhibit the undesirable characteristics that the Forest Service said would occur. The court observed that differences in stream channels above and below diversions were for the most part "subtle" and did not constitute a threat to the integrity of the channels. The court found no evidence of increased flooding downstream from diversion points, which further weakened the Forest Service claims.[90]

Though unconvinced by the application of the Forest Service methodology in these particular circumstances, the court did not reject the general theory or line of reasoning employed by the United States. Neither did the court adopt the theory and reasoning in their totality. The United States argued that Congress had intended stream channels in the national forests to be maintained in an unimpaired condition. The objectors argued that the maintenance of channels was totally irrelevant to national forest purposes. The court reasoned that the truth lay somewhere between, that channel maintenance was required to secure favorable conditions of water flows but only "to a reasonable degree."

The quest to identify and claim flows necessary to secure favorable conditions of water flows in the national forests continues. The Forest Service has several channel maintenance claims pending in general adjudications throughout the West, including adjudications of the Klamath River and its tributaries in Oregon, the Snake River and its tributaries in Idaho, and the three water divisions that have not yet adjudicated Forest Service claims in Colorado. The channel maintenance methodology has been revised to eliminate problems made evident in Colorado's Division 1 adjudication, and the Forest Service remains hopeful of using the methodology to obtain federal reserved water rights for channel maintenance flows in the national forests.

Indian Reservations

There is no doubt that there are reserved rights associated with Indian reservations. However, as with water rights on federal lands, a number of controversies have surrounded Indian reserved rights. Though most Indian treaties, and hence the priority dates of most Indian reserved water rights, typically predate most if not all other water right priorities in the West, much of the water subject to Indian reserved rights has been used by others. The development of water on Indian lands has lagged well behind the development of water on non-Indian lands,[91] and western water law allows junior water rights holders to use any water that seniors aren't using.[92] Much of the opposition to Indian reserved rights has come from junior water rights holders who fear that assertion of the full measure of the Indians' reserved rights will deprive the juniors of the water they have been using for years, even generations. Opposition has also come from the states, which, in addition to representing the interests of state water rights holders, fear the loss of sovereignty that may result if substantial quantities of water are administered through federal or tribal law and agencies.

Controversies over Indian reserved water rights concern the methods by which such rights are quantified, the ability of tribes to transfer water subject to reserved rights off the reservation, and the purposes for which the water will be used. Of primary interest here is the latter issue, the purposes for which water subject to Indian reserved rights can be used. This issue can be further broken down into two questions: (1) Do Indian reserved water rights exist for instream purposes? (2) Can Indian reserved rights based on other purposes be transferred to instream use?

Do Indian reserved rights exist for instream purposes? The answer to the first question is that reserved rights exist for instream purposes only if the instream purposes constitute one or more of the original purposes for which an Indian reservation was created. In general, the courts follow a three-stage process when evaluating and quantifying Indian reserved rights. First the court must determine the intent of the parties in creating the reservation. Then the court must decide what water uses are necessary to accomplish the intended purposes. Finally, the court determines how much water is needed for those uses.[93]

The purposes of a reservation are culled from the treaty or proclamation that created the reservation. Most of these documents are terse, employ very general terms, and were written long ago, so they leave much room for interpretation.[94] In many cases the courts have determined that reservations were created for the purpose of establishing agricultural economies. The standard used to quantify the amount of water necessary to accomplish this purpose, "practicably irrigable acreage," or PIA, was established by the U.S. Supreme Court in its 1963 *Arizona v. California* decision. PIA is based solely on the amount of water needed to irrigate the practicably irrigable land on a reservation, and does not contemplate any instream needs.[95]

However, the purposes of some treaties have been interpreted to include instream water uses, especially fisheries.[96] Some tribes, particularly in the Pacific Northwest, have been successful in getting reserved water rights for these purposes. Among the court cases verifying reserved rights for fishery purposes are *Colville Confederated Tribes v. Walton* in 1981, *United States v. Adair* in 1983, and *Muckleshoot Indian Tribe v. Trans-Canada Enters.* in 1983.[97] The *Colville* case was particularly interesting. First, the reserved right of the Colville Confederated Tribes for fishery purposes was given a priority date of "time immemorial" rather than the date that the treaty establishing the reservation was signed, in recognition of the fact that the tribes had relied on salmon and trout for food since long before the reservation was created. Second, the court recognized and quantified the reserved right for a fisheries purpose even though the historical fishery had since been destroyed by the construction of dams on the Columbia River. The tribes were subsequently able to use the water covered by the reserved right to develop and maintain a replacement fishery.[98]

Can Indian reserved rights based on other purposes be transferred to instream use? This question is important given that most Indian reserved rights are based on agricultural purposes.[99] Examination of relevant federal law implies that the answer is yes. The special master in the 1963 *Arizona v. California* case that created the PIA concept stated:

> This does not necessarily mean, however, that water reserved for Indian Reservations may not be used for purposes other than agriculture and related uses. . . . The measurement used in defining the magnitude of the

> water rights is the amount of water necessary for agriculture and related purposes because this was the initial purpose of the reservation, but the decree establishes a property right which the United States may utilize or dispose of for the benefit of the Indians as the relevant law may allow.[100]

The Interior Department solicitor subsequently advised the secretary of the interior that Indian reserved water rights could be used for any beneficial purpose, and the federal Ninth Circuit Court adopted this reasoning with respect to the *Colville* case described above.[101] However, actual court decisions have not always followed the lead of the *Arizona* court and the Interior Department solicitor. The idea of allowing Indian reserved rights—particularly the portion of the rights that has not previously been put to use by the Indians—to be transferred to new uses is frequently resisted by other water users and by the states. Though the state courts apply federal law when adjudicating Indian reserved rights, conclusions reached by the state courts sometimes constrict the options available to the tribes for using their water.

The Wind River Instream Flow Claims. A recent Indian reserved water rights case involving the Wind River Reservation in the Big Horn River basin of west central Wyoming illustrates many of the issues and conflicts surrounding the transfer of Indian reserved water rights to instream purposes. The Wind River Reservation, home to the Eastern Shoshone and Northern Arapaho tribes, was established by the Second Treaty of Fort Bridger in 1868. Water development in the region proceeded in an uneven fashion, with non-Indian water users, aided by the Bureau of Reclamation, eventually irrigating larger tracts of land than the tribes, who received more limited aid from the Bureau of Indian Affairs.[102] In the 1970s the tribes began to consider possible water policy options available to them under the reserved water rights doctrine, and in 1977 the state of Wyoming joined the tribes in a general adjudication to clarify the status of all water rights in the Big Horn River basin.[103]

The special master appointed to hear the case by the Wyoming district court determined in 1982 that the reservation had been created for the purpose of establishing a permanent homeland for the Indians, and that the tribes owned a reserved water right for irrigation, stock watering, fisheries, wildlife, aesthetic, mineral, industrial, domestic, commercial, and municipal water uses.[104] The special master quantified the tribe's water right using the U.S. Supreme Court's PIA standard, which resulted in a total award of approximately 500,000 acre-feet per year. The district court subsequently agreed with the special master that the tribes had a reserved right but rejected the special master's "permanent homeland" reasoning and held that the purpose of the reservation was solely agricultural.[105] The district judge did state, however, in an amended decree in 1985, that the tribes could subsequently use the water covered by the reserved water right for any purpose, so long as the use took place on the reservation and did not increase consumptive use beyond the amount quantified by the PIA standard.[106]

In 1988 the Wyoming Supreme Court affirmed the decision to award the tribes a reserved water right and agreed that the Wind River Reservation had been created solely for the purpose of establishing an agricultural community.[107] Though noting that this purpose might include municipal, domestic, livestock, and commercial water uses, the court also held that these uses were subsumed within the agricultural purpose and did not entitle the tribes to water in quantities beyond that determined through the PIA standard.[108]

The Wyoming courts awarded the tribes a reserved water right of 500,017 acre-feet a year. The large size of the right, and its priority date of 1868, which predated all other water rights in the basin, caused great consternation among non-Indian water users and led the state to appeal the decision to the U.S. Supreme Court. However, the U.S. Supreme Court in 1989 affirmed the Wyoming court's use of the PIA standard to quantify the right.[109] This decision was considered a substantial victory for the Wind River Tribes.

Though owning the most senior water right in the basin, the Wind River Tribes were not yet using all of the water specified in the right. A substantial portion of the award was based on the amount of water needed to irrigate acreage on the reservation that had not yet been developed. In 1990 the tribes enacted a water code that listed among its beneficial uses "fisheries, wildlife, and pollution control, aesthetic and cultural purposes"[110] and created a Water Resources Control Board to administer the tribes' water rights.[111] The Board then issued an instream water permit for 252 cfs in the Wind River (a tributary of the Big Horn River) in the name of the tribes, specifying the "future uses" part of its reserved water right as the source of the water. The instream flow was to be used to establish and maintain a fishery, and for other instream purposes[112] would have amounted to approximately 80,000 acre-feet of water per year, or about 16 percent of the tribes' total reserved water right.[113]

Much of the water used for irrigation on and near the reservation is diverted by both Indian and non-Indian water users at points above and within the reach protected by the tribes' instream water permit. When flows in the river fell below 252 cfs in April 1990, the tribes asked the Wyoming state engineer to close headgates on the diversions of upstream junior water users to protect the instream right. Because the Indian diversions were included in the "existing uses" portion of the reserved right and thus shared the senior 1868 priority date, the request, if honored, would have led the state engineer to close down the diversions of the non-Indian water users. However, the state engineer refused to take any action against the upstream diverters, stating that until the entire Big Horn adjudication was completed—resolution of the Indian reserved rights in the basin was just one phase of the three-phase general adjudication of the Big Horn basin[114]—he intended to administer all rights in the basin as they had been administered in 1977, when the adjudication began.[115]

The tribes then filed suit in Wyoming district court to force the state engineer to enforce the tribes' instream permit against junior water rights holders upstream. The special master appointed by the district court concluded that the tribes could use their reserved rights for any purpose.[116] This finding was adopted by the district court, which held in its 1991 decision that "the Tribes may change their reserved water right to instream flow without regard to Wyoming state water law."[117] The court also ruled that the tribes' Water Resources Control Board, rather than the Wyoming state engineer, should administer and enforce all water rights, both state and federal, within the boundaries of the reservation.

The tribes' victory was, however, short-lived. The state appealed the decision to the Wyoming Supreme Court, which stayed enforcement of the district court's judgment and decree pending the appeal.[118] Then in June 1992, the Wyoming Supreme Court handed down a decision that completely reversed the district court. The Wyoming Supreme Court held that the tribes may use their water rights "solely for agricultural and subsumed purposes and not for instream purposes,"[119] and that the Wyoming state engineer was the appropriate officer to administer water rights on the reservation. Following consultation with a number of legal experts, the tribes decided not to appeal the decision to the U.S. Supreme Court, and so the decision of the Wyoming Supreme Court was allowed to stand.[120]

The decision by the Wyoming Supreme Court was hard to interpret. All five justices wrote separate decisions and appeared to use five different lines of reasoning to support their decisions. The vote on both issues—conversion of the right to instream purposes, and administration of tribal water rights—was 3–2, with different majorities on each issue. As such, the decision may not have much precedent in other cases.[121] However, it did result in the frustration of the Wind River tribes' efforts to change the use of part of their reserved water right to instream purposes. Administration of the reserved right by the state engineer may further limit the tribes' use of their reserved right.[122] Westwide, the issue of whether Indian reserved rights can be changed to instream purposes is probably best characterized as "unsettled."[123] The outcomes of similar cases may well vary with the individual state, reservation, and physical, legal, and institutional characteristics in which they arise.

Federal "Nonreserved" Rights

The legislation, executive orders, and proclamations used to reserve federal lands generally state the purposes for which the land is being reserved and will subsequently be managed. Congress often adds to or amends these purposes through subsequent legislation, such as the Multiple-Use Sustained-Yield Act of 1960, the Federal Lands Policy and Management Act of 1976, and the National Forest Management Act of 1976. Given the supremacy of federal law and the probability that the accomplishment of at least some of the mandates specified in the additional legislation will require the use of water, some have argued that these subsequent statutes also constitute a basis for creating federal water rights. These rights would not technically be "reserved" rights, because the rights would be created at the time that the federal land management agencies actually appropriate and put to use the water necessary to achieve statutory objectives, rather than at the time the land in question was first reserved from the public domain.

The suggestion that such rights exist has been hotly contested, as illustrated by a series of opinions issued by the Interior Department solicitor and the Office of Legal Counsel in the Department of Justice. The Office of the Solicitor is administratively independent of the Interior Department's individual resource management agencies but has a substantial impact on the agencies' activities. Strictly speaking, the solicitors' opinions, issued as "Interior Decisions," are not law, but actions taken by the affected agencies are made in accordance with the opinions. For example, if a solicitor's opinion states that there are no federal reserved water rights associated with wilderness areas, then neither the Park Service, the BLM, nor the Fish and Wildlife Service, all of which are agencies of the Interior Department, would be allowed to assert claims for federal reserved water rights in wilderness areas. Opinions issued by the attorney general affect all federal agencies, not just those within the Department of the Interior.

Interior solicitor Leo Krulitz issued a 1979 opinion in which he argued that, barring a specific congressional directive to the contrary, virtually any statutory assignment of land management responsibilities to federal agencies gave those agencies the authority to appropriate water necessary to meet their responsibilities.[124] The solicitor referred to these appropriations, dependent on statutory land management goals rather than on the primary purposes of the initial land reservations, as "federal nonreserved rights." The solicitor's opinion relied heavily on property clause and supremacy clause powers and stated that appropriations of water could take place without regard to the laws of the state in which the land is situated. For example, nonreserved rights would not be subject to state beneficial-use requirements and could be used to claim instream flow rights on federal lands in states that require physical diversions to establish water rights, or that allow only the state to hold instream flow rights.[125]

Federal statutes encompass a great many goals that might require the use of water. Given the substantial number and breadth of the mandates under which nonreserved rights could potentially be claimed, the Krulitz Opinion was extremely unpopular with the states and with many others in the water resources community. Nonetheless, in early January 1981, a new solicitor, Clyde Martz, issued a supplement to the Krulitz Opinion in which he essentially left intact the concept of federal nonreserved rights. Martz did decide, however, that there was no authority in either of two specific federal statutes, the Federal Lands Policy and Management Act of 1976 and the Taylor Grazing Act of 1934, for federal agencies to appropriate water.[126]

Later that same January a new administration, and a new solicitor, took office. The new solicitor, William Coldiron, quickly issued an opinion focusing on the desire of Congress to yield sovereignty over water resources to the states and concluding that federal nonreserved water rights did not exist.[127] This opinion was followed by a memorandum, referred to as the Olson Opinion after its author, Assistant Attorney General Theodore Olson, issued by the Office of Legal Counsel of the United States Department of Justice.[128] The Olson Opinion completely reversed the premise of the original Krulitz Opinion. Rather than holding that nonreserved rights exist unless specifically denied by Congress, Olson said the presumption is that the federal government will acquire water rights in accordance with state law unless there is specific congressional intent to preempt state water laws.

However, the Olson Opinion stated that the presumption of deference to state water law is a rebuttable one. Close analysis of individual land management statutes, delegated authorities, and federal water needs may provide evidence of congressional directives or primary purposes that would justify the exercise of federal rights arising outside of state water rights systems:

> Congress could establish "primary purposes" for the management of public domain lands that could be the basis for federal water rights. [It is] . . . the specification of particular purposes for which the lands should be maintained and managed and the implicit intent that water be available for these purposes that give rise to those rights. It may be possible to argue, therefore, that in the relevant statutes Congress intended that the public domain be managed for specific purposes that cannot be accomplished without implication of federal water rights.[129]

The Olson Opinion essentially states that congressional intent should be construed quite narrowly, because of the presumption that Congress intended the states to maintain control over water allocation. In this respect the Olson Opinion on federal nonreserved rights mirrors the Supreme Court's *New Mexico* decision on federal reserved water rights, which in fact was cited in the Olson memorandum. Unlike the *New Mexico* court or Solicitor Martz, however, the Olson Opinion did not analyze specific statutes to determine whether they actually contained evidence of congressional directives or primary purposes that would justify the assertion of federal nonreserved rights.[130] A look at some of these specific statutes may help clarify the theory, opportunities, and limitations of federal nonreserved rights.

The Federal Lands Policy and Management Act

The Federal Lands Policy and Management Act of 1976 (FLPMA) is the BLM's primary source of land management authority. Most people view the primary purpose of FLPMA as the assignment of particular *land* management functions to BLM. However, Congress specifically identified thirteen different policies to be furthered by FLPMA, including at least a few that could be read to imply a need for instream flows. In addition to inventory and planning, multiple-use and sustained-yield goals, FLPMA mandates that

> the public lands be managed in a manner that will protect the quality of scientific, scenic, historical, ecological, environmental, air and atmospheric, water resource, and archaeological values; that, where appropriate, will preserve and protect certain public lands in their natural condition; that will provide food and habitat for fish and wildlife and domestic animals; and that will provide for outdoor recreation and human occupancy and use.[131]

Protecting the environment, water resources, and public lands in their natural condition would almost certainly require the protection of instream flows, as would the goal of providing food and habitat for fish and wildlife. In addition, FLPMA requires that "regulations and plans for the protection of public land areas of critical environmental concern be promptly developed." This is important because of the way that "areas of critical environmental concern" is defined:

> The term "areas of critical environmental concern" means areas within the public lands where special management attention is required (when such areas are developed and used or where no development is required) to protect and prevent irreparable damage to important historic, cultural, or *scenic values, fish and wildlife resources, or other natural system or processes,* or to protect life and safety from natural hazards [emphasis added].[132]

The BLM has identified 524 areas of critical environmental concern (ACECs) totaling over 8.7 million acres. ACECs are found in all of the non-Plains western states except for Washington. The largest number of critical areas are on BLM lands in California (112) and Oregon (104), but the 27 critical areas in Alaska account for almost half of the total acreage.[133]

It can certainly be argued that these passages make fish, wildlife, and other instream uses primary purposes of FLPMA, and that to accomplish these purposes the BLM will need to obtain instream flow rights, even in the face of conflicting state law. This, in short, is the argument for federal nonreserved rights on the BLM lands. However, it is important to emphasize that the BLM has not actually tried to assert such rights and does not, so far as is known by the authors, have any plan to do so in the future. At this point the argument is purely theoretical. As noted by one commentator:

> Three things must happen to make this a live controversy: (1) the BLM must be willing to assert claims for instream flows under the management authority of FLPMA; (2) the BLM must have the proper fact situation—a significant fish and wildlife habitat that requires minimum instream flows for proper management, and a watercourse that contains unappropriated

water in danger of upstream consumptive appropriation; and (3) the BLM must be unable to establish instream flows under state law.[134]

Any observer of BLM water resources activities will readily note an abundance of appropriate fact situations where BLM is unable to establish instream flows under state law, but BLM's willingness to assert claims for instream flows under the authority of FLPMA has been nowhere in evidence, leaving the argument in its speculative state.

Federal Nonreserved Rights in the National Forests

The National Forest Management Act of 1976 (NFMA) would be the most likely source of federal nonreserved rights in the national forests.[135] NFMA was motivated primarily by concerns about timber cutting, but the Act also contains provisions affecting water management. The provision most likely to spur federal nonreserved rights claims is a requirement that the Forest Service maintain viable populations of all vertebrate species—which includes fish—currently existing in the national forests. The argument would be that water is necessary to achieve the purposes that Congress, through passage of NFMA, has directed the Forest Service to achieve. If water is necessary to fulfill these purposes, and Congress has delegated responsibility for achieving these purposes to the Forest Service, then the supremacy clause of the Constitution would enable the agency to assert rights—based on federal, rather than state, law—for the quantity and timing of water necessary to achieve these purposes.

As proposed by two legal scholars, the process of claiming such rights would begin with the issuance of notice to the public, including relevant state agencies.[136] The Forest Service would then hold public hearings, so that current water users, state officials, and other members of the public would have an opportunity to present their points of view. The right would be established only following these hearings, if a determination was made that the flows were necessary. The establishment of water rights through this method would have some advantages, for both sides. The Forest Service would be able to claim water rights even in states where laws do not recognize water rights for instream purposes, or where such rights exist but only state agencies are allowed to hold them. Concerned members of the public would have an opportunity to present their views in an official hearing and would be able to challenge an unfavorable decision if proper administrative procedures and judgment were not exercised. Existing rights would not be preempted, because the priority date of the right would be the date that notice was first given to the public rather than the date at which the national forest was first established.

It is far from certain that the Forest Service would ever actually claim federal nonreserved rights. It does not appear that the assertion of such rights has ever been tried. None of the Forest Service officials interviewed by the authors of this book were aware of a situation in which it had been tried or of any plans to make this argument in the future. Forest Service officials are aware that attempts to assert rights on this basis would likely draw extreme opposition, despite the advantages described above. States and other water users might well be upset by a process in which a federal agency seeks to establish water rights through its own procedures, based on federal law, rather than through state procedures and law. Even federal reserved water rights are adjudicated through state courts, and there would likely be numerous objections to attempts by the Forest Service to circumvent the state water rights system.

Federal Use of State Water Rights

Federal agencies may claim water rights under state law. The scope and diversity of federal activities that affect water in the West are enormous, and whether seeking water for traditional purposes such as irrigation and hydropower, or "newer" purposes such as instream flows, the potential impact of federal activities on other water users is correspondingly large. For this reason the federal government's attempts to establish water rights are often contested by other water users or potential water users. Nonetheless, the courts have consistently upheld the right of the federal government to seek water rights from the states just like any other water user, and federal agencies have routinely sought such rights.[137] Federal agencies have sometimes been directed by federal statutes, such as the Reclamation Act of 1902, or by the courts, as in the *New Mexico* case, to seek the water rights they need through the states rather than to try to establish rights under federal law.

Two of the more recent and well-known cases protecting the right of the federal government to seek water rights under state law both concerned instream flow issues. One, *In Re Water of Hallett Creek Stream System,*[138] decided by the California State Supreme Court in 1988, concerned claims for common law riparian rights by the federal government. The U.S. Forest Service had sought enhanced instream flows for the use of wildlife in the Plumas National Forest. The rights were based on the assertion that the federal government had riparian rights because of its ownership of the land on which the spring and stream in question were situated. The ability of the federal government to claim riparian rights in California was of great concern to the state and to other water users, given that 46 percent of the state (over 45 million acres) is owned by the federal government. The federal government's claims were originally rejected by the state Water Resources Control Board but were later upheld by the California Supreme Court, which ruled that the federal government must be treated the same as any of the state's other landowners.

The availability of riparian rather than appropriative rights to protect instream flows in California has several advantages for the federal government. Riparian rights attach to land ownership, do not depend on use, and do not require a state permit. Though the California State Water Resources Control Board has some administrative control over riparian rights, the courts, not the Board, retain ultimate jurisdiction. And, perhaps most important for instream flow protection, riparian rights can be applied to any use deemed "reasonable" by the state, whether or not appropriative rights can be granted for the same purposes.[139]

The California Supreme Court's *Hallett* decision was not a complete success for the federal government, however, as certain restrictions on its riparian rights were also defined. First, the court drew a distinction between lands "reserved" by the federal government for specific purposes (such as national forests, national parks, and wilderness areas) and those that remain in the "public domain." The court ruled that riparian rights could be claimed on reserved lands but not on the public domain. This ruling had the effect of reducing the federal lands on which riparian rights could be claimed by approximately half. Second, relying on an earlier decision, which clarified the relation between the state's riparian and its appropriative rights,[140] the court ruled that it was constitutionally permissible for the state to give previously "unexercised" riparian rights a lower priority than appropriative rights granted prior to the exercise of the riparian right.

The second major case protecting the right of the federal government to establish water rights under state law took place in 1988 in Nevada, which subscribes to the prior appropriation doctrine. At issue in *Nevada v. Morros* was the granting of a permit to the Bureau of Land Management by the Nevada state engineer for appropriative rights to protect fish, wildlife, and recreational values in Blue Lake, located on BLM land in northwestern Nevada.[141] In addition to ruling that water rights could be granted for an in situ purpose, the court rejected objectors' arguments that the state should not grant permits to the federal government. The court cited California's *Hallett* case to the effect that states may not discriminate against the federal government or its agencies. Nevada law states that "any person" may apply to the state engineer for a water permit, and it defines "any person" to include individuals, the state of Nevada, and the United States. Another of the objectors' arguments, that granting a permit to the United States violated Nevada requirements that permits not be issued if against the public interest, was also rejected; the Nevada Supreme Court found that because the permit would be held by a public agency, was for a nonconsumptive use, and would give wildlife and stock access to water, the appropriation was clearly not against the public interest.[142]

The federal government has other means of obtaining state water rights. One is to invoke its power of eminent domain to condemn existing rights. The Constitution does, however, require the government to pay just compensation to the owner of the condemned right when using this power.

Though the authority of the government to use the power of eminent domain is undisputed, the power is seldom exercised. Far more common is an exchange of rights or their outright purchase.

A prominent example of the use of purchase and exchange procedures comes from the Bureau of Reclamation's Central Valley Project (CVP) in California. The Bureau of Reclamation is required by Section 8 of the Reclamation Act of 1902 to obtain water rights for its development projects from the states—a primary example of the way in which Congress has typically deferred to the states in the matter of water allocation. For many Reclamation projects, water rights necessary for successful implementation of the projects were already held by other water users. In California the Bureau was able to exchange water pumped through its Sacramento–San Joaquin Delta facilities for rights on the San Joaquin River that it needed to build the CVP's Friant Dam. Similarly, the Bureau was able to supply supplemental water to farmers in the Sacramento Valley in dry years in return for quantification and limitation of the farmers' monthly diversions, thereby making water available for other project purposes. In a normal year the Bureau delivers approximately 4 million acre-feet of water to these CVP "exchange contractors" and "water rights settlement contractors," which amounts to approximately half of all CVP water deliveries.[143]

Advantages and Disadvantages of Rights Under State and Federal Law

There are several advantages of having federal agencies use state water rights to protect instream flows. Perhaps the most important is that it helps to minimize conflicts between state and federal agencies over the control of local water resources. Congress has repeatedly expressed the desire to have federal agencies minimize conflicts with state water allocation systems, and state officials uniformly prefer to retain control over the administration of all the waters within their boundaries. Oversight by a single level of government may also simplify and reduce the cost of administration and lead to more consistent application of water policies. State administrators may be more responsive to local needs and concerns, and placing authority in the hands of the several states may encourage management diversity and innovation. In those states that utilize public interest criteria when evaluating permit applications, submitting federal claims to state water allocation agencies may force an evaluation of the trade-offs between the benefits of the federal uses and the losses caused to private and state interests;[144] such an evaluation might not otherwise occur.

There are also several disadvantages, at least from the perspective of the federal government, in submitting claims for instream flow rights to the states. First among these, as discussed in previous chapters, is that many states do not recognize instream flow purposes as beneficial uses of state waters. Another problem is that most states that have instream flow programs allow only a specified state agency to administer the rights, so that a federal agency applying for state instream rights would lose control over the administration of the right if the right were granted. The interests of a state in achieving federal land management objectives are not likely to be as extensive as those of the federal agency managing the land,[145] so it is possible that state administration would impair federal objectives. Federal agencies are also concerned that water rights held by the state may not prove to be as permanent, given the possibility of changes in state water policy or law.

The necessity of acquiring and maintaining rights under the laws of many different states would place additional administrative burdens on federal agencies. Though similar in many ways, especially in their general adherence to prior appropriation principles, each of the states has its own particular methods and criteria of granting and administering water rights. This seems particularly true with respect to instream flow rights. Federal agencies have difficulty achieving their missions when forced to operate separate programs in each of the different states.

An additional problem is that, even in the states where they exist, instream rights are relatively new and therefore have very recent priority dates. Because many western streams are already fully allocated, instream rights with junior priority dates will not have much, if any, effect on actual streamflows. This problem, together with those listed above, has frequently been cited as a justification for pursuing water rights based on federal law and authority rather than on state law.

State Rights and the Federal Land Management Agencies

The degree to which federal agencies seek state water rights to protect instream flows varies widely. As described above, the National Park Service is in a strong position to claim federal reserved water rights and has often done so. This makes the Park Service a notable exception to the Department of Interior's other agencies, which find themselves bound much more often by the department's general policy of relying first, if possible, on the acquisition of state rather than federal rights to achieve federal purposes. The Fish and Wildlife Service is not yet very active with respect to seeking any water rights to protect instream flows but it has sought instream flow rights for national wildlife refuges in Alaska and Arizona, and more recently in Idaho and Utah.

The Bureau of Land Management has relied almost exclusively on state water rights. The BLM is authorized to develop and use water by several statutes, including FLPMA, the Taylor Grazing Act, and the Fish and Wildlife Coordination Act, but private parties are the most significant water users on BLM lands. With respect to water rights, FLPMA mandated maintenance of the status quo, which encompassed (1) the appropriation by private persons of previously unappropriated water, under state law, for activities taking place on BLM lands; (2) the appropriation of previously unappropriated water under state law by the BLM; and (3) the assertion of federal reserved water rights by the BLM for certain limited and specified purposes, as described above. BLM policy recognizes the primacy of the states in the allocation and management of water resources. With regard specifically to instream flows, the BLM manual, Section 7250.1(A)(4), states that "the Bureau may file for State appropriative water rights needed to maintain instream flows on public lands where State water laws recognize instream flows."

The Forest Service has put substantial energy into the effort to obtain water rights based on federal law but has, in some situations, also applied for state water rights. Forest Service efforts with respect to riparian rights in California were mentioned above, and the agency is also free to apply for appropriative rights in any of the western states. So far as is known by the authors, the Forest Service has applied for instream flow rights under state law only in Arizona, where the agency has been fairly active in working with the state to protect instream flows in the national forests.

Federal Administrative Protection of Instream Flows

The federal government can also affect streamflows through mandates that may or may not be directly related to the allocation of water. Congress has authorized federal agencies to undertake a variety of actions that, while aimed primarily at other goals, affect the way that water resources are used or prohibited from being used. Primary among these are land management activities that affect the status of water development. For example, as mentioned earlier, strong land management controls are a significant factor in reserving streamflows in the national parks. Although designed primarily to protect features of the land and natural environment rather than streamflows per se, prohibitions on development have been an effective means of protecting streams from diversion or other development within park boundaries.

Administrative controls are distinguished from the nonreserved rights discussed above in that administrative controls don't create water rights. The effectiveness of administrative controls stems not from the ownership of a right to use water, but from prohibitions on water development that are either a direct or indirect consequence of the attainment of other property management goals. Land management mandates affecting streamflows on federal lands are particularly effective in the national parks but have also been used on wildlife refuges, BLM lands, and the national forests.

The Forest Service has probably been the most active agency in using administrative controls to protect streamflow. The national forests, among all federal lands, have been the most heavily used for water development. The primary administrative tool used by the Forest Service to protect water resources is the issuance or denial of special-use permits. Anyone wishing to develop water within the boundaries of a national forest must first obtain a permit from the Forest Service. When the permits expire, they must be reissued to allow continued use of the water development. The authority of the Forest Service to issue permits dates from the 1897 Organic Act, which authorized the secretary of the interior to make rules and regulations governing the "occupancy and use" of the forest reservations. Legislation in 1901 explicitly extended this general permitting authority to the construction of water developments.

The secretary of agriculture became responsible for issuing permits on the national forests when management authority for the national forests was shifted to the Department of Agriculture in 1905. The secretary of the interior retained authority to issue permits for the rights-of-way involved in transporting water. This bifurcation of responsibilities was eliminated by passage of the Federal Lands Policy and Management Act in 1976. FLPMA consolidated all regulatory functions for the national forests within the Forest Service and authorized the secretary of agriculture to "grant, issue or renew rights-of-way over, under, or through such lands for reservoirs, canals, ditches, flumes, laterals, pipes, pipelines, tunnels, and other facilities and systems for the impoundment, storage, transportation, or distribution of water."[146] No water development within the national forests is allowed to occur without a permit, and the Forest Service will often stipulate conditions within the permits that will ensure attainment of national forest purposes.

The purposes to be protected through the permitting process have been expanded over the years to include many values related to instream flows. For example, Section 505 of FLPMA requires that:

> Each right-of-way shall contain—(a) terms and conditions which will . . . (ii) minimize damage to scenic and esthetic values and fish and wildlife habitat and otherwise protect the environment . . . (v) require location of the right-of-way along a route that will cause least damage to the environment . . . and (vi) otherwise protect the public interest.

Terms and conditions specified in the permits issued to water developers can—and frequently do—require permit holders to allow water to pass through their facilities at specified times and in specified amounts necessary to maintain instream resources. When necessary, the Forest Service will deny permits if even these "bypass flows" would be insufficient to protect the purposes for which the national forests are managed. The courts have consistently supported these regulatory activities of the Forest Service, ruling that while Forest Service decisions must be supported by the administrative record and may not be arbitrary or capricious, the agency otherwise has been delegated the authority to exercise its judgment in the issuance, denial, or conditioning of permits.[147]

The authority to administer permits is a powerful tool. As noted by the court in Colorado's Water Division No. 1, exercise of this permitting authority has successfully protected stream channels from degradation even in the absence of water rights held by the federal government.[148] There are, however, several possible reasons the Forest Service has pursued federal reserved water rights in addition to the exercise of its permitting authority. First, the ability to condition or deny permits for water development occurring on national forest land does not enable the Forest Service to protect streams from development occurring on private land located above or within the boundaries of the national forest. This is not usually a problem for national forests that encompass virtually the entire headwaters of a river system, but it is a problem for the Forest Service in other settings.

Second, the establishment of water rights for instream purposes provides more effective notice to other existing and potential water users about the status of the stream and the government's flow requirements than does the permit evaluation system. With the permit system, proposed developments are evaluated individually, based on their own merits, and the process for each permit starts from scratch. Instream-flow water rights allow a developer to prospectively determine whether it is possible to construct additional developments or alter existing facilities, because the legal availability of water would already have been specified and confirmed by the courts.

Third, the administration of the special-use permit system in the absence of water rights requires additional agency resources. Reliance on the permit system forces the Forest Service to evaluate applications each time they arise, a problem compounded by the fact that the scheduling of permit evaluations is driven by events outside the agency's control. The permitting process is initiated each time a developer applies for a new permit or for the renewal of an existing permit, rather than at times that best fit agency priorities and the availability of administrative resources. The Forest Service would probably prefer to have its instream needs confirmed as water rights and then be able to call on the state to administer and enforce its claims, just like any other water user, rather than having to do all of the monitoring and administrative work itself.[149]

There is one more problem for the Forest Service in relying on its permitting authority to protect instream values. Attempts by the Forest Service to exercise its permitting authority have sometimes been met with the same strenuous objections as have its assertions of federal reserved water rights. A recent and striking example of this opposition occurred in some of the same national forests that were adjudicated in Colorado Water Division No. 1, when the permits for several water projects operated by the cities of Boulder, Fort Collins, Greeley, and Loveland, as well as by two private companies, came up for renewal in the late 1980s and early 1990s. No instream flow conditions had previously been attached to the permits, but, based on initial consultations with Forest Service personnel, the permit holders feared that the agency would condition the new permits with a requirement to allow bypass flows to protect aquatic habitat downstream. The project owners feared not only the loss of stored water, but also the cost of retrofitting dams to allow winter bypass of water in the small amounts necessary to preserve instream resources.

Some of the permittees took their concerns to members of Colorado's congressional delegation, who in turn initiated a series of meetings with Forest Service and Department of Agriculture officials.[150] In response to the concerns expressed by members of Congress, the secretary of agriculture wrote a letter October 6, 1992, to one of Colorado's senators stating that he would have the Forest Service issue the new permits without conditioning the permits to require bypass flows.[151] Forest Service rules state that the decision to issue, deny, or condition permits is to be made by the appropriate forest supervisor, so the chief of the Forest Service forwarded a copy of the letter to officials of the Arapaho-Roosevelt National Forests, where the projects were located.

Arapaho-Roosevelt officials then sought the advice of the agency's Office of General Counsel, which determined that a decision on the permits would need to be accompanied by environmental impact documents to comply with requirements of the National Environmental Policy Act. Arapaho-Roosevelt officials postponed a final decision on the permits in order to complete the required statements.[152] During the time it took to complete the statements, a new administration took office in Washington, and soon thereafter both the secretary of agriculture and the chief of the Forest Service were replaced.

As explained in greater detail in chapter 10, the Endangered Species Act requires that preparation of the relevant environmental documents include consultations with the U.S. Fish and Wildlife Service. The Fish and Wildlife Service issued its first document, a Biological Assessment, in May 1993, stating that operations of the projects might negatively affect six endangered species. This evaluation made it clear that there were two separate environmental problems to be resolved in the permitting process: (1) negative impacts on aquatic habitat immediately below the projects; and (2) negative impacts to endangered species much farther downstream, in Nebraska. To address the first problem, three of the permittees operating projects on tributaries of the Poudre River proposed a voluntary measure, known as the Joint Operations Plan (JOP), in which they would cooperatively manage projects under their control in such a manner as to ensure greater streamflows in the mainstem Poudre River during the winter.[153] In return, they asked the Forest Service to eliminate bypass flow conditions from their permits. The permittees reasoned that the JOP would be cheaper, because modifications to dams that would be necessary to release small amounts of water through the winter would be very expensive. They also believed the JOP would do more to improve aquatic habitat in the national forest than would the provision of bypass flows on the relatively short segments of the tributary channels on which the dams were located.

Water users and others brought substantial pressure to bear on the Forest Service. The conflict drew extensive coverage in the local media, much of it unfavorable to the Forest Service. City officials, irrigators, and members of Congress portrayed the Forest Service not only as overly rigid, but also as attempting to unfairly "change the rules" and deprive water users of their vested water rights through a kind of bureaucratic subterfuge. They asserted that the Forest Service was trying to interject new criteria and values—the maintenance of instream flows—into established water-use systems, at the expense of existing water users and holders of vested water rights. A United States senator publicly accused the Forest Service of "federal blackmail,"[154] and others asserted that the actions of the Forest Service with respect to the renewal of these permits constituted yet one more battle in the federal "war on the West."

The first of what later became a series of three-month extensions on the permits was issued when the original permits expired. Water project owners were able to operate under the terms of their old permits while awaiting a final determination of the conditions, if any, to be imposed on the new permits. By August 1994, environmental statements had been completed and the forest supervisor had issued decisions on five of the permits, while two others were postponed. Each permit application was addressed and decided separately, though in all five Records of Decision the Forest Service explained in identical terms the numerous authorities under which the supervisor had based his decision.[155] In response to the controversy surrounding the permitting process, the supervisor judiciously noted only: "the issue of whether or not the Forest Service could require bypass flows in a land-use authorization was controversial. I tried to consider all points of view and address the many divergent and strongly held opinions."[156]

The forest supervisor required bypass flows on one of the projects, operated by the city of Fort Collins. He reasoned that even with implementation of the JOP, flows below the dam would leave the tributary dewatered for 8 months of the year: "In the half-mile section immediately below the Reservoir, where at times no surface flows occur, habitat potential for all life stages of all fish species is zero. These losses are directly attributed to the operation of Joe Wright Reservoir."[157] The forest plan for the Arapaho and Roosevelt National Forests, developed under the National Forest Management Act and in conformance with the Federal Lands Policy and Management Act, requires that fish populations in the national forests be maintained at a level of at least 40 percent of their natural potential.[158] The Supervisor argued that this standard must apply to each watercourse individually, not to the forest as a whole. Although some degradation in aquatic habitat below the dam would exist even with the bypass flows, the supervisor determined that this would be acceptable given that "the establishment of minimum streamflow improves the quantity and quality of aquatic habitat," and that "there are significant public benefits to allowing the continued occupancy and use of NFS lands for Joe Wright Reservoir."[159]

Three other projects located on tributaries of the Poudre were not required to provide bypass flows but were required to participate in the Joint Operations Plan. The permit for the fifth project, operated by the city of

Loveland, was conditioned to require compliance with the terms of an existing Memorandum of Agreement between the city of Loveland and the Colorado Department of Wildlife that called for a specified, minimal level of flow in the stream channel below the project. The final statement on the projects by the U.S. Fish and Wildlife Service, a Biological Opinion issued in June 1994, indicated that operation of all five of the projects—even with bypass flows and implementation of the JOP—was likely to jeopardize the endangered whooping crane, least tern, piping plover, and pallid sturgeon farther downstream in the Platte River. The forest supervisor therefore decided that permits for all five projects would be amended, following completion of a Central Platte River Recovery Implementation Program by the federal government and the states of Colorado, Nebraska, and Wyoming, to require additional actions on the part of the project operators that will be necessary to protect these species. Until then, the project operators will be required to make annual financial contributions,[160] which would be used to restore riverine habitat and supply additional water to the Platte River below North Platte, Nebraska.

One thing that became clear during this drawn out and complicated course of events was that the process of making a determination on the renewal of the permits—just a few of over 100 such permits on the Arapaho and Roosevelt National Forests alone—had been anything but easy or routine for the Forest Service. Personnel involved in the process included biologists and other professionals who prepared the environmental impact statements and conducted public meetings, the forest supervisor and other staff members from the Arapaho and Roosevelt National Forests, the regional forester of Forest Service Region 2, attorneys in the agency's Office of General Counsel, Forest Service Washington Office personnel, the Forest Service watershed director, and the chief of the Forest Service. Other government officials participating in the process included the assistant secretary of agriculture, the secretary of agriculture, and members of both the United States Senate and the House of Representatives.

Forest Service officials repeatedly explained that they were not trying to eliminate the water projects or to disrupt legitimate expectations; the Forest Service was merely seeking to have water projects in the national forests operated in a balanced manner consistent with the preservation of values that the agency is obligated to protect. Forest Service opponents had repeatedly accused the agency of trying to "steal" water, but, as pointed out by a Forest Service attorney, "Holders of water rights do not hold by virtue of their water rights the right to use land on the National Forest."[161] In the Records of Decision for all five projects the forest supervisor stated, "Water uses continue to be a welcome and legitimate use of the public's lands, just as they have been for decades," but said that the Forest Service must "find a balance between the use of NFS lands for water facilities and protection of aquatic resources."[162] One Forest Service official noted that public comments received by the agency were running 4 to 1 in favor of requiring bypass flows to maintain ecosystems,[163] effectively making the point that the Forest Service had much more support for their activities than the objectors often claimed. But there is no doubt that the entire process was enormously expensive both in terms of its impact on agency morale and in the amount of administrative resources necessary to make a decision on the permits.

The final word on the use of special-use permits by the Forest Service to require bypass flows in the national forests is not yet available. An amendment to the 1996 Farm Bill, signed into law by the president on April 4, 1996, establishes a Water Rights Task Force to examine and report on the bypass flow issue. In the meantime, the Act stipulates that there will be an 18-month moratorium on any Forest Service decision to require bypass flows or any other relinquishment of the unimpaired use of a decreed water right as a condition of renewal or reissuance of a land-use authorization permit.[164]

Additional Streamflow Protection Activities

The land management agencies have undertaken a number of instream flow activities in addition to those described above. Agencies have created new initiatives and policies, brought together personnel into groups focused on instream flow issues and activities, and created new procedures for quantifying and protecting instream flows.

The National Park Service

For example, the Park Service has created two administrative units that spend a substantial amount of time working on instream flow protection issues. The first is the Water Rights Branch of the Water Resources Division, a national office based in Fort Collins, Colorado. The Water Rights Branch was created in 1984 to centralize and coordinate water management support activities among the individual park system units, using personnel specifically trained in water resource management. Efforts to protect park system water resources were previously carried out by Park Service personnel in individual park system units or regional offices, many of whom did not have expertise in water matters. Staff members for the new branch, primarily hydrologists and other technical experts, were largely drawn from other federal land management agencies and have a great deal of experience in dealing with federal water issues. Branch efforts are focused on the determination of legal water entitlements and the quantification of water flow needs.[165]

Threats to streamflows in the parks are brought to the attention of the Water Rights Branch in several ways, most commonly as a result of the Park Service being made a party to a water rights adjudication. There are general adjudications or other large water rights determination processes in progress in most of the western states, and the Park Service is served notice by the courts when park system waters are affected by the proceedings. On occasion the Park Service is made a party to an adjudication brought by private interests or initiates a suit in response to a request from either an individual park system unit or a private interest group.[166]

The Water Rights Branch has developed an instream flow needs quantification method known as "departure analysis."[167] The basic approach of this method is to start with the assumption that 100 percent of a river's flow is needed to preserve natural conditions, and to then incrementally subtract—through modeling—small units of water until the decreased streamflow results in an observable and/or quantifiable impact on the environment. At that level of streamflow a "departure" is said to have occurred. It does not matter whether the impact is construed as "positive" or "negative," because the Park Service has been directed to manage resources in such manner as to keep them "unimpaired." Any deviation from natural conditions is considered to be an impairment. In practice, Park Service modelers usually continue to subtract units of water until the existence of significant impacts is evident enough to be widely accepted and confirmed by other members of the scientific community. The Park Service believes that this approach enhances the credibility of the method, but so far departure analysis has not been tested in court. The method has been used in educational presentations, however, and in support of Park Service positions in negotiations, where it has been favorably received. Presentations based on the method have even resulted in satisfactory water rights settlements in several situations.[168]

The Park Service's Rivers, Trails, and Conservation Assistance (RTCA) office, essentially an advisory and assistance program, also undertakes activities related to instream flows. The RTCA office was created as an umbrella organization for Park Service programs having a national or regional scope, including the many projects and activities taking place under the heading of the National Rivers Program.[169] Among these projects is the responsibility for aiding and overseeing Wild and Scenic River designations that occur through the process of gubernatorial petition and secretarial approval, a process explained in more detail in chapter 10. The RTCA office has been particularly active in working to designate rivers that run through private lands as components of the Wild and Scenic Rivers system. In 1990 Congress directed the Park Service to serve as an advocate for recreation interests during hydropower facility licensing and relicensing procedures by the Federal Energy Regulatory Commission (FERC). This resulted in the creation of the Riverwatch program—now known as the Hydropower program—under the auspices of the RTCA. The RTCA has played an active role in directing FERC applicants to commission studies and provides other data related to the impacts of hydropower facilities on the recreational potential of affected waterways. The RTCA has also sponsored a variety of educational activities, often in cooperation with other groups and agencies. These activities have included seminars, workshops, and the publication of documents related to instream flow issues, particularly those associated with recreation.

The Bureau of Land Management

The Bureau of Land Management has created two policy initiatives and a process that have the potential to affect instream flows. The first of the initiatives began in 1991 with an effort to systematically identify water resource strategies available to the agency. The overall objective was "to ensure that BLM has a consistent, comprehensive water availability strategy" driven by, and based on, resource management needs. The process resulted in a November 1992 report containing recommendations and selected water rights data for the BLM lands.[170] Strategies identified in the report include, among others, the recommendation that the Bureau issue an Instruction Memorandum and revisions to the BLM manual that will "encourage vigorous adjudication and protection of existing water rights" and "clarify the purposes for which BLM seeks instream flow protection or other water rights." The report calls for the establishment of priorities within each state "for streams needing instream flow protection based on agreed-upon national criteria" and the setting of top priority for the pursuit of BLM water needs "where there is a risk of precluding management options through inaction, with special emphasis on wetlands and waterfowl habitat, riparian areas, watershed and water quality, wilderness values, grazing management, and in-situ wildlife and recreation needs." The report also called for development of a handbook of techniques and methodologies that agency personnel could use in resolving conflicts over instream flow; identified a need to incorporate water concerns into BLM's planning process; and recommended development of "a position paper/guidance for consistency in dealing with state governments, especially state legislative proposals affecting water rights, emphasizing instream flow, definition of beneficial use, and filing procedures."[171]

Instream flows on BLM lands may also be affected by the agency's riparian and wetlands program, known as the BLM Riparian-Wetland Initiative for the 1990s.[172] The general goal of the initiative is to restore, maintain, and protect riparian and wetland areas on BLM land. The primary methods for accomplishing this task fall more under the heading of land-use management than water policy, but the initiative also calls for more active protection of instream resources: "Using the procedural requirements of the various state laws, obtain on a case-by-case basis the rights or cooperative agreements for water necessary to sustain riparian-wetland areas and their associated uses. Site-specific studies will be conducted to determine water amounts, including instream flows, needed to support healthy riparian-wetland values."[173] The impact of such protection could be significant, given BLM's estimate that the agency manages almost 24 million acres of wetlands and 182,320 riparian stream miles.[174] Over 95 percent of the wetlands and almost 75 percent of the riparian stream miles are on BLM land in Alaska, but the initiative might also be important for instream flow protection in many of the other western states, particularly those in the northern Rockies and northern Plains.

One of the BLM's greatest streamflow-related successes has been the development of a unique interdisciplinary process for evaluating and quantifying instream flow needs and identifying protection strategies. The process has been designed to operate within the many constraints imposed on the BLM by law, policy, and the scarcity of administrative resources.[175] Rather than pursuing a narrow focus on the quantification of instream flow needs necessary to support a water rights claim, the agency seeks to study an instream flow situation from a variety of perspectives. The ultimate goal is to provide decision makers with information that can be used to formulate and implement any of a variety of actions that have the potential to protect the broadest possible spectrum of instream flow needs.

BLM's current study process is a result of steady evolution. The agency's first instream flow quantification studies, in the early 1970s, were generally performed by professionals from a single discipline, usually hydrology, and were often focused on a single resource value, such as fisheries. By the early to mid-1980s, BLM had developed a more interdisciplinary approach, in which multiple instream values were evaluated by interdisciplinary teams, with a focus on interaction between the disciplines rather than just separating the study into different parts to be handled by the different members. The most recent evolution of the process has been to make it not just interdisciplinary, but also multi-institutional. Rather than pursuing its studies by itself, the BLM has pursued cooperative efforts with other public agencies and even private interest groups, making an effort to include any person or group that has a stake in the management of a basin's water resources. Most of this effort to identify stakeholders and build cooperative working groups takes place prior to the initiation of the actual study. BLM makes extensive efforts to identify and gain the cooperation of other affected parties, who then also participate in the study process. This is not a trivial enhancement of the process; pre-study efforts to gain the cooperation of other entities lead to solutions that are more likely to work in the long run, and may take years to complete.[176]

The BLM's interdisciplinary study process is best illustrated by example. One of the agency's most recent studies was done on the Rio Chama, in New Mexico; the study served as the source of information about the Rio Chama described in chapter 3. The first step in the process is to complete a preliminary assessment of the situation, including field assessments and literature reviews.[177] The primary goal at this stage is to identify the values and/or uses for which the river is to be managed and to define project objectives. The identification of values is a key to the entire process, because BLM's goal is to relate the study's final management recommendations directly to the values identified during this stage of the process. Rather than placing their own values into the process, project participants seek to discover the values that have already been defined by other agencies and interests, paying special attention to existing management plans and mandates, from any source.[178] BLM took its values for the Rio Chama Wild and Scenic River primarily from the management plan prepared by the BLM, Forest Service, and Army Corps of Engineers.[179] BLM also took direction from the New Mexico Department of Game and Fish, which desired maintenance of a valuable brown trout fishery in the river and requested that the brown trout be identified as the target management species.

The next stage of the process is to describe the ways in which the identified values are affected by streamflow. For the Rio Chama the managing agencies sought to describe the flow-related needs of all stages of the target fish species' life cycles: flushed gravel and sufficient area for spawning, fresh water supplies for incubating eggs, sufficient cover for fish fry, acceptable levels of turbidity, etc. Every attempt is made to describe not only these short-term flow needs, but also long-term needs associated with such things as channel structure and form.

Step three in the process is to analyze the hydrology and geomorphology of the river system to quantify existing and probable instream flow regimes and their associated hydraulic and geomorphic attributes. Rather than attempting a broad-scale description of all hydrologic and geomorphic attributes, the study is focused on those characteristics that are of relevance to the particular values identified and described in steps one and two.

Step four relates the flow-dependent values established in step two to the hydrologic and geomorphic regimes defined in step three. The goal at this stage is to describe—and quantify, if possible—the effect of alternative flow regimes on the values that are to be protected. The goal is to produce information that can be used to evaluate the effects of incremental increases or decreases in flow on resource values. A variety of analytical tools is used to accomplish this goal. For example, fisheries professionals on the Rio Chama project employed the Physical Habitat Simulation System and Instream Flow Incremental Methodology to analyze data previously collected by the U.S. Fish and Wildlife Service.

Step five takes the information collected in step four one step further by identifying recommended flows that will protect the established values. Study team members evaluate both optimum and minimum flows necessary to maintain resource values, once again using analytical tools such as the Instream Flow Incremental Methodology whenever appropriate. The goal, however, is not to rigidly apply a given methodology to the situation at hand, but to use whatever methods are most likely to yield insight about the flow regimes necessary to protect resource values. There is great emphasis on interdisciplinary interaction during this stage, and, in the end, recommended flow levels depend more on the application of professional judgment than on the results of any particular methodology. Flow recommendations are expressed as some fixed discharge rate, together with the necessary timing of flows. Because the recommended flow levels for the Rio Chama were not all compatible, the BLM also analyzed the trade-offs inherent in two different scenarios, one designed to maximize ecological values and one designed to maximize recreational values.

The final step in the process is to develop flow protection strategies likely to protect identified values. The BLM makes every effort to identify realistic options, taking into account prevailing legal, institutional, and attitudinal constraints. The BLM's study process is based on the philosophy that successful instream flow protection measures are based not only on technically sound quantification procedures, but also on the development of equally sound legal and administrative strategies. The BLM recognizes that solutions that are to be effective in the long run must be both administratively feasible and attractive. For the Rio Chama, the primary constraint around which the BLM had to fashion protection strategies was the multitude of laws, compacts, agreements, regulatory agencies, facilities, and water rights holders associated with the storage and distribution of water in the Rio Grande basin. For example, the Rio Grande Compact of 1929 places numerous conditions on the amounts and types of water (natural versus imported) that can be stored at different facilities and at different times. At least two major dams in the basin are operated for flood control purposes by the Corps of Engineers, under authority of several flood control acts, and other dams are operated to supply irrigation and municipal water by the Bureau of Reclamation, or are operated by private entities. And water from the San Juan–Chama project is treated differently from water originating within the basin. Given these numerous complexities and constraints, the BLM determined that the most feasible option for protecting instream values in the near term would be some sort of cooperative interagency management of imported San Juan–Chama Project waters.

The BLM process incorporates the use of several methodologies and analytical tools, but the agency believes that the study process itself is just as important as the individual methods. The process is designed to be inclusionary and the participation and agreement of all affected parties are critical to meeting the BLM's goal of developing broad-based agreement for managing river systems. Some of the early studies done by the BLM focused on quantifying instream needs to support claims for federal reserved water rights in court, but more recent studies have been less tied to specific legal outcomes. The studies may be used to support federal or state water rights claims in some cases but are just as frequently used—as in the Rio Chama study—to support a variety of cooperative agreements, water exchanges, or other methods that have the potential to "solve" an instream water problem. This cooperative, interdisciplinary method is perhaps most likely to lead to the implementation of practical, flexible, and effective flow protection strategies that will work in the long run.

The Fish and Wildlife Service

The U.S. Fish and Wildlife Service is one of the nation's premier conservation agencies. The mission of the Service is "to conserve, protect, and enhance fish and wildlife and their habitats for the continuing benefit of the American people."[180] In addition to managing the national wildlife refuges, described earlier, and protecting endangered species, described in chapter 10, the Service has had two other responsibilities that affect instream flows. First, it protects wildlife habitat through monitoring, study, and consultations with other federal agencies regarding the impacts of federal projects on wildlife. Second, it was actively involved in instream flow research.

The Fish and Wildlife Service has no organic act of its own, but it is called upon by numerous statutes to evaluate proposals and make recommendations regarding impacts on wildlife. Many of the Service's responsibilities stem from the Fish and Wildlife Conservation Act of 1934. This act was designed to ameliorate negative impacts to fish and wildlife arising from the construction and operation of water development projects. The act required federal agencies to consult with the Service when constructing or permitting water development projects. The act authorized federal agencies to modify future water development projects to "accommodate the means and measures for such conservation of wildlife resources as an integral part of such projects. . . ."[181] These means and measures have great potential to affect instream flows. However, the impact of the Service in its consulting role has been limited in practice by a lack of agency resources to adequately evaluate proposed developments.

Research was, until recently, a strong component of the Fish and Wildlife Service's mission. Instream flows became a particular focus of the Service with the creation of an instream flow research team in 1976. The Cooperative Instream Flow Service Group, known more widely as the Instream Flow Group (IFG), was headquartered in Fort Collins, Colorado. The original focus of the group ran heavily toward the development of methods for quantifying instream flow needs and to the synthesis and dissemination of existing information. Specifically, the group intended to (1) develop improved methods for assessing and predicting instream flow requirements; (2) establish an effective communication network for disseminating instream flow information; and (3) develop and improve guidelines for implementing instream flow recommendations. "It is intended that IFG will become recognized as the place where the latest information on instream flow matters is readily available."[182]

The efforts of the Instream Flow Group were highly successful. The synthesis and extension of existing research on the relationship between streamflow levels and habitat suitability led to development of the Instream Flow Incremental Methodology (IFIM), described in more detail in chapter 3. The IFG consulted widely with a variety of federal and state agencies and other practitioners and became a clearinghouse for information on a range of technical and policy issues related to instream flows. The Group succeeded in coordinating many of the previously disparate efforts of professionals in the field. In doing this, the IFG helped forge a more unified, effective, and technically credible movement to quantify and protect streamflows for instream purposes.

The Instream Flow Group is still actively involved in instream flow protection issues. The IFG continues to support use of the IFIM, but its focus has otherwise changed. In the early 1980s the mission of the IFG shifted toward more basic research and less consultation. More effort was put into establishing the scientific credibility of the concept of habitat suitability curves, including the testing of existing relationships between flow levels and resource status and the development of new relationships. Research was also done in the social sciences related to instream flows, including social, economic, and institutional analysis.[183]

The most recent changes for the Instream Flow Group came in November 1993, when IFG team members were transferred to the new National Biological Survey (NBS),[184] and then in October 1996, when the NBS became the Biological Resources Division of the U.S. Geological Survey. The IFG has been somewhat dispersed, as team members have been assigned to three different sections of the U.S. Geological Survey Midcontinent Services Center in Fort Collins.

Variation within Agencies

Everything presented in this chapter so far reveals that there is substantial variation among agencies in the ways and degree to which they address instream flow protection problems. But attempts to briefly characterize the instream flow protection efforts of individual agencies is frustrated by the fact that there is also substantial variation within those agencies.

Though all of the federal agencies are guided by national policies, on-the-ground management is usually decentralized. For example, individual park superintendents and regional agency officials have a great deal of control over management initiatives in the Park Service. District rangers, forest supervisors, and regional foresters all have substantial control and discretion over management activities in the national forests. The BLM is managed by the state, and there is substantial variation in the emphasis given to instream flow issues in the different BLM state offices. At least some of this variation is explained by differences in the degree to which individual states give federal agencies the opportunity to protect instream flows. Differences in the philosophy and training of personnel in the individual BLM state offices also play a significant role.[185]

Instream flow protection activities by the Fish and Wildlife Service on the national wildlife refuges are a good example of the variation found within agencies. Though instream flow protection has not been a priority for the refuge system nationwide, the attention devoted to instream flow issues varies substantially among the different regional offices. Management authority for the national wildlife refuges is concentrated at the level of individual refuges and the regional offices.[186] Since 1990 there has been a Western Water Rights Coordination Group that advises the Fish and Wildlife Service director on water rights issues, but most water-related policy action takes place at the regional level.[187]

Very little emphasis has so far been placed on instream flow protection activities for the refuges in Region 1, which includes the states of California, Oregon, Washington, Idaho, and Nevada[188] and Region 6, which covers refuges in the states of Colorado, Kansas, Montana, Nebraska, North Dakota, South Dakota, Utah, and Wyoming.[189] Water resources personnel in Region 1 are focusing their attention on a more basic task, the collection of baseline data for water quantity and quality in each of the refuges. This data will eventually be used to create water budgets, determine water needs, and acquire rights, including instream flow rights where necessary, but the process was not started until 1990 and is expected to take decades to complete. Personnel in both regions believe that very few of the refuges require instream flows to accomplish the purposes for which they were created.

State willingness to grant federal agencies water rights for instream flow purposes has led to somewhat more instream flow protection activity in the southwestern states. The Service has been active in instream issues in Arizona, which along with New Mexico, Texas, and Oklahoma forms U.S. Fish and Wildlife Service Region 2.[190] Arizona allows federal agencies to hold water rights for instream flow purposes. The Service has already acquired an instream flow right in Arizona for the San Bernardino National Wildlife Refuge; has applied for an instream right in the Bill Williams National Wildlife Refuge; and is investigating the possibility of applying for a right in the Buenos Aires National Wildlife Refuge. The Service is investigating the possibility of asserting a federal reserved right to protect instream flows on the Bill Williams River, but, as in other regions, it generally prefers to work through the state water rights system. The Service has not applied for instream rights in other states of the region, though it is considering an application for instream rights on the lower Rio Grande River in Texas.

The Fish and Wildlife Service is quite active in instream flow protection activities in Alaska, which comprises Region 7.[191] Region 7 has had a substantive instream flow program in place since the mid-1980s. For example, in December 1993, regional personnel were planning to submit applications for instream flow rights on 13 stream segments and 150 lakes in the Arctic National Wildlife Refuge during the winter of 1993–94 and applications for instream rights on other refuges following collection of the appropriate data.[192] Because the Service had not yet submitted an instream flow rights application to the state Department of Natural Resources, it was not yet known whether the methodology used by the Service for quantifying instream flow needs would be accepted, but Service personnel felt that the prospects were good.

An advantage for those wishing to protect instream flows in Alaska is that, so far, there has been very little diversion of water from stream channels and only limited competition for water rights. In the words of one Fish and Wildlife Service employee, it gives the Service—and others—an "opportunity to do it right the first time."[193] Most of the other Fish and Wildlife Service regions are forced to concentrate on efforts to ameliorate the negative effects of past practices, but in Alaska it is more often possible to protect flows before they are gone rather than trying to recoup minimal flows in streams that had previously been allocated for other uses.

One disadvantage of trying to protect streamflows in Alaska is that there is much less basic streamflow data available in Alaska than in most other states. There is also a potential problem for instream flow reservations granted by the state, because Alaska's instream flow statute allows the state to review instream flow reservations every 10 years.

Availability of Resources for Instream Flow Protection

Important factors influencing the level of involvement by federal agencies in instream flow protection activities include the purposes for which particular classes of federal land are managed, methods available to the agencies for protecting instream flows, and the degree to which instream values are threatened. The availability of funds and personnel necessary to engage in protection activities is also important. Though some of the activities described above, particularly the protracted legal battles of the Forest Service, may give the impression that federal agencies have virtually unlimited funds to devote to instream flow protection, the reality is often much different.

Instream flows are just one of many needs on the federal lands that must be met with limited human and financial resources. The process of quantifying instream flow needs and legally acquiring or protecting instream flow rights requires a substantial investment of technical and legal expertise and can be very expensive. Compounding the problem is the fact that water rights acquisition procedures are different in each of the western states and often must be undertaken separately for each national park, monument, forest, wildlife refuge, and BLM Resource Area.

National park superintendents are often forced to give priority to their efforts to cope with the phenomenal growth in visitation to the parks, growth that usually has not been matched by increases in park personnel or other administrative resources. Park staff preoccupied with efforts to accommodate additional visitors at overburdened facilities often have little time to deal with any but the most pressing threats to natural resources and are not likely to add labor- and dollar-consuming water protection efforts to their slate of management activities unless the need is critical. Water concerns—among others—are often forced onto the back burner.[194]

Many of the recommendations found in the BLM's Water Availability Strategy, described above, have yet to be implemented, primarily because of a continuing lack of agency resources. The report itself was intended at least in part to make a case within the agency for better funding of water rights activities; but it was not successful. Other commentators have noted that "throughout its relatively short history, the BLM has been assigned difficult management tasks without the wherewithal to accomplish them."[195] This situation applies to the agency's water-related goals just as much as it does to the agency's seemingly intractable problems with administering and restoring the public rangelands. The BLM has been given many mandates and has put in place significant policies and strategies to deal with water issues, but it has not been given the resources necessary to accomplish its mandates. Implementation of water rights policies has fallen well short of agency goals.

The BLM's policy of seeking state water rights to protect instream flows has not yet gone very far. Data compiled for the strategies report indicated that BLM district offices had identified 529 streams needing some level of instream flow quantification and protection. The report stated that quantification and protection of slightly over half of these "could" be completed by the end of 1997, especially if the BLM were to be successful in finding partners to share the heavy workload.[196] This projection, however, was based on the assumption that the BLM would receive more funding for water rights activities, which did not happen. It is also important to realize that the 529 streams identified in the strategies report as being in need of instream protection do not include all streams on BLM lands that might need protection. The BLM does not have a formal process for identifying streams needing instream flow work, so there has been no comprehensive attempt to systematically identify all such streams. Most of the streams identified in the report are larger, more noticeable streams and do not include many of the smaller headwater tributaries that may also be in need of protection.

The lack of agency resources may be particularly pronounced with respect to the national wildlife refuges. It is necessary to identify existing water rights, develop water rights databases, conduct water management engineering studies, and develop water-budget models if refuges are to be protected from injurious water diversions.[197] However, a 1993 study by the Interior Department's Inspector General's Office reported that the refuge system was facing a $325 million maintenance backlog and was running a $200 million operations deficit.[198] Even if the Fish and Wildlife Service were to concentrate solely on water issues, the process of quantifying water needs and acquiring rights would take many years. There are many other needs competing for agency resources. Even the assertion of basic water rights has not yet been done for many of the refuges. Of the 224 refuges in the West, the Service identified only 57 that have quantified water rights; only 60 that have fully documented their water needs; and of the 91 refuges that may have federal reserved water rights only 11 that have quantified those rights.[199] This state of affairs cannot be attributed solely to a lack of threats to the nation's wildlife refuges; the Service also discovered that conflicts with other water users have been reported in 150 of the 224 western refuges.[200] Faced with a situation in which even basic water needs have not been met, the acquisition of instream flow rights has rarely been made a priority.

Summary Comments

Conflicts between the federal and state governments over the control of water resources are frequent and widespread and are much in evidence with respect to instream flow protection activities. When considering the material in this and the following chapters, which describe the efforts of federal agencies to protect instream flows, it is interesting to consider the question of which level of government has been most responsive to the changing values and needs of the public. It is not clear that there is a single or simple answer to this question. Some of the recent efforts of the federal agencies seem quite protective, whereas others have been criticized by river advocates as being based on outdated conditions and values. No matter which level of government seems most responsive to instream flow concerns in any given situation, it is worthwhile to look for situations in which the states and federal government either have, or could have, worked together to solve problems related to instream flow protection issues.

CHAPTER NINE

Federal Water Development Programs Affecting Instream Flows

The federal government has been heavily involved in the development of western water. Federal agencies manage hundreds of flood control, irrigation, hydropower production, and navigation facilities in the West, including most of the region's largest dams. The operation of these facilities has substantial influence over the timing and magnitude of river flows downstream.

Federal Reclamation Program

The Interior Department's Bureau of Reclamation has been a major developer of western water since passage of the Reclamation Act in 1902.[1] The Bureau has played an extremely important role in settling the West and in establishing productive agricultural economies. The Bureau's focus has been almost exclusively on the supply of water to agricultural and municipal users, and the Bureau is now the West's largest water supplier. As of 1993, the Bureau delivered water to over 30 million people and to nearly 10 million acres of land on over 137,000 farms in the western states. The agency had constructed 348 storage dams and reservoirs, numerous diversion structures, and over 69,000 miles of canals and other water conveyance and distribution facilities.[2] Though accomplishing many of their intended objectives, the facilities have also had a negative impact on natural streamflows, altering, severely reducing, or even eliminating instream flows on rivers throughout the West. The Bureau continues to operate its myriad facilities to achieve historical goals, but the Bureau's mandate—and possibly its mentality—may now be changing to better accommodate new goals, including the provision of instream flows.

First Changes in the Reclamation Program

A paucity of remaining high-quality sites for dam construction, a reluctance on the part of Congress and the American public to continue funding the heavily subsidized projects typical of the reclamation program, and a new emphasis on preserving environmental values have combined to halt major dam construction by the Bureau.[3] As a result, the Bureau in 1987 proclaimed the arrival of a "new era," an era in which the emphasis would shift from construction of new reclamation projects to the management of existing projects, with the new management to occur in more economically efficient and environmentally sensitive ways.

The announced change in policy was followed by development of a Strategic Plan to guide Bureau actions in the new era. Implementation plans were written for specific components of the Strategic Plan, including a draft implementation plan for instream flows.[4] The draft instream flow plan—though never succeeded by a final plan or formally adopted by the agency—appears to be a serious attempt at promoting the integration of instream flow needs and opportunities into the Bureau's operations and decision-making processes.[5] It describes the history of Bureau involvement in instream flow issues, delineates the scope of instream flow needs and opportunities at Bureau facilities, lists several factors that need to be considered and acted on when pursuing such opportunities, and proposes strategies for enhancing instream benefits at Reclamation facilities.[6]

Existing attitudes within the Bureau are identified in the plan as "perhaps the greatest impediment to implementing instream flows." The draft is clearly intended to educate Bureau personnel about the benefits of enhanced instream flows. Interspersed throughout the document are examples of ways in which the Bureau has previously been successful—even before the start of the new era in 1987—in creating or enhancing instream benefits at a variety of Reclamation facilities. Methods of implementation identified in the plan include: changes in the operations of water users and/or dam operations; storage buyback, in which water currently being stored for offstream purposes is purchased for use instream; use of currently uncontracted reservoir storage capacity for meeting instream flow needs; water exchanges and conjunctive use of surface water and groundwater; subordination of hydroelectric generation goals to instream goals; dry-year options in which offstream water users are compensated for using less water in dry years so that more water is available for instream needs; increased reservoir storage capacity; structural modifications to dams and outlet works that will allow releases in quantities suitable for meeting instream needs; and water conservation incentives. It is interesting to note that modifications to existing dams and operations play a prominent role in this list of methods, as befits an agency that will continue to play an important role in the operation of the thousands of dams already existing on the rivers of the West.

Because of the sheer number and size of Reclamation facilities throughout the West, it is worth looking more closely at how the facilities have sometimes been managed, and might be more extensively managed in the future, to enhance instream uses.[7] The following three examples also illustrate that the "old" Bureau was not immune to instream flow concerns.

One of the first efforts by the Bureau to enhance instream uses took place on the North Platte River below Kortes Dam, in Wyoming. A six-mile reach of river below the dam known as the Miracle Mile Recreation Area contains a blue-ribbon trout fishery. Prior to the changes described here, water releases from Kortes Dam were scheduled to produce peaking power from the dam's generators and to meet the water-supply needs of downstream irrigators. Flows in the river below the dam ranged from zero to 3,000 cfs and were quite variable; both extremes sometimes occurred within the course of a single day. In 1971, Congress authorized the Bureau to modify dam operations in order to provide a minimum streamflow of 500 cfs in the river. This minimum flow was to be maintained at all times, except when there was not enough water available in the reservoir to produce electricity. No other restrictions were placed on flow rates, or on the rate of change of flows, but the Bureau did undertake efforts to ensure that flows would not be reduced during periods of trout spawning or incubation.[8]

Another example of the Bureau of Reclamation's efforts to accommodate instream uses occurred at Island Park Dam on Henry's Fork of the Snake River, Idaho. The river below the dam contains a blue-ribbon trout fishery as well as important wintering habitat for trumpeter swans.[9] Island Park Dam's original operating procedure was to pass inflows through the dam until November 15, and then close the dam gates and retain all flows until the reservoir filled. But in 1984 the Bureau signed a memorandum of understanding with two downstream hydroelectric facilities—which held water rights senior to those of the Bureau—that allowed the Bureau to start storing water prior to November 15, and then release enough water to maintain a minimum streamflow in the river during the time that the reservoir was filling. Minimum flows have helped maintain the downstream fishery but so far have fallen short of the 700 cfs flows thought optimal for the wintering trumpeter swans. The Bureau is continuing negotiations with hydropower interests to provide adequate flows for the swans.

A third example of the Bureau's efforts to enhance instream uses took place on the Yakima River in Washington. Salmon runs on the river have been decimated, declining from estimated runs of 500,000–600,000 in the late 1800s to just 1,000 fish in 1981. There are numerous causes of the decline, one of the most significant being the construction and operation of numerous large dams on the Yakima and on the Columbia River downstream from the confluence of the two rivers. The Bureau's first efforts to improve fish runs in the Yakima took place in the 1970s and focused on constructing yet more storage facilities so that more stored water would be available to maintain flows in the river during critical periods of the year. This solution—building more dams to ameliorate problems caused by building dams—was strongly criticized by some environmentalists. Opposition to the proposal inspired the Bureau to initiate additional, nonstructural streamflow protection measures in the 1980s. These new methods included the coordinated management of releases from multiple dams, additional water releases during fish incubation periods, and the partial curtailment of diversions to two hydropower facilities to leave more water in stream channels for fish migration. The Bureau is currently investigating the possibility of making structural improvements to on-farm water conveyance and distribution systems and other, nonstructural changes that would increase the amount of water available for instream flow—and for additional irrigation. In recent years salmon and steelhead runs in the Yakima have increased to 8,000–12,000 fish per year.

Continuing Changes at the Bureau

New life was breathed into reform efforts after the 1992 presidential election. New Bureau commissioner Daniel Beard seemed firmly committed to the goal of managing water resources to meet the contemporary needs of American society, as was his boss, Secretary of the Interior Bruce Babbitt.[10] Commissioner Beard declared that there would be no new federally funded irrigation projects, that the Bureau would work with other agencies and the Congress to change the budget appropriations structure to better reflect new program goals, and that Bureau operations would be decentralized, in large part to promote the integration of public participation into Bureau decision-making processes.[11]

The agency had been realigned to effect many of the changes pushed by Beard and Babbitt by the time Beard left the Bureau in 1995. Numerous employees in construction-related occupations had been laid off, and, although agency-wide policy goals were still being established in Washington, D.C., greater program authority had been delegated to the Bureau's 25 field offices, known as area offices. The Bureau's Denver Service Center had been relieved of many of its "line of command" responsibilities and was reorganized with a mission of providing technical support to the area offices. Area managers were being encouraged to think of their areas as entire watersheds rather than as specific reclamation projects, and responsibility had been placed on area managers to embrace the changes occurring in the agency's mission. As part of this mission, area managers were being strongly encouraged by the commissioner's office to incorporate instream flow benefits into ongoing decisions about dam operations and contract renewals.[12]

The federal reclamation program is also being changed from outside the Bureau. Reclamation activities historically have been undertaken in close association with directives from Congress, and Congress too has been facing demands for change in the ways that western water resources are being managed. One of its most notable actions in response to these demands, and one that had a significant impact on instream flows, was passage of the Reclamation Projects Authorization and Adjustment Act of 1992, an omnibus bill that authorized funding for numerous water projects.[13] Of interest with regard to instream flow protection policy were Titles 2–5, the Central Utah Project Completion Act, and Title 34, the Central Valley Project Improvement Act.

The Central Utah Project Completion Act

The Central Utah Project (CUP) was first authorized by Congress in 1956. The CUP is a massive engineering project designed to divert water from streams in Utah's Uinta Basin—streams that are tributaries to the Green River, itself a tributary of the Colorado River—and transport the water over and through the mountains for delivery to cities and farms along the Wasatch Front, where most of the state's population lives. Most of the project remained unbuilt, a victim of escalating costs, growing awareness of the environmental damages associated with the project, and, at least some allege, poor design and management of the project by the Bureau of Reclamation.[14] To get more funding for the project, proponents got together with environmentalists and others to work out a compromise bill. The rewritten bill, passed in 1992 by Congress as the Central Utah Project Completion Act (CUPCA), deleted several of the project's planned construction starts and added substantial funding for environmental mitigation. The bill also removed the Bureau from further involvement with the project, turning over responsibility to the Central Utah Water Conservancy District.

Environmental features of the CUPCA provide substantial protection for instream flows. Section 301 created a presidential commission, the Utah Reclamation Mitigation and Conservation Commission, to coordinate implementation of the mitigation and conservation provisions of the act among federal and state fish, wildlife, and recreation agencies.[15] The Commission, a federal agency, was directed to acquire some existing water rights in the Utah Lake drainage basin and transfer the rights to the Utah Division of Wildlife Resources for the purpose of maintaining instream flows in the Provo River.[16]

Section 303 of the Act directs the Central Utah Water Conservancy District to acquire, with federal funds, a number of existing water rights and use the rights to increase minimum streamflows in the Strawberry River and its tributaries.[17] Because development of the Central Utah Project is expected to increase peak flows in the Provo River, to the detriment of a number of instream values, the District was also directed to undertake a comprehensive study of alternative means to avoid or mitigate such flows.

Section 303 also requires the District to manage deliveries of water in several specified streams affected by the project in such manner as to protect instream flows necessary to maintain fish and wildlife habitat. The intent was to require the District to provide the flows as a contractual obligation, rather than to create new water rights.[18] The District is to provide at least 44,000 acre-feet of water per year to maintain 50 percent of the historic adult trout habitat in the Strawberry River, Rock Creek, West Fork Duchesne River, and Currant Creek in the Uinta Basin. The water provided by the District is to be available for diversion after flowing through the designated river reaches and so is not permanently lost to the District.

Other sections of the CUPCA, although not directly affecting instream flows, provide related benefits. Such sections authorize funding for fish habitat restoration,[19] improving stream access and riparian development,[20] and fish hatchery production.[21] Funding for the Commission's activities comes from a mixture of state and federal sources. In each fiscal year from 1994 through 2001, $3 million will be provided by the state of Utah, $0.75 million from the conservancy district, $5 million from the secretary of energy, using funds appropriated to the Western Area Power Administration, and $5 million from amounts authorized by Congress to be appropriated for completion of the CUP and other components of the Colorado River Storage Project.[22]

The Central Valley Project Improvement Act

The Central Valley Project (CVP), far and away the largest of the Bureau's projects, was originally launched by the State of California near the start of the Great Depression with the goal of promoting agricultural development in the state's immense but largely arid Central Valley. The valley is drained by the Sacramento River and tributaries in the north and the San Joaquin River and tributaries in the south. The natural flows in both rivers were highly variable, running extremely high in late spring and early summer with snowmelt from the surrounding mountains, and very low in the late summer, fall, and winter. The CVP was designed to harness the flows in these rivers and their tributaries to provide water year-round for irrigation and cities.

Depression-era difficulties in financing the massive project led the state to transfer responsibility for construction and operation to the federal government. The Bureau of Reclamation subsequently built 20 large storage dams and 500 miles of water conveyance and distribution facilities as part of the project. The project's 12 million acre-feet of total storage capacity and 8 million acre-feet of water delivered annually to 3 million acres of farmland and over 2 million urban residents helped convert the largely arid valley into an agricultural cornucopia.[23]

Development of the valley's water resources also wreaked havoc with the natural environment. Riparian areas that had been dependent on the periodic cycles of flooding and low flows were largely destroyed, miles of rivers and streams were inundated by reservoirs, wetlands were drained to expand irrigable acreage, and problems with drainage from agricultural lands created severe water-quality problems. The alteration and reduction of river flows and the placement of large dams in river channels drastically reduced suitable habitat for salmon and the other anadromous fish that had played an important role in the economic and cultural life of the region.[24]

Though Congress had sporadically authorized the Bureau to mitigate environmental damage arising from construction and operation of various CVP facilities, it was not until passage of the historic Central Valley Project Improvement Act (CVPIA) in 1992 that these environmental considerations became prominent.[25] Section 3406(a) of the CVPIA makes mitigation, protection, restoration, and enhancement of fish and wildlife official purposes of the CVP. The act creates a $50 million annual habitat restoration fund and firms up deliveries of CVP water to several national wildlife refuges and state wildlife management areas that had previously been dependent on surplus CVP water available only in wet years.[26] The Act also gave the secretary of the interior three years to develop a plan to roughly double the number of natural anadromous fish in Central Valley rivers by the year 2002.

Another prominent feature of the act, the one most relevant to instream flow protection, is the allocation of roughly 10 percent of CVP water for environmental purposes. Specifically, Section 3406(b) states that 800,000 acre-feet of water—over and above the amount allocated to the wildlife refuges discussed above—will be dedicated and managed for fish, wildlife, and habitat restoration and to assist the state of California in protecting the waters of the Sacramento–San Joaquin Delta, where the Sacramento and San Joaquin rivers merge as they flow into the San Francisco Bay.[27] Provisions of the CVPIA are to be administered jointly by the Bureau and the U.S. Fish and Wildlife Service, in compliance with state law. At approximately the same time as the CVPIA was passed, the California State Water Resources Control Board, acting on new water-quality standards established by the U.S. Environmental Protection Agency, called for enhanced freshwater flows through the delta. The flows were to be used to prevent the encroachment of saline waters; eliminate "reverse flows," which caused fish to be sucked into CVP pumps; and create "pulse flows" to help juvenile salmon migrate to the ocean. The agencies administering the CVPIA agreed that for the first three years the entire 800,000 acre-feet allocation would be used for these purposes, although the act allows for lower allocations depending on water conditions.[28] In essence, the CVPIA provided the water necessary for a massive instream flow enhancement program.

The water allocated for environmental purposes was previously allocated to other water users. Many of these users will be forced to cut back their use of CVP water to meet CVPIA requirements. If climatic conditions are such that CVP water yields are high, all CVP water uses, including the environment, can be satisfied. But in years in which CVP water yields are low, those who contract with the Bureau for CVP water deliveries most likely will have their deliveries cut back to accommodate the new water deliveries to the environment.[29]

The cutbacks in water deliveries are not intended to be permanent. The CVPIA directs the secretary of the interior to develop a plan for increasing the yield of the CVP, and the new water is supposed to offset the water allocated to the environment. The act is not explicit about how this new water will be developed, though conservation, structural modifications to CVP facilities, land fallowing, conjunctive use of groundwater and surface water, and increased system efficiencies have all been discussed. The new water is to be available within 15 years of the CVPIA's passage.

The CVPIA directed the Bureau to allocate the 800,000 acre-feet "upon enactment," and the Bureau has done so, despite some protests and legal wrangling that remains unresolved. The water was allocated to meet water-quality standards at the delta in water years 1993 through 1995, thus completing the original three-year agreement. A final plan for distributing the annual 800,000 acre-feet allocated to the environment was to be defined in a programmatic environmental impact statement (EIS) being prepared by the Bureau to cover all aspects of CVPIA implementation.

The Future of the Reclamation Program

The potential for the federal government to effect the status of instream flows through the reclamation program is enormous. The Bureau operates most of the truly large western storage reservoirs, and water storage is the key to much of what happens with water in the West. Dams and reservoirs have many negative effects on instream flow values, but they also are a fact of life. With limited exceptions, virtually no one has seriously proposed that the West's reclamation facilities be removed.[30] However, there are opportunities to operate reclamation facilities to enhance instream flow uses. To date, reservoir releases designed to enhance instream values, as described in several of the examples above, are clearly the exception. Dam operations can of course be altered to mitigate some adverse impacts of their presence. Though the very notion may be anathema to many enthusiasts of free-flowing rivers, reservoir operations may actually be used to "improve" on nature for some river uses, because natural river flows tend to be highly variable, and both floods and drought have been responsible for destroying fish and wildlife habitat and endangering rather than enhancing recreational opportunities. Reservoir releases could be managed to facilitate different stages of the fish life cycle, to create different kinds of recreational opportunities, to influence channel morphology in desired ways, and to enhance the aesthetic qualities of rivers. In short, the power of existing reclamation dams might be harnessed to improve conditions for some values.[31]

A discussion of recent trends in the federal reclamation program must include a few reservations. First, political developments since the congressional election of 1994 may significantly change the future of the reclamation program. Congressional representatives and others have been discussing the possibility of privatizing the Bureau of Reclamation, and possible repeal of the Central Valley Project Improvement Act.

Second, changing the mission and operations of an entrenched bureaucracy is notoriously difficult. Bureau leadership has established an ambitious course of action for the future, but it is too early to judge the success of these efforts. So far the Bureau seems to want to enhance instream flow values without eliminating services to existing clients. This goal is certainly laudable but may not be realistic. Improvements in efficiency and awareness may enable the Bureau to provide some new benefits without corresponding sacrifices from existing users, but without sacrifice somewhere these gains are likely to be limited.

The Bureau's hands are also somewhat tied, because the agency's authority is limited by existing contracts, federal and state law, and congressional mandates, which are not always conducive to instream flow protection. A definitive evaluation of the effectiveness of the "new" Bureau of Reclamation in meeting contemporary needs will have to wait. But the Bureau is positioned to have a large impact on the future of the American West whether or not changes recently espoused by Bureau leadership and the Congress actually occur.

Federal Hydropower Programs

The federal government has played a major role in hydroelectric energy generation since the first hydropower plants came on-line in the late 19th century. That role has included the development and operation of federal facilities and the regulation of facilities built and operated by others.

Federal Power Agencies

A discussion of federal hydropower facilities in the West must start with the Bureau of Reclamation. Although hydroelectric generation was originally considered to be little more than an add-on to major reclamation projects, hydropower later became a major component of virtually every storage dam built by the Bureau. One reason for this shift was the adoption of river basin accounting methods by the Bureau. Viewed individually, proposed irrigation projects—the primary mission of the Bureau—often fared poorly when subjected to benefit-cost analysis. With river-basin accounting, the costs and benefits accruing to all portions of a project were combined, rather than viewed individually on their own merits. Unlike many other kinds of water development, the generation of electricity often yields revenues far in excess of costs, so under a river-basin accounting format, surplus revenues resulting from the generation of electricity could be used to justify construction of a facility that otherwise lacked a favorable benefit-cost ratio. Further, because Bureau projects typically included more than one major facility, surplus revenues from electricity generation at one facility could even be used to justify construction of separate facilities that, judged on their own merits, would not be economically efficient. Hydroelectric facilities thus became the "cash registers" of large multipurpose projects built by the Bureau and were used to justify an array of water development activities.[32]

Most other federal hydropower facilities in the West have been constructed and operated by the U.S. Army Corps of Engineers. The civil projects branch of the Corps is involved in all manner of water development projects throughout the United States and beyond, with purposes ranging from flood control to navigation to the construction of bridges and marinas and other facilities. Most projects serve multiple purposes. Flood control is often the primary purpose, but the generation of electricity has also constituted a major component of Corps water development projects. The Corps has not been as active in the West as the Bureau of Reclamation, but Corps development of hydropower has been substantial, particularly on the Columbia and Missouri rivers. Hydroelectric generation made the Columbia River the world's largest producer of electricity and transformed the economy of the entire Pacific Northwest. Projects on the Missouri were intended to have, but fell short of achieving, a similar effect on the economy of the Plains states in the upper Missouri River basin.

The electricity generated at federal facilities is marketed through agencies within the Department of Energy. The Bonneville Power Administration (BPA) and Western Area Power Administration (WAPA) are the most prominent of these agencies in the western states. BPA markets electricity from facilities in the Columbia River basin, and WAPA markets electricity from facilities in most of the remaining 18 western states. The Alaska Power Administration markets electricity produced by two facilities in that state, and the Southwestern Power Administration markets electricity produced by federal facilities in Oklahoma and parts of Kansas and Texas. The focus here is on the two primary western agencies, BPA and WAPA.

The Bonneville Power Administration is the oldest, and, in terms of generating capacity, the largest federal hydropower marketing agency. Created in 1937 to market and transmit the power produced at Bonneville Dam on the Columbia River, the agency currently markets the electricity generated at 30 federal dams in the Columbia River watershed, which includes Oregon, Washington, Idaho, western Montana, and small parts of Wyoming, Nevada, Utah, and California.[33] BPA markets about half of all electricity consumed in the Northwest. Eighty-five percent of BPA's electricity is generated by hydroelectric facilities. The total generating capacity of the 30 hydropower plants, which include many of the largest hydropower facilities in the United States, is 20,095 megawatts (MW). The largest of these facilities are listed in table 9.1.[34] The Bureau has nine hydropower facilities in the region, and the Corps has the remaining 21.

WAPA was established on December 21, 1977, to take over power marketing responsibilities previously managed by the Bureau of Reclamation. WAPA sells 15 percent of the entire nation's hydropower, marketing electricity from 54 power plants with a total capacity of 10,082 megawatts.[35] All of these 54 facilities are operated by the Bureau except for six dams on the mainstem Missouri operated by the Corps and two dams on the Rio Grande operated by the International Boundary and Water Commission. The facilities with the greatest installed generating capacity are listed in table 9.2.[36] The other facilities under the jurisdiction of WAPA are considerably smaller.

Operating Rules for Federal Hydropower Facilities

General principles for operating federal hydropower facilities are found in umbrella statutes and in authorizing or supplementary statutes for specific facilities and projects. Examples of the former include the Reclamation Act of 1902 and the Flood Control Act of 1944; examples of the latter include the Boulder Canyon Act of 1932, Colorado River Storage Project Act of 1956, and the Grand Canyon Protection Act of 1992. Operations are also

TABLE 9.1
The Bonneville Power Administration's Largest Facilities

Dam	River	Operator	Generation capacity (MW)
Grand Coulee	Columbia	Bureau	6,998
Chief Joseph	Columbia	Corps	2,614
John Day	Columbia	Corps	2,484
The Dalles	Columbia	Corps	2,074
Bonneville	Columbia	Corps	1,147
McNary	Columbia	Corps	1,127

TABLE 9.2
The Western Area Power Administration's Largest Facilities

Dam	River	Operator	Generation capacity (MW)
Hoover	Colorado	Bureau	2,074
Glen Canyon	Colorado	Bureau	1,346
Oahe	Missouri	Corps	786
Shasta	Sacramento	Bureau	578
Garrison	Missouri	Corps	546
Big Bend	Missouri	Corps	538

affected by federal statutes guiding other federal programs, such as the National Environmental Policy Act and Endangered Species Act.

Within these statutory guidelines, however, the operating agencies have substantial discretion to operate the facilities in such manner as to best achieve broadly-worded project purposes. Project purposes are often listed in project authorization statutes. Purposes commonly include flood control, irrigation, municipal and industrial water supply, navigation, generation of electricity, fish and wildlife, and recreation. Operating procedures that would maximize any one use are often in conflict with those needed to maximize other uses, so the operating agencies must strike a balance among competing uses. Congress has usually not assigned priorities among listed uses, leaving to the operating agencies a determination of the precise balance of uses for which dams will be operated.

Hydropower rarely if ever receives top priority. Federal projects are primarily operated to meet other needs, especially flood control, irrigation, water supply, and, where applicable, navigation. Although there is no formal list of water-use priorities, and virtually all uses receive at least some water, an informal, generalized priority list might look like this: first flood control, then water supply (for agriculture, cities, and industry), navigation, hydropower, fish and wildlife, and recreation. The presence of endangered species usually elevates fish and wildlife purposes to a place near the top of the list, though fish and wildlife are not likely to supplant flood control in the top position.

None of the foregoing means that hydropower is anything other than a major water use at federal facilities, however. Water releases for other uses can usually be run through the turbines, and enormous economic value is produced by using falling water to produce electricity. Hydroelectric generation at federal facilities often produces much greater economic benefits than do higher-priority irrigation, navigation, or water-supply purposes. Hydropower is frequently the only water use that fully pays for itself through repayment of construction, operation, and maintenance costs to the federal treasury. The only other purpose that consistently shows a comparable level of benefits is flood control, for which benefits are measured as damages avoided rather than as revenue received. The proportion of project costs attributed to flood control are "nonreimbursable," which is to say that they are financed by the federal government without expectation or requirement of repayment by project beneficiaries. This leaves hydropower as the primary revenue producer.

To the agencies in charge of hydropower operations and marketing, hydropower receives undeservedly low priority. To some, it often seems that western water projects are operated for everything except power production. It would be a mistake, however, to understate the role that hydropower plays in dam operations. Although the total volume of water released annually or monthly from dams might be designed to accommodate downstream water uses, the timing of daily water releases, and sometimes even of releases over longer periods, is heavily influenced by electric energy production goals. As described in chapter 3, hydropower facilities are extremely valuable because of their responsiveness and ability to quickly meet peak power demands. As a result, the timing of daily releases at many federal dams reflects the fluctuating level of demand for electricity more than it does the demands of other water users. The potent economic benefit derived from using federal dams to produce electricity keeps hydropower from being overlooked and contributes to the substantial political and economic clout of agencies such as BPA, WAPA, the Bureau of Reclamation, and the Corps of Engineers.

Historically, little emphasis was placed on operating dams and reservoirs for fish, wildlife, recreation, or other instream purposes. Dams were constructed to facilitate economic development, and most mention in project authorization statutes of fish, wildlife, and recreation focused on opportunities available at the reservoirs created by the dams rather than on instream uses impaired by the dams. However, changes in public knowledge, attitudes, and valuation of the environment have brought pressure for changes in dam operating procedures. Operating agencies have been urged to use their often substantial discretion in setting dam operation parameters to accommodate instream uses. The following two examples of the emerging importance of instream uses in the operation of federal hydroelectric facilities, Glen Canyon Dam on the Colorado River and the dams of the Columbia River, may help clarify the ways in which such operations can be altered to accommodate instream values.

Glen Canyon Dam

The conflict between hydropower and downstream water uses is the focus of an ongoing evaluation of operating procedures at Glen Canyon Dam. The Dam is located in north central Arizona, just south of the border with Utah. The huge reservoir created by the dam, Lake Powell, extends for many miles into southern Utah and has a storage capacity of almost 25 million acre-feet, twice the average annual flow into the lake. Of particular interest here, Glen Canyon Dam is upstream from the Grand Canyon of the Colorado.

Glen Canyon Dam was authorized by the Colorado River Storage Project Act of 1956. Its operation is affected by the terms of a multitude of federal and state laws, court decisions, interstate compacts, and treaties collectively referred to as the Law of the River. The dam was designed primarily to help the four upper basin states of the Colorado River basin—Wyoming, Utah, Colorado, and New Mexico—meet their water delivery obligations to the three lower basin states—Arizona, Nevada, and California—by providing enough capacity to store large inflows and regulate outflows on the mainstem river.

The Colorado River Storage Project Act authorized Glen Canyon Dam for the purposes of regulating flows, water storage, reclamation, and flood control. Hydroelectric power generation was listed "as an incident of the foregoing purposes."[37] Reservoir operations have been conducted in the past to produce the greatest amount of firm capacity practicable while adhering to releases required under the Law of the River.[38] Generating capacity of 1,356 megawatts makes Glen Canyon more than an incidental supplier of electricity. Glen Canyon is by far the largest supplier of hydroelectric power in the upper Colorado River watershed, accounting for 75 percent of the total federal hydropower capacity of the projects authorized by the Colorado River Storage Projects Act of 1956. Hydropower operations at Glen Canyon Dam have produced about $55 million in revenue during minimum-release years, more in higher-release years.[39] The electricity is distributed to customers in Arizona, Colorado, Nevada, New Mexico, Utah, and Wyoming.[40]

The upper basin states are required to release a minimum of 8.23 million acre-feet of water per year from Glen Canyon Dam. The actual volume of water released is based on many factors, including forecasted inflow, existing storage levels, monthly storage targets, annual release requirements, efforts to reduce the risk of spilling, and the goal of equalizing storage between Lake Powell and Lake Mead downstream from the Grand Canyon. Releases for hydropower, fish and wildlife, and recreation are accommodated to the degree that they are not inconsistent with delivery obligations, water-supply needs, and flood control. Because power demand and power prices are highest during summer and winter, greater volumes of water are released during those months whenever possible.

The total volume of water released annually from Glen Canyon Dam is not much affected by efforts to maximize hydropower revenues, but hydropower concerns have been dominant with regard to the daily and weekly timing of reservoir releases. The power plant at Glen Canyon has been operated to produce peaking power, as noted by the Bureau:

> To the extent possible within higher priority operating constraints, the following guidelines are used in producing hydroelectric power: maximize water releases during the peak energy demand periods, generally Monday through Saturday between 7 A.M. and 11 P.M.; maximize water releases during peak energy demand months and minimize during low demand months; minimize and, to the extent possible, eliminate power plant bypasses.[41]

Flows have fluctuated widely in response to daily changes in power demand. Minimum and maximum flow constraints prior to 1992 were set at 1,000 cfs and 31,500 cfs, respectively, and there were no restrictions on "ramping rates," which are the rates at which the flow rate is changed. Fluctuating releases resulted in large, rapidly occurring changes in river stage.

This timing of releases to accommodate hydroelectric goals has been a source of great concern to downstream recreationists, environmentalists, fish and wildlife agencies, downstream Indian tribes, and others. Widely fluctuating flows have lowered the safety and quality of fishing, boating, and camping experiences and impaired aquatic and riparian habitat, cultural and archaeological sites, and endangered species. The role of streamflows in carrying sediment has been at the heart of many of these concerns. Fluctuating flows were thought to be detrimental to the creation and erosion of beaches, terraces, and sandbars along the river, which in turn affect fish and wildlife habitat, the availability of camping sites, and the preservation of archaeological and cultural artifacts.

These concerns led the Bureau in 1982 to initiate research into the impacts of water releases for power purposes on downstream resources. These studies were known as the Grand Canyon Environmental Studies (GCES). As a result of data from these studies and continuing political pressure, the secretary of the interior in 1989 ordered preparation of an Environmental Impact Statement (EIS) to evaluate and recommend alternatives for operating the dam.[42] The Bureau was specifically given the mandate to develop "a range of alternative Glen Canyon Dam operations designed to protect downstream resources and Native American interests in Glen and Grand Canyons, as well as to produce hydropower" and "to determine specific options that could be implemented to minimize—consistent with law—adverse impacts on the downstream environmental and cultural resources. . . ."[43]

This mandate was endorsed by Congress in the Grand Canyon Protection Act of 1992, passed as part of the same umbrella piece of legislation as the Central Utah Project Completion Act and Central Valley Project Improvement Act described above. Section 1802(a) of the Grand Canyon Protection Act requires the secretary of the interior to operate Glen Canyon Dam "in such manner as to protect, mitigate adverse impacts to, and improve the values for which Grand Canyon National Park and Glen Canyon National Recreational Area were established, including, but not limited to natural and cultural resources and visitor use."[44] The EIS was completed in March 1995.[45]

Decisions on the future operating procedures at Glen Canyon Dam will be made by the secretary of the interior. For the timebeing, the dam has been operated under the terms of interim operating criteria developed by the Bureau of Reclamation and Western Area Power Administration and implemented in November 1991. The criteria substantially changed previous patterns of water releases, placing new limits on maximum and minimum total daily releases, on total daily fluctuations in releases, and on ramping rates. Annual release rates were not changed.

Alternatives developed during the EIS process have been of three basic types: (1) "unrestricted fluctuating flows," including the no-action alternative (historical release rates) and a maximum power production alternative (which differs only slightly from the no-action alternative, making use of additional generating capacity installed in 1987 but not yet used); (2) "restricted fluctuating flows," designed to provide a range of downstream resource protection measures while offering varying amounts of flexibility for power operations (within constraints, maximum water releases would be scheduled for times of peak electrical demand); and (3) "steady flows," designed to provide a range of downstream resource protection measures by minimizing daily release fluctuations. Common to all alternatives are additional "flood frequency reduction" measures (reserving more reservoir capacity for flood control), beach/habitat-building flows (scheduled high releases beyond power-plant capacity but of short duration, to occur approximately once every five years), and emergency exception criteria.

The Bureau of Reclamation's preferred alternative is the "modified low fluctuating flow."[46] This alternative would reduce daily flow fluctuations well below those of the "no-action" level. It is designed to "provide high steady releases of short duration with the goal of protecting or enhancing downstream resources while allowing limited flexibility for power operations." It would not change the volume of annual releases from the dam and would retain essentially the same volume of monthly releases as the no-action alternative but would place strong limitations on daily fluctuations.

Changes to the historical operating procedures encompassed by the preferred alternative are shown in table 9.3. The preferred alternative specifies a sharply increased minimum release rate and moderately diminished maximum release rate. The rate at which flow rates could be changed—up and down ramps—was not previously constrained but under the preferred alternative would be limited to 4,000 cfs/hr on the up ramp and 1,500 cfs/hr on the down ramp. Compared to previous operating criteria, the preferred alternative would significantly restrict the production of peaking power.

The preferred alternative includes both "habitat-maintenance flows" and "habitat/beach-building flows" to compensate ecologically for the absence of high natural flows. Habitat-maintenance flows will consist of high, steady releases, still within power-plant capacity of 33,200 cfs, for one to two weeks in the spring. They are designed to rebuild sandbars, reduce accumulation of sediment at low elevations, and maintain camping beaches and return-current channels free of vegetation. Beach/habitat-building flows, mentioned earlier as a component of all proposed alternatives, are much higher—above power-plant capacity—and are scheduled to occur less often, approximately once every five to ten years. Habitat-maintenance flows would not occur during years in which beach/habitat-building flows are scheduled, and neither flow would occur in years when there was concern for endangered fish or other sensitive resources.[47]

TABLE 9.3
Historic and Proposed Operating Criteria for Glen Canyon Dam

Rate	Historic operations criteria	Modified low fluctuating flow (preferred) alternative
Minimum release rate	1,000 cfs Labor Day to Easter 3,000 cfs Easter to Labor Day	8,000 cfs 7 a.m.–7 p.m. 5,000 cfs 7 p.m.–7 a.m.
Maximum release rate	31,500 cfs	25,000 cfs
Maximum daily fluctuation	30,500 cfs Labor Day to Easter 28,500 cfs Easter to Labor Day	5,000 cfs low vol. months[a] 6,000 cfs med. vol. months 8,000 cfs high vol. months
Ramping rates	no restrictions	4,000 cfs/hr up 1,500 cfs/hr down
Beach/habitat-building flow	none	Greater than 33,200 cfs One every 5–10 years
Habitat maintenance flow	none	30,000–33,200 cfs One every year, except years in which beach/habitat-building flows are scheduled

[a]Low volume months are those in which less than 600,000 acre-feet are released from the dam; over 800,000 acre-feet are released during a high volume month, and an amount between the two extremes is released during a medium volume month.

The Bureau also intends to provide specific flows requested by researchers to aid in the study of the linkages between endangered fish, their habitat, and Colorado River flows. The Bureau has committed itself to implement additional changes in dam operations necessary to protect endangered species, identified during the course of the studies, to comply with the Endangered Species Act.

The first beach/habitat-building test flow was released from Glen Canyon Dam in March 1996. On March 22, flows were released into the river at the rate of 8,000 cfs for four days.[48] Starting March 26 flows were gradually increased to 45,000 cfs and held constant at that rate for seven days. The high flow was achieved by running all eight turbines at the Glen Canyon power plant at full throttle and opening all four jet tube outlets in the dam.[49] Starting April 2, flows were gradually reduced to 8,000 cfs in a manner designed to mimic the reduction of flow after a natural flood. The constant flow rates of 8,000 cfs before and after the flood were designed to permit aerial photography and on-site evaluation of sedimentation patterns and effects on other downstream resources.[50]

Over 100 scientists were stationed along the flood route between Glen Canyon and Lake Mead, including the length of the Grand Canyon, to monitor the flood. Full evaluation of the data collected during the flood will likely take years, but initial results seem promising. Initial reports indicate that the test flood was successful in stirring up sediments from the channel bed and redepositing the sediments to create beaches and sandbars and in creating backwater areas that scientists believe will be conducive to the propagation of the endangered humpback chub. Scientists expect that results from the data gathered during and after the test flood will be used to help determine future Glen Canyon dam releases and might well be applied to other dams elsewhere.[51]

The Bureau expects implementation of the preferred alternative to improve fish habitat, lead to the expansion of riparian vegetation and wildlife habitat, and improve the safety and quality of recreational experiences on the river below the dam. All of these improvements come at a substantial cost, however, including the cost of the studies themselves. A total of $5.3 million was spent on Phase I of the Grand Canyon Environmental Studies. The cost for Phase II of the GCES, scheduled to be completed by October 1995, was estimated to be $46.4 million, and the Bureau expected to spend an additional $7.3 million preparing the EIS. The cost of monitoring interim operations was expected to be a further $11.1 million.

The figures cited above do not include the cost of foregone power revenues resulting from decreased production of peaking power. The cost to the WAPA of replacing peaking power previously produced at the dam but not produced during Phase II of the GCES was $15.4 million. Replacement costs under the interim operating criteria were expected to total an additional $20.9 million through fiscal year 1995. Altogether, the cost of studies, monitoring, and the replacement of peaking power was expected to total $106.4 million by the end of fiscal year 1995.[52] Revenue foregone when water released from jet tubes bypassed power plant turbines during the beach/habitat-building test flow in March 1996 totaled an additional $1.5 million.[53]

Expenses associated with the adoption of new operating procedures will continue well beyond 1995. Implementation of the preferred alternative will significantly reduce the flexibility with which over 4 billion kilowatt hours of electricity per year are produced at Glen Canyon Dam. The preferred alternative will cause a reduction in energy production during peak demand times and consequent increase in production during nonpeak (base-load) demand times. The drop in production during peak demand times must be made up at other plants, especially at relatively expensive gas-fired plants.[54]

This will result in electricity rate increases. In addition, the change may create a regional need to build additional power plants and associated power lines 5 to 10 years sooner than would otherwise have been necessary.[55] Adoption of the preferred alternative is estimated by the Bureau to result in an economic cost to society of from $15 million to $44 million per year relative to adoption of the no-action alternative.[56]

The Grand Canyon Protection Act of 1992 states that "all costs of preparing the environmental impact statement . . . , including supporting studies, and the long-term monitoring programs and activities . . . shall be nonreimbursable." This means that taxpayers nationwide, rather than the consumers of electricity and water produced from the dam, will ultimately bear the costs.[57] The portion of the dam's capital costs attributable to fish, wildlife, and recreation, like flood control and certain other kinds of benefits, is also nonreimbursable. Because the proportion of the dam's remaining unpaid construction costs attributable to these purposes will be increased, the repayment obligation arising from other uses, including hydropower, will be reduced correspondingly.

The Grand Canyon Protection Act directs the secretary of energy, in consultation with the secretary of the interior, power and water customers, environmental organizations, and the Colorado River basin states, to "identify economically and technically feasible methods of replacing any power generation that is lost through adoption of long-term operational criteria for Glen Canyon Dam. . . ." The secretary is directed to investigate the feasibility of adjusting operations at Hoover Dam to replace all or part of such lost generation, and to present a report of the findings, and implement draft legislation, if necessary, within two years after adoption of long-term operating criteria.[58] To the degree that these replacement sources represent more expensive energy—which they probably will, because the most likely substitute for the peaking power capacity of hydropower plants is the more expensive gas-fired plants—the price of electricity will probably increase.

Opponents of the proposed restrictions on peaking power capacity at Glen Canyon Dam have sometimes argued that the restrictions will cause an increase in air pollution, because replacement power would have to come from polluting fossil-fuel plants. However, the Bureau's modeling efforts have predicted an improvement in air quality if the restrictions are adopted. This is because the shift of hydropower production to off-peak periods will displace an equivalent amount of production from existing base-load energy producers, which are usually coal-burning plants. Because coal-burning plants are more polluting than the gas-fired plants used to produce peaking power, the reduction in pollution resulting from lowered production at coal-burning plants will more than offset the increase in air pollution resulting from increased peaking power production at gas-fired plants.

In summary, hydropower operations at Glen Canyon Dam will be significantly constrained to produce greater instream benefits below the dam, and possibly some air-quality benefits. The trade-off has been, and will continue to be, many millions of dollars in increased electricity costs borne by homes, businesses, and other users in the service area and by taxpayers nationwide.

The Pacific Northwest Power Planning and Conservation Act

As at Glen Canyon, the operation of dams to produce electricity has been at the center of debates over the future management of water in the Columbia River and its tributaries. The conflict in the Columbia River watershed is usually characterized as "hydropower versus fish," or more informally as "volts vs. smolts." The fish referred to are anadromous salmon and steelhead. The federal government is heavily involved in this conflict, because, as described above, it is the owner, operator, and energy marketer of the largest and most significant dams on the mainstem Columbia and Snake rivers.

The Columbia is one of the most highly developed river systems in the world. Federal dams have transformed the river as well as the economy of the Pacific Northwest. Of the 1,214 miles of Columbia River from Bonneville Dam (the dam nearest the river's mouth) to the Canadian border, only 50 miles remain free-flowing.[59] The rest of the river has been turned into a series of enormous slackwater pools backed up by the dams. Dams on many of the tributaries have been built by nonfederal entities, but the federal role has been dominant on the mainstem Columbia and its major tributary, the Snake. There are currently eleven dams on the mainstem Columbia and ten on the Snake.

The construction and operation of these dams, in conjunction with heavy commercial fishing pressure (mainly at sea and in the lower reach of the river) and development in the upper reaches of the watershed, have devastated the annual runs of salmon and steelhead:

> From estimated runs of 10–16 million salmon and steelhead before non-Indians arrived, today's numbers are around 2.5 million (only about one-quarter of which are native stock). Entire stocks have been wiped out, and others are nearly gone. Dams make access difficult to much of the salmon's original spawning habitat. An estimated one-third to one-half of the habitat is now completely inaccessible to migrating fish. But the adult fish swimming upstream may have it easier than the smolts heading for the sea. The highly regulated flows in the Columbia River fail to flush those young fish at anywhere near the speed of the unharnessed river. It now takes young chinook salmon an additional forty to fifty days [relative to 1–2 weeks under natural conditions] to reach the ocean during low-flow years, increasing its susceptibility to predation and impeding its ability to convert from a freshwater to a saltwater environment.[60]

The conflict between hydropower and fish is unfortunate, as both fish and hydropower have been extremely important to residents of the Northwest. The federal government has not attempted to resolve the conflict on its own. Instead, Congress, through passage of the Pacific Northwest Power Planning and Conservation Act of 1980 (known more simply as the Northwest Planning Act, or NPA), authorized the creation of a regional entity to solve a whole range of issues associated with hydropower and fish in the Columbia River basin.

There were numerous motivations for the NPA. First among these was the desire to reduce conflict over the allocation of federal energy by the Bonneville Power Administration. After years of abundance, the Northwest was facing a potential energy shortage. BPA was producing and selling federal power at rates much lower than those that would be available from newer, alternative sources and consequently was deluged by customer attempts to procure a larger share of the power produced by the federal dams. The solution proposed by BPA and others was to give BPA the authority to purchase additional supplies, thereby expanding the available pool of "federal" power. Meanwhile, environmentalists, Native Americans, and others saw the looming crisis as an opportunity to address their own concerns with respect to the impact of federal hydropower facilities on fish. The resulting legislation accommodated both parties. The NPA gave broader authorities to BPA with respect to the acquisition and marketing of power, but also called for the creation of a regional program "to protect, mitigate and enhance the fish and wildlife."[61]

The means established for achieving these goals was creation of a regional entity known as the Northwest Power Planning Council.[62] The Council is composed of eight members, two appointed by each of the four governors of the affected states (Washington, Oregon, Idaho, and Montana). Its activities are financed by BPA, using revenues from the sale of electricity. The Council provides BPA with advice and direction on a number of different issues. In addition to its energy-related and public participation mandates, the Council was required to implement a program for meeting NPA's fish and wildlife goals. This program was to go beyond the typical admonition to "consider" fish and wildlife when making decisions about power generation. The NPA mandates that the Council and BPA give fish and wildlife equal *treatment*. Congress also required other federal agencies to cooperate with the Council's plans, at least to the extent that the agencies' own legal obligations would not be impaired. The Federal Energy Regulatory Commission, Corps of Engineers, Bureau of Reclamation, and other federal agencies must take the Council's fish and wildlife program into account to the "fullest extent practicable."

The fish and wildlife program implemented by the Council consists in large part of efforts to construct and operate fish passage facilities around dams, restore and improve fish habitat throughout the watershed, and operate hatcheries. More relevant here is NPA Section 4.(h)(6)(E)(ii), which calls for the Council to include in its program measures that will "provide flows of sufficient quality and quantity between such [hydroelectric] facilities to improve production, migration, and survival of such fish as necessary to meet sound biological objectives." These measures, essentially an enhanced instream flow known as the "water budget," were duly implemented by the Council.

The water budget makes additional water available during specific times of the year when it would be most useful in aiding salmon smolt migration. Each year the water budget is allocated a specific quantity of water, which is stored in upper basin reservoirs in Washington, Idaho, and Montana. In 1994, the budgeted amount was 3.4 million acre-feet, twice the amount allocated to the water budget just a few years previously.[63] The water is released at times and in quantities that are requested by fish and wildlife agencies from each of the four states, and by a collection of tribal fish and game agencies acting collectively as the Columbia River Inter-Tribal Fish Commission.[64] The requests of these agencies are coordinated by the Council's Fish Passage Center, formerly known as the Water Budget Center. Flow requests are based on extensive monitoring of fish migration in the watershed. The Council adopts, subject to some conditions, the measures proposed by the fish and wildlife managers, and then passes them along to the facility operators for implementation.[65] The extra flows are used to create stronger currents for pushing smolts downstream, sometimes all the way to the ocean but more often to collection sites from which the smolts are captured and barged to the river's mouth.

Requirements imposed by the Endangered Species Act have sometimes resulted in the release of water in quantities beyond that allocated to the water budget. The U.S. Fish and Wildlife Service's Biological Opinion on endangered species in the Columbia-Snake watershed, issued in January 1994, resulted in release of a "flow augmentation" of approximately 3 million acre-feet in 1994. This augmentation included spills—water goes over the spillway rather than through a dam's power-plant turbines—at certain facilities.[66] Low-water flow conditions resulting from the ongoing drought in the Northwest resulted in the release of an additional 10.5 million acre-feet later that year. Between 15 and 20 percent of the additional flow was spilled.[67] These additional flows are managed by all of the relevant agencies—BPA, the Corps, Bureau of Reclamation, National Marine Fisheries Service, U.S. Fish and Wildlife Service, state fish and wildlife agencies, and tribes. The agencies continually review dam operations and make adjustments throughout the year in a procedure known as the in-season management process.

The NPA and Endangered Species Act have given fish protection a higher priority in the management of federal reservoirs in the Columbia River watershed. Indeed, the operations of some federal projects in the watershed are now dedicated almost entirely to fish.[68] Because flood control continues to receive the highest priority, and releases for irrigation and navigation are largely fixed as a result of statutory and contractual obligations, the elevated priority of fish has come primarily at the expense of hydropower.

Fish protection costs have been enormous. Fish and wildlife programs have cost the BPA a total of $1.662 billion since implementation of the NPA in 1981.[69] Not all of this cost, by any means, has been for instream flow programs. Habitat improvement, monitoring, enforcement, trucking and barging fish, controlling predators, and research accounted for $320 million. Repayments to the federal treasury for construction of hatcheries and fish passage facilities, plus operations and maintenance of the facilities, cost $662 million. And expenditures for operating the Council have totaled $40 million. However, expenditures directly related to operation of the water budget and other augmented spills have totaled over $660 million, virtually all of it a result of reduced power revenues. Some of the reduction in revenues is due to spills; because water released through spillways does not go through power-plant turbines, it cannot be used to generate electricity. But most of the revenue reduction is due to the altered timing of flows. Electricity is most valuable in the summer and winter, when demand is highest and flows are lowest, whereas releases for the water budget and augmentation flows occur primarily in the spring, when flows are already high and the price for electricity is relatively low. Shifting releases from summer and winter to spring has necessitated the purchase of over $400 million worth of replacement power by the BPA.[70]

It is difficult to assess the effectiveness of the Council's fish recovery programs. Some have criticized extraordinarily large expenditures made on fish protection, particularly given the absence of effective monitoring programs that could be used to evaluate cost effectiveness:

> More than $150 million is spent annually on the recovery of the degraded salmon and steelhead runs in the Columbia River, yet a monitoring program that would enable the measurement of the major sources of mortality at key points in the river and ocean ecosystem does not exist. With little or no formal peer review, this spending constitutes well over twice the annual budget of the Environmental Biology Program at the National Science Foundation (NSF), which is the primary source of competitive funding for basic research in freshwater ecology in the United States.[71]

Runs of most salmon and steelhead stocks have continued to decline, though probably by less than would have occurred in the absence of the recovery programs. Protection, let alone recovery, of the remaining fish stocks will continue to require extraordinary effort. Drought may be the biggest obstacle in the near term, for there is not enough water storage in the system to sustain the recent pattern of releases on a regular basis. Drought conditions have persisted in the Northwest for eight of the last ten years, and without additional rainfall there will not be enough water in storage to be released for fish augmentation flows. Recent releases, together with other expenses related to the drought, have depleted BPA's financial resources. A contingency fund maintained by the agency is nearing exhaustion.[72] Given the wealth of factors affecting fish populations and recovery efforts, only time will tell whether salmon and steelhead will continue to be an important part of the Northwest.

Federal Regulation of Nonfederal Hydroelectric Facilities

The federal government is heavily involved in the regulation of hydropower facilities owned and operated by nonfederal entities. This involvement has its origins in the Federal Water Power Act of 1920, the terms of which were later incorporated within the Federal Power Act of 1935. The two pieces of legislation are collectively known as the Federal Power Act (FPA).[73] The FPA firmly embraced the many desirable characteristics of hydropower and established a national program to promote the development of hydroelectric facilities on the nation's rivers. The FPA also created a new federal agency, the Federal Power Commission, now the Federal Energy Regulatory Commission (FERC), to regulate all nonfederal power facilities. Virtually all nonfederal entities, whether public or private, wanting to construct and operate a hydroelectric facility are required to apply to FERC for a license. FERC is authorized to issue licenses for the construction and operation of hydropower facilities. Licenses, if issued, are valid for up to 50 years. When a license expires, the facility owner must apply for relicensing, following a process essentially the same as that for a new license. FERC is required to evaluate all sorts of factors during the licensing process, including environmental impacts of the proposed development. The effects of facility construction and operation on instream values are a part of these considerations.

The licensing process for a hydropower facility can be quite involved. The applicant is first required to share information about the proposed project with all relevant resource agencies, including agencies of affected states and Indian tribes. The applicant must then undertake studies to obtain information requested of the applicant by the resource agencies. Finally, the applicant must solicit the resource agencies' license recommendations. The application, together with supporting study documentation and agency recommendations, is then submitted to FERC. The process can be time consuming and expensive even for relatively small projects. For large projects, especially those involving multiple facilities, the process can take several years and cost millions of dollars.

Once submitted, FERC must evaluate applications for their suitability. Most of the criteria used in the evaluation process are concerned with engineering and other power-related features of the proposed projects, but FERC is also required to evaluate the effects of the proposed projects on nonpower concerns. In particular, Section 4(e) of the FPA, as amended by the Electric Consumers Protection Act of 1986, requires FERC to give conservation interests "equal consideration" with power and development interests in the agency's efforts to determine the public interest. Equal *consideration* is not the same as equal *treatment*—and wasn't intended to be—but this provision was intended to make FERC pay more attention to the environmental ramifications of its actions.[74]

Section 4(e) also requires FERC to adopt the recommendations of affected federal agencies if a project is located within a federal reservation, such as a national forest. FERC must adopt such recommendations, including recommendations to deny the license, without question in such situations.[75] This gives federal land managers substantial control over the development of nonfederal hydroelectric facilities on federal lands. On other lands, FERC must consider the recommendations made by federal, state, and tribal entities that may be affected by the proposed project. FERC must give "due weight" to the expertise of the agencies making the recommendations but is not necessarily obliged to follow the recommendations. However, FERC is required to state its reasons, in writing, for rejecting or modifying such recommendations. This process and statement provide the basis for subsequent legal challenges.

FPA Section 10(j) gives FERC authority to write conditions into licenses. Conditions are recommended by state and federal fish and wildlife agencies to adequately protect, mitigate damages to, and enhance fish and wildlife. Typically, conditions imposed by FERC include specified minimum flows that must be left in stream channels to ensure the protection of fish and wildlife. FERC conditions mandating minimum flows do not have the same strength as an instream flow right, because the bypassed water is not legally protected from appropriation by others. However, minimum flows mandated by FERC are important in ameliorating instream damages resulting from hydropower development.

Environmental resources, including instream flows, are potentially protected through FPA Section 10(a). Section 10(a) requires FERC to ensure that hydropower projects are "best adapted" to "comprehensive" plans for improving or developing a waterway. This provision does not refer to any particular plan, nor does it require FERC to develop such plans. Instead, FERC is required to consider any existing plans produced by state, federal, tribal, or other agencies that have jurisdiction over some part of the project or region. If these plans call for the protection of environmental values, Section 10(a) should result in greater protection of the environment from the potentially negative impacts of hydroelectric development.

There are approximately 2,350 conventional[76] hydroelectric facilities operating in the United States.[77] Of these, 1,825 (78 percent) are licensed by FERC, and the rest are either federally owned or privately owned but are exempt from licensing. The 1,825 plants licensed by FERC account for just under half of all developed hydropower capacity in the United States.[78]

The proportion of all plants licensed by FERC, and their corresponding proportion of total licensed capacity, is shown in table 9.4.[79] The high proportion of licenses issued from 1978 to 1985 reflects a boom in hydroelectric facility construction resulting from a renewed emphasis on the development of renewable energy sources. Most of the facilities licensed during that boom, however, were relatively small. Though comprising 54 percent of all licensed facilities, they represent only 22 percent of all licensed capacity. The largest facilities, for the most part, were built earlier. Note that the 28 percent of facilities licensed prior to 1978 account for 73 percent of total licensed generating capacity.

The environmental protection features written into the Federal Power Act imply a strong role for FERC in protecting the environment. However, FERC is more widely known for its aggressive promotion of hydropower development and its attempts to minimize instream flow requirements. This agency posture was born of the national attitudes existing at the time the agency and its programs were created. The clear national emphasis at that time was on the need to develop the nation's resources, including the rivers that had so much potential for generating electricity. The development orientation of the agency has been slow to change. The agency has been chided by the courts and by Congress on numerous occasions for failing to adequately consider nonpower concerns.[80] Passage of the Electric Consumers Protection Act in 1986, which added many of the environmental protection features discussed above, was in large measure a reaction by Congress to FERC's single-minded pursuit of development objectives.

The philosophy and actions of FERC are extremely important. As a federal agency operating under federal law, FERC can override conflicting state law. The authority of FERC vis-à-vis state agencies has been the subject of several major cases before the U.S. Supreme Court. The first of these, in 1946, was *First Iowa Hydroelectric Cooperative v. Federal Power Commission*.[81] *First Iowa* held that the Federal Power Act authorizes FERC to exercise exclusive jurisdiction over hydropower projects, even if exercise of the

TABLE 9.4
Proportion of Licenses Issued and of Licensed Capacity at Different Time Periods

License issue date	Percent of all licensed facilities	Percent of total licensed capacity
pre-1970	24	63
1970–1977	4	10
1978–1985	54	22
1986–1993	18	5
TOTAL	100	100

agency's authority were to entail the preemption of state laws. The case arose in Iowa, where a proposed hydroelectric project would have diverted water from the Cedar River and transported it several miles to a power plant above the Mississippi River. The Federal Power Commission denied the project a license because the project developer failed to obtain a permit from the state. The project developer appealed the denial to the courts, and the U.S. Supreme Court overturned the FPC's denial. The Supreme Court decision turned on an interpretation of FPA Section 9(b), which requires applicants to verify compliance with the requirements of state law. The court ruled that this requirement was an "informational provision" only, stating that to rule otherwise would give the state veto power over projects, thus creating an unworkable system of dual control.[82]

First Iowa was for many years the definitive case on federal-state relations in hydropower development. Its continued relevance came into question, however, after the Supreme Court *California v. United States* decision in 1978.[83] *California v. United States* involved the New Melones dam project built by the Bureau of Reclamation on the Stanislaus River. The Bureau had applied to the state for water rights sufficient to operate the project, as required by Section 8 of the federal Reclamation Act. But conditions on the permits subsequently issued by the California State Water Resources Control Board were considered unacceptable by the Bureau. The Bureau took its objections to the courts and received favorable rulings from the federal district and appellate courts. But these decisions were reversed by the U.S. Supreme Court, which ruled that the Bureau had been explicitly directed by Congress, in Section 8 of the Reclamation Act, to obtain all necessary permits from the state. This ruling seemed to be in direct conflict with the Supreme Court's ruling in *First Iowa,* because Section 9(b) of the Federal Power Act is virtually identical to Section 8 of the Reclamation Act.

Many observers expected this apparent contradiction to be settled in another case, *California v. Federal Energy Regulatory Commission.*[84] The central issue in this case, also known as the Rock Creek case, was the relative authorities of FERC and the state to set minimum flow levels for hydroelectric projects.[85] The Rock Creek Hydroelectric Project was designed to divert water from Rock Creek, a tributary of the American River in California, and return the water to the stream channel at a point located approximately one mile below the point of diversion. In 1983 FERC issued a license for the project that included a condition for the provision of a minimum flow of 11 cfs from May to September and 15 cfs from October through April to protect resident fish. These conditions were imposed on an interim basis pending completion of studies by the California Department of Fish and Game. In 1984 the State Water Resources Control Board issued water right permits for the project that adopted the same interim flow levels specified by FERC, though the Board reserved the opportunity to set permanent flow levels at a later date.

The State Water Resources Control Board in March 1987, following completion of the study by the Department of Fish and Game, set the permanent flow levels at much higher levels. Levels were set at 60 cfs March–June and 30 cfs July–February, because the agency had decided that the lower levels were not sufficient to protect the fish. At that point the project developer applied to FERC for an order declaring FERC's exclusive authority to set flow levels. FERC complied with this request, issuing its own order, also in March 1987, directing the project operator to comply with the lower flow levels set by FERC rather than the higher flow levels set by the state. The State of California filed a petition for review by the Ninth Circuit Court.

The Ninth Circuit Court issued a decision in June 1989 upholding FERC's exclusive authority to regulate water use at hydroelectric facilities, including the setting of minimum flow levels. The court stated that FERC, not the states, has the authority to set flow releases—that only FERC has the authority to do the necessary balancing between conflicting needs. If FERC's actions put the agency in conflict with the states, then the state agencies or statutes must give way. In justifying its decision with respect to the apparent contradiction between *First Iowa* and *California v. United States,* the court drew a distinction between the federal reclamation and the hydropower programs, arguing that they "address two different and entirely separate water use programs."[86]

The State of California, and many observers, found this explanation unconvincing. The state, supported by 43 other states that filed an *amicus curiae* brief, appealed the decision to the U.S. Supreme Court. The Supreme Court, however, affirmed the Ninth Circuit Court's decision, stating:

> As Congress directed in FPA sec. 10(a), FERC sets the conditions of the license, including the minimum stream flow, after considering which requirements would best protect wildlife and ensure that the project would be economically feasible, and thus further power development [citations omitted]. Allowing California to impose significantly higher minimum stream flow requirements would disturb and conflict with the balance embodied in the considered federal agency determination.[87]

Hydropower developers generally applauded this ruling, the court's most recent statement on the relationship between states and the federal government in the licensing of hydropower facilities and the setting of minimum flows. The hydropower industry favors the exclusive jurisdiction of FERC, at least in part on the grounds that state licensing would be impractical for facilities tied together in grids that supersede state boundaries.[88] There is no doubt, however, that the hydropower industry believes FERC is less sympathetic to minimum instream flow requirements than are the states, and prefers to deal solely with FERC for this reason. From the perspective of the hydropower industry, minimum flows, though often politically or environmentally necessary, raise costs. Minimum flow requirements lead to reduced power production, additional record keeping requirements, and increased costs for studies prior to the filing of license applications. Minimum continuous flow requirements may require that peaking plants essentially be converted to run-of-the-river projects, thus degrading the value of electricity sold by the projects.[89] The Rock Creek decision was viewed as a defeat by environmentalists, who in recent years have believed that state governments are often more sensitive to environmental problems resulting from power development than are some federal agencies.[90]

FERC's activities will continue to play a great role in the status of the West's instream flows. Hydropower facilities have had an enormous impact on flow regimes throughout the West. New facilities continue to be built, and there has been a recent surge in relicensing applications as numerous 50-year licenses issued in mid-century have started to expire. Environmentalists, recognizing the huge potential long-term impacts of FERC's activities on the environment, have succeeded in drawing renewed attention to the licensing process and in bringing pressure to bear on FERC. Many new members were appointed to the Commission following the change of presidential administrations in 1992, and the "new" Commission has recently been giving more deference to environmental goals—at least verbally. Environmentalists are encouraged by recent pronouncements from the Commission but remain cautious.[91] No one expects major changes in the agency to occur overnight.

CHAPTER TEN

Federal Environmental Protection Legislation and Programs Affecting Instream Flows

A variety of environmental features, from wildlife to vegetation to channel form, are dependent on streamflow. The federal government has played an active role in efforts to protect the environment, so it should not be surprising that environmental legislation has had a substantial impact on instream flows.

The relationship between environmental legislation and instream flows may be indirect. For example, the Endangered Species Act directs the Fish and Wildlife Service to protect certain threatened or endangered species from extinction. But because some of these species live in or rely on water for their survival, accomplishment of this mandate may require that a certain quantity of water be left in streams. The Constitution's supremacy clause gives precedence to federal laws if federal laws are in conflict with state laws, so the net effect of some legislated federal programs has sometimes been to force other water users to leave a certain amount of water in streams that otherwise would be diverted, just as if the federal government were exercising a right to use the water. Because the effects are so similar to those achieved through the exercise of water rights, some people now refer to the restrictions placed on other water uses by these federal programs as "regulatory property rights,"[1] though they are not truly property rights, or water rights, at all.

This chapter focuses on five pieces of federal legislation: the National Environmental Protection Act, the Clean Water Act, the Endangered Species

Act, the Wild and Scenic Rivers Act, and the Wilderness Act. All have had a substantial impact on instream flows.

The National Environmental Policy Act

The National Environmental Policy Act of 1969 (NEPA), signed into law January 1, 1970, by President Nixon, was one of the first major national pieces of environmental legislation.[2] NEPA modifies administrative procedure rather than mandating specific actions to protect the environment, but it is ubiquitous. The act applies to all major actions taken by federal agencies that have the potential to significantly affect the quality of the human environment.

NEPA made environmental protection a national policy goal. Section 2 of the Act states that the purposes of NEPA are:

> To declare a national policy which will encourage productive and enjoyable harmony between man and his environment; to promote efforts which will prevent or eliminate damage to the environment and biosphere and stimulate the health and welfare of man; to enrich the understanding of the ecological systems and natural resources important to the Nation; and to establish a Council on Environmental Quality.

NEPA requires federal agencies to consider the environmental effects of their proposed actions before actually performing those actions. "Action" is defined broadly to include such things as rule making and the adoption of plans, in addition to the more physical acts usually associated with that term. The requirements of NEPA have been extended to actions taken by nonfederal agencies or entities if those actions are federally funded, controlled, permitted, or approved. This broader interpretation brings a great many additional actions within the purview of the act.

NEPA requires federal agencies to inform the public about proposed actions and environmental effects and to consider comments and evidence submitted by the public. NEPA does not supplant existing agency mandates or statutes or in any other way change the primary missions of the federal agencies, but it does supplement those mandates.

NEPA requires all federal agencies to utilize a systematic, interdisciplinary approach in planning and decision making that affect the environment. The Act requires federal agencies to identify and develop procedures and methods that will ensure that unquantified environmental values are given appropriate consideration, and to prepare a detailed statement to accompany every recommendation for federal actions that significantly affect the environment. These detailed statements are of two primary kinds: Environmental Assessments (EAs) and Environmental Impact Statements (EISs).

Environmental Assessments

Environmental Assessments are used to determine whether a proposed project involves significant environmental effects. Preparation of an EA is less formal and rigorous than preparation of an EIS. The Council on Environmental Quality recommends that an EA not exceed ten to fifteen pages, though in practice this recommendation is often exceeded. The EA may be prepared by a consultant, or by the project applicant if the federal agency is involved as a licensor or permitting agent. The agency is required to review, assess, and verify statements prepared by other parties. The result of the EA preparation process is either a "finding of no significant impact" (FONSI) or a conclusion that a more comprehensive statement (an EIS) must be prepared. An agency may determine that the effects of a proposed action are significant and proceed straight to the preparation of an EIS, without first preparing an EA. Or an agency may mitigate the effects of a proposed action or otherwise change a proposal so that only an existing EA, rather than an additional EIS, will be required.

Environmental Impact Statements

Preparation of an Environmental Impact Statement can be, and usually is, a much more involved process.[3] It starts with agency issuance of a notice of intent, and a "scoping" process to identify key issues. This process usually involves a series of meetings that is open to the public and any other interested parties. The agency prepares and issues a draft EIS after the key issues have been identified. Following an appropriate period of time for public review of the draft, the agency then collects comments and responses from concerned agencies and the general public. NEPA requires that the "responsible federal official"—the agency proposing the action and/or preparing the EIS—consult with other federal agencies that have legal jurisdiction or special expertise relevant to the area or issue of interest. The comments from both the public and the consulting agencies are published in the final EIS, together with the lead agency's responses to the comments.

The EIS itself has an invariant format, because guidelines for the statements issued by the Council on Environmental Quality are binding on all federal agencies. An EIS must describe: (1) the environmental impact of the proposed action; (2) any adverse environmental effects that cannot be avoided should the proposal be implemented; (3) alternatives to the proposed action, including a no-action alternative; (4) the relationship between local short-term uses of the human environment and the maintenance and enhancement of long-term productivity; and (5) any irreversible and irretrievable commitments of resources that would be involved in the proposed action should it be implemented.

The agency considering a proposed action does not have to follow the advice or recommendations received during the commenting process. The agency is, however, required to respond to the comments, and the process enters evidence introduced by the public into the official record. The agency is required to prepare a "record of decision" (ROD), a concise statement issued by the agency to identify: (1) the alternatives that were considered when making its decision; (2) the environmentally preferred alternative; and (3) the agency's preferred alternative. The agency must identify and discuss all factors that were relevant in making the decision. The draft and final environmental impact statements and the ROD are distributed as widely as possible, so that information about federal actions affecting the environment is available to all interested parties. The processes mandated by NEPA allow the public to participate in the decision-making process and to monitor agency decisions, actions, and results.

Effectiveness of the Act

NEPA has had a real effect on administrative procedure. Agencies have put forth a considerable effort in compiling the required information and distributing it to the public. Virtually every federal agency has changed procedures to comply with the Act. As of the late 1980s, federal agencies were preparing approximately 10,000–20,000 environmental assessments annually, plus about 425 draft and final environmental impact statements.[4]

It is difficult to say exactly how effective this effort has been in protecting the environment. The processes mandated by NEPA have been successful in identifying the potential environmental impacts of proposed projects, and in involving the public in agency decision-making processes. However, NEPA does not mandate results that protect the environment nor require that agencies change their decisions. The Act merely requires that agencies follow a specified process in making decisions.

NEPA has been the subject of much criticism. The degree to which NEPA goals are accomplished varies by agency and by the type of project involved. Some observers have claimed that federal agencies treat the process as just another bureaucratic hoop that must be jumped rather than allowing the process to affect the way decisions are made. Many of the environmental statements prepared under NEPA have failed to live up to the promise of the Act, and the amount of public participation in the process has often been minimal.[5] NEPA requirements have been abused by those who wish to delay projects—perhaps permanently—for reasons that may or may not be related to environmental quality.

On the positive side, processes mandated by the Act have led to the accumulation of a substantial database on the effects of proposed projects on the environment.[6] This increases our knowledge about both environmental processes and their susceptibility to damage by human actions. NEPA procedures allow the public and other interested agencies and parties to introduce evidence about the effects of proposed projects into the agency decision-making process and to have this evidence included in the administrative record. Information developed during the preparation of environmental statements and the comments and evidence introduced by the public and other agencies are often considered by the agencies making decisions. And if an agency is taken to court, the EIS provides the administrative record used by the courts to determine whether a good-faith effort has been made to take environmental values into consideration during the decision-making process.

NEPA and Instream Flows

The NEPA does not directly address instream flow issues. However, to the extent that the environmental effects of a proposed action include changes to river or stream flow regimes, instream flow issues may be among those that receive greater attention from agency officials when making decisions. Heightened awareness may well have led to increased protection.

The establishment of a federal policy that seeks "harmony" with the environment, though not enforceable, may steer some agency decisions in the direction of leaving additional water within streams. It would be a mistake to place too much emphasis on this aspect of the act, however. Probably the greatest impact that NEPA has had on instream flows has been a result of the act's application in conjunction with more substantive environmental laws. For example, the preparation of environmental statements should raise and evaluate issues associated with the Endangered Species Act and Clean Water Act, discussed in greater detail below.

The Clean Water Act

The stated goal of the Clean Water Act is "to restore and maintain the chemical, physical, and biological integrity of the Nation's waters."[7] This includes providing for "the protection and propagation of fish, shellfish, and wildlife" and for "recreation in and on the water."[8] Though the act is clearly related to instream values—fish, wildlife, and recreation all use water within stream channels—it was aimed primarily at the improvement and protection of water quality rather than water quantity. For example, the heart of the Clean Water Act is Section 402, which establishes a system requiring polluters to obtain permits to discharge wastes. Though the type and quantity of pollutants discharged into the nation's waterways have an obvious impact on the uses that can be made of that water, Section 402 has not had an impact on the provision of instream flows per se. The situation is similar with other major sections of the Act, such as Section 208, which governs water-quality planning and the control of nonpoint sources of pollution. There are, however, three sections of the Act, Sections 404, 401, and 303, that have had a substantial impact on instream flows and are likely to have a larger impact in the future.

Section 404

Section 404(a) of the Clean Water Act authorizes the Army Corps of Engineers to issue permits for the discharge of dredged or fill material into navigable waters. Involvement of the Corps in water-quality regulation dates to the 1899 Rivers and Harbors Act, which made unlawful the excavation or placement of fill in navigable waters without authorization from the U.S. Army Corps of Engineers.

The Corps issues 404 permits within guidelines established by the Environmental Protection Agency (EPA). During the permitting process the Corps undertakes a comprehensive public interest study to determine whether the benefits of the proposed project outweigh the environmental costs. Permits may be issued for individual projects or may be issued on a statewide, regional, or nationwide basis for certain types of projects determined to have only a minimal impact on the environment. The U.S. Fish and Wildlife Service is required to comment on all individual and general permits issued by the Corps.

Amendments to the Clean Water Act in 1977 made it possible for the states to establish and administer permit programs for nonnavigable waters. The states must follow extensive guidelines set by the EPA, and the Corps remains in charge of navigable waters. The EPA retains authority to comment on permits issued by the states, as does the Corps. Significantly, Section 404(c) allows the EPA to veto a permit, whether issued by the Corps or the states, if the EPA believes the permitted discharge would have "an unacceptable adverse effect on municipal water supplies, shellfish beds and fishing areas (including spawning and breeding areas), wildlife, or recreational areas."

The scope of Section 404 is very broad. It applies to all "waters of the United States," which the Corps and the courts have interpreted to include wetlands and virtually any natural surface water body.[9] Regulations promulgated by the EPA define "unacceptable adverse effect" as "an impact on an aquatic or wetland ecosystem which is likely to result in significant degradation of municipal water supplies (including surface or ground water) or significant loss of or damage to fisheries, shellfishing, or wildlife habitat or recreation areas."[10] Section 404(c) states that "any discharge of dredged or fill material into the navigable waters incidental to any activity having as its purpose bringing an area of the navigable waters into a use to which it was not previously subject, where the flow or circulation of navigable waters may be impaired or the reach of such waters be reduced, shall be required to have a permit under this section." This means that Section 404 applies to any activity related to dredging or to the placement of fill or the erection of structures in a natural watercourse or wetlands—to virtually any water development whatsoever.

The potency of Section 404 may be seen in its application to two separate water development proposals, Wildcat Reservoir and Two Forks Dam, both located in the state of Colorado. The Endangered Species Act also played a role in the evaluation of the projects, but the focus here is on the Clean Water Act.

Wildcat Reservoir. Colorado's Riverside Irrigation District proposed construction of a dam on Wildcat Creek, a tributary of the South Platte River in eastern Colorado, in the late 1980s. Riverside applied for a nationwide permit to discharge sand and gravel as part of its construction activities. The Corps refused to issue the permit, largely because of the impact that operations of the completed dam would have on habitat for the endangered whooping crane downstream on the Platte River. The District took the Corps to court, where it argued that Section 404 applies only to water-quality impacts occurring as a result of the dredge and fill itself, not to downstream water-quantity impacts. But the 10th Circuit Court of Appeals, in *Riverside Irrigation District v. Andrews,*[11] strongly supported the Corps' decision. The court ruled that:

> both the statute and the regulations authorize the Corps to consider downstream effects of changes in water quantity as well as on-site changes in water quality in determining whether a proposed discharge qualifies for a nationwide permit. . . . The guidelines for determining compliance with section 404(b)(1) . . . require the permitting authority to consider factors related to water quantity, including the effects of the discharge on water velocity, current patterns, water circulation, and normal water fluctuations. Thus, the statute focuses not merely on water quality, but rather on all of the effects on the "aquatic environment" caused by replacing water with fill material.[12]

The *Riverside* ruling validated a potent extension of the scope of Section 404. However, the decision itself did not determine the outcome of this project, as Riverside District could still have applied for an individual project permit.[13] Whether such a permit would have been granted is not known, because the project was subsequently abandoned on other grounds.

Two Forks. Section 404 regulatory activities were central to the outcome of the proposed construction of Two Forks Dam. Two Forks was to be financed, built, and operated by the Denver Water Department and a consortium of approximately 40 suburban governments and special districts known as the Metropolitan Water Providers. The dam was to be built 24 miles upstream from Denver, just downstream of the South Platte's confluence with the North Fork of the South Platte. The resulting reservoir would have inundated the valleys of both streams, hence the name of the dam and project.

The construction of Two Forks had been contemplated since at least 1890. The Denver Water Department had long since established the necessary water rights (dating to 1902), purchased land that was in private ownership, and obtained rights of way for portions of the dam and reservoir on federal lands. As proposed, the dam would have been 615 feet high and would have stored over 1 million acre-feet of water. Most of the water stored in the reservoir was to come from across the continental divide, in the watershed of the Colorado River, delivered to the eastern side of the Rockies by way of a tunnel. The firm yield of the project was estimated to be 98,000 acre-feet of water per year.

The Corps became involved in the project when it was asked by the Denver Board of Water Commissioners to prepare a systemwide EIS. The EIS was in part required by an earlier agreement to study water-supply alternatives before constructing additional water projects, but it was also required as part of the 404 permit application process.[14] Work on the EIS started in October 1982 and was supposed to be finished by August 1984. Completion of the EIS actually took the better part of six years and cost the permit applicants almost $40 million, making the EIS one of the most expensive NEPA documents ever prepared.

Preparation of the EIS was a boon to the hundreds of consultants who worked on the document, but it was the focus of much controversy and strife. The entire Two Forks application process was at the center of a bitter fight, with both the merits and the demerits of the project hotly contested by environmentalists, developers, city officials, and others. Very much at issue were the existing instream uses of the river. The river supported a gold-medal trout fishery and was heavily used by kayakers and other recreationists. The proposed reservoir would have inundated habitat of a rare butterfly, the Pawnee montane skipper, as well as several homes and other structures. The draft EIS was released by the Corps in 1986, the final EIS in 1988.

The final EIS projected an annual water shortage of 166,000 acre-feet for the Denver metropolitan area by the year 2035 if no new water supplies were developed. This finding, together with a "no jeopardy" opinion issued by the U.S. Fish and Wildlife Service and the Corps' belief that any environmental damage from the project could be mitigated, was a key reason that the Corps decided to issue a 404 permit.[15] Another factor was that Colorado's governor, while not endorsing the project, chose not to oppose it; the Corps typically has not issued construction permits over the objection of the governor of a state in which a project is to be built.[16]

Construction would have soon followed the Corps' announcement in

January 1989 that it intended to issue the 404 permit had the EPA not intervened. The EPA was under considerable pressure from an array of environmental and recreational groups to review the permit and consented to do so pursuant to its authority under Section 404(c) of the Clean Water Act. The EPA had actually intervened earlier in the process, rejecting the Corps' draft EIS on the grounds that it was environmentally unacceptable and technically inadequate.[17] Staff at the EPA's regional office in Denver reviewed the subsequent, final EIS and also found that document inadequate. The staff reasoned that there were reasonable, less damaging alternatives available and recommended a veto to the EPA administrator.

The administrator did indeed veto the 404 permit. The veto was based on the administrator's judgment that the Two Forks project held the potential to significantly degrade the environment while doing little to replace the area's existing recreational value. The administrator specifically cited four grounds for his veto: (1) inundation of a prized trout stream; (2) elimination of a major recreation area around the dam and reservoir site; (3) destruction of valuable wildlife; and (4) the availability of less damaging alternatives.[18] Following eight years of formal study and $40 million in expenses, the Two Forks project was denied by application of Section 404.

Sections 303 and 401

The EPA's veto of Two Forks, a major water development project, was widely noted throughout the United States and sent shock waves through the West's water development community. By comparison, Sections 303 and 401 of the Clean Water Act, which govern the setting of water-quality standards and procedures for state certification of certain water developments, have received much less attention. These two sections may, however, eventually have more impact on instream flows.

The Water Quality Act of 1965 directed the states to classify their waters by intended use and to adopt other water-quality standards. This requirement was strengthened by amendments to the Act in 1972 and now comprises Section 303 of the Clean Water Act. Standards officially consist of two parts: "designated uses" and "water-quality criteria." Designated uses are the specific use or uses for which a given body of water must be protected. Examples include irrigation and livestock, domestic water supply, water contact recreation, and maintenance of cold-water or warm-water fisheries. Criteria are the characteristics of the water necessary to allow the designated uses to occur. Criteria may be either numeric or narrative. Numeric criteria represent the maximum or minimum concentrations of various constituents that could be present in the water that would still allow the designated uses to be met. For example, a minimum of 6.0 mg/l dissolved oxygen or a maximum of 0.05 mg/l arsenic. Narrative criteria may be necessary to describe water characteristics necessary to support uses such as preservation of the environment, or aesthetics. The EPA encourages states to include narrative as well as numeric criteria in water-quality standards.

The establishment of water-quality standards is the responsibility of the states, and the states have some latitude in setting both designated uses and water-quality criteria. However, the EPA is required to set standards for any state that fails to do so on its own. Section 303 also requires the states to establish antidegradation programs to prevent further degradation of water quality. Antidegradation programs are to be applied to any body of water already meeting or exceeding water-quality standards, and to waters of exceptional recreational or ecological significance that the states may designate as Outstanding National Resource Waters.

Standards set under Section 303, while protecting water quality for various instream uses, would, by themselves, have very little impact on the provision of adequate flows for those uses. But the Section 303 standards may be used in conjunction with Section 401 to have a very strong impact on instream flows.

Section 401 gives states the authority to certify water projects for compliance with state water-quality requirements. Section 401 applies to all water development activities that require a federal license or permit, including activities regulated by the Federal Energy Regulatory Commission under the Federal Power Act, or by the Army Corps of Engineers under Section 404 of the Clean Water Act. Virtually any activity involving the construction or operation of reclamation, hydropower, or other facilities that might result in a discharge into intrastate navigable waters or otherwise affect the quality of the aquatic environment is covered by Section 401. Presumably Section 401 also applies to special-use permits for water management projects located on federal land.

Regulations promulgated by the EPA require activities as well as discharges to comply with the standards. Section 401 states that "no license or permit shall be granted until the certification required by this section has been obtained or has been waived." A state may issue a certification that includes limits, monitoring requirements, standards, or "any other appropriate requirement of State law." These conditions must be incorporated within the relevant federal license or permit. State denial of a certification is binding on federal officials, though project applicants may—and often do—seek review of state actions in the state courts. Section 401 essentially gives states an opportunity to veto or condition projects, even those viewed favorably by the responsible federal agency, if the project requires a federal license or permit and would be detrimental to the achievement of state water-quality standards.

Despite the urgings of some commentators,[19] very little effort has been made by the states to apply Section 401 to the protection of minimum streamflows. This situation may now be changing, largely as the result of a 1994 ruling by the U.S. Supreme Court. In *PUD No. 1 of Jefferson County v. Washington Department of Ecology,*[20] the court firmly drew together the concepts of water quantity and quality protection under the single umbrella of Clean Water Act Sections 303 and 401. The case originated in the state of Washington, where a local utility district wanted to build a hydroelectric facility on the Dosewallips River in the Olympic National Forest just outside of Olympic National Park. The project was designed to divert water from the stream channel, route it to an offstream power plant, and then discharge the water back into the channel 1.2 miles below the diversion point. The Dosewallips River had been assigned water-quality "Class AA" by the state. Class AA is for waters of "extraordinary" quality, waters that markedly or uniformly exceed the requirements for virtually all uses, including fish migration, rearing, and spawning.[21]

The Washington State Department of Ecology issued a 401 certification imposing a minimum stream flow of 100–200 cfs, depending on the season, within the 1.2-mile reach. This flow level was designed to protect the habitat and needs of anadromous salmon and steelhead trout, but it was much higher than what the utility had anticipated. The utility took the state to court and won, but the Washington Supreme Court upheld the action of the Department of Ecology on appeal. The court ruled that antidegradation provisions of the state's water-quality standards required the imposition of minimum stream flows. The utility then appealed the case to the U.S. Supreme Court.

The U.S. Supreme Court held that a state may indeed impose conditions on 401 certifications as necessary to protect a designated use under state water-quality standards. The court specifically stated that Section 303 does not require states to protect such uses solely through implementation of water-quality *criteria,* as the utility had argued. The court stated that water development projects must be consistent with both water-quality criteria *and* designated uses. The court also stated that the action by the state of Washington was reasonable in light of the state's water-quality antidegradation policy, which required that existing instream uses be "maintained and protected." The court declared that the utility's assertion that the Clean Water Act is relevant only to water quality, not water quantity

> makes an artificial distinction, since a sufficient lowering of quantity could destroy all of a river's designated uses, and since the act recognizes that reduced stream flow can constitute water pollution. . . . In many cases, water quantity is closely related to water quality; a sufficient lowering of water quantity in a body of water could destroy all of its designated uses, be it for drinking water, recreation, navigation or, as here, as a fishery. In any event, there is recognition in the Clean Water Act itself that reduced stream flow, i.e. diminishment of water quantity, can constitute water pollution.[22]

The ruling should be equally applicable to recreation, aesthetics, and other instream uses, as explicitly mentioned in the decision, so long as these uses appear within a state's water-quality standards. Unlike the *First Iowa* and *Rock Creek* cases discussed earlier in conjunction with the federal regulation of private hydropower facilities, the conflict in this case was between two federal statutes—the Clean Water Act and the Federal Power Act—rather than between conflicting state and federal policies. The state action was allowed to stand because the actions of the state were authorized by the federal Clean Water Act rather than solely by state law. The opinion, written by Justice O'Connor, was signed by six other justices.[23] The decision represents strong court support for the authority of the states to implement minimum flow standards on hydropower and other facilities through use of Section 401.

Effectiveness of the Clean Water Act in Protecting Instream Flows

From the discussion above, it appears that these three sections of the Clean Water Act provide comprehensive protection of instream uses and values, even in the absence of other protection programs. However, this is usually not the case. The authorities described above, although very real and potent, are not consistently applied, have numerous loopholes and exceptions, and are not as pervasive as might first appear.

For example, Section 404(f) specifies numerous exemptions to the 404 permitting process. Permits are not required for whole categories of activities with the potential to substantially alter instream flows. Exemptions are made for normal farming, ranching, and logging activities; construction or maintenance of farm or stock ponds or irrigation ditches; maintenance of structures such as dams, levees, and bridges; use of temporary sedimentation basins at construction sites; and construction of roads for farming or forestry purposes. This array of activities may have significant effects on the instream use of water, but it isn't covered under Section 404. Perhaps the largest exemption to Section 404 is the hundreds of thousands of water projects that were already in existence when it became law; existing projects are not required to have permits.

General permits issued on a state-, region-, or nationwide basis allow many other activities to escape significant regulation under Section 404. Such permits are supposed to be issued only in situations in which the activities would have minimal environmental effects, but the Corps lacks the administrative resources necessary to adequately review all of the applications that it receives and so must rely to a large extent on the data supplied by permit applicants. Though the Riverside case may give the contrary impression, the Corps issues permits to most applicants. Some have argued that the relatively stringent review requirements imposed by the EPA are applied more as the exception than as the rule.[24] Enforcement of permit conditions is weak, because the Corps does little surveillance or investigation once permits have been issued. Civil and criminal remedies specified in the law are rarely applied.[25]

It is too early to assess the impact that Section 401 will have on instream flow protection. Clearly the potential is very large. The Supreme Court's decision in *PUD No. 1 of Jefferson County v. Washington Department of Ecology* was widely noted and reported throughout the West. All of the states have water-quality standards and could use them for 401 certifications if they chose to do so. But will they choose to do so? There is significant variation in the degree to which states set and enforce standards and monitor water-quality programs.[26] The states hadn't done much to utilize their 401 certification authorities prior to this decision,[27] and very few states had designated state waters as Outstanding National Resource Waters under the antidegradation requirements of Section 303.[28]

Other factors may prevent wide application of Section 401. The Clean Water Act declares that certification of pending applications will be waived if a state "fails or refuses to act on a request for certification, within a reasonable period of time (which shall not exceed one year) after receipt of such request." In practice, applications have frequently been waived, whether due to lack of administrative resources or to lack of interest. Few states have been reviewing permits, let alone denying them, and there has been so little activity under Section 401 that the EPA hasn't even maintained systematic records or other information on state 401 programs.[29] The Federal Energy Regulatory Commission is working to determine whether the agency is required to accept conditions that are not, in the agency's opinion, strongly enough related to water quality. There is the possibility that Congress will weaken state authority under Section 401 if the states take action deemed detrimental to national interests.[30] The stance of the EPA is another unknown. The agency has previously been reluctant to deal with water-quantity issues, both because the subject seemed removed from the agency's mission and because of the potential for strong opposition. The EPA did not immediately issue any comment on the Supreme Court decision.[31]

In summary, provisions of the Clean Water Act that have the potential to affect instream flows have not yet been consistently or uniformly applied. When applied, they have been potent. The recent Supreme Court 401 decision may well stimulate new interest in the Clean Water Act as a tool to protect minimum flows and may eventually prove to be one of the most significant developments in the history of federal and state efforts to protect instream flows.

The Endangered Species Act

Wildlife policy in the United States has primarily been left to the states. States even have authority to manage wildlife on federal lands, notwithstanding the obvious impact of federal land management activities on wildlife. Nevertheless, the federal government now plays a substantial role in wildlife and fish management, especially in its enforcement of the Endangered Species Act.

The Endangered Species Act (ESA), passed with almost no opposition and signed by President Nixon in 1973 during the height of the environmental movement, is one of the strongest pieces of environmental legislation ever enacted by the United States Congress.[32] The ESA was intended to preserve biodiversity by protecting individual species from extinction. Many of the species protected under the ESA are full- or part-time inhabitants of aquatic environments, so the ESA has had an important impact on instream flows.

The ESA requires federal agencies to identify species that are susceptible to extinction and to perform actions necessary to protect and enhance remaining populations of these species and their habitats. The ESA extends to more than just wildlife, as all plant and animal species, excluding only certain insects considered pests, are eligible for consideration under the ESA. Specific provisions of the ESA that have the potential to affect instream flows include: (1) the listing of species; (2) prohibitions on the taking of a listed species; (3) the prohibition of agency actions that might jeopardize endangered species; and (4) agency consultation requirements.

Listing of Species

The ESA authorizes the U.S. Fish and Wildlife Service and the National Marine Fisheries Service to identify and list species threatened with extinction. The Marine Fisheries Service has responsibility for most marine species, including salmon. The Fish and Wildlife Service has responsibility for all other plants and animals, by far the largest share of responsibility under the ESA. Species that the agencies find to be in imminent danger of extinction throughout all or a significant portion of their range are listed as "endangered." Species likely to become endangered in the foreseeable future are listed as "threatened." The ESA requires that criteria for listing species be based solely on biological evidence and the best scientific and commercial data available. Economic or political considerations are not supposed to influence the decision to list a species. Additions or deletions to the list of threatened or endangered species may be proposed by any person or organization that presents adequate evidence about the status of a species. Once listed, there is little difference in the management of species listed as threatened and those listed as endangered.

The two agencies had listed approximately 500 U.S. species as threatened or endangered as of 1994.[33] Over 3,000 additional species had been identified for possible listing.[34] Amendments to the ESA in 1978 required that additions to the lists of threatened and endangered species be accompanied by a listing of the species' "critical habitats." Though subsequent amendments eliminated the requirement to list critical habitat at the same time that a species is listed, the two agencies are still authorized to list critical habitats and have done so for several of the threatened and endangered species.

Prohibition of Takings—Section 9

ESA Section 9 prohibits the "taking" of listed species by any individual or entity under the jurisdiction of the United States. The word *take* is defined broadly, to include "harass, harm, pursue, hunt, shoot, wound, kill, trap, capture, or collect, or attempt to engage in such conduct." The word *harm* has been interpreted in regulations as "an act which actually kills or injures wildlife. Such act may include significant habitat modification or degradation where it actually kills or injures wildlife by significantly impairing essential behavioral patterns, including breeding, feeding, or sheltering."[35] Essentially, no one is allowed to pursue any action that might harm a listed species, even if the harm is but an indirect result of the action. For example, operating a reservoir in such manner that the habitat used by an endangered species is lost or degraded is prohibited under Section 9. Exceptions are possible, primarily for species that are part of officially designated "experimental populations" or for takings designated as "incidental."

Protection by Federal Agencies—Section 7

ESA Section 7 has had the most impact to date. Section 7 states that:

> The Secretary [of the Interior] shall review other programs administered by him and utilize such programs in furtherance of the purposes of this chapter. All other Federal departments and agencies shall, in consultation with and with the assistance of the Secretary, utilize their authorities in furtherance of the purposes of this chapter by carrying out programs for the conservation of endangered species and threatened species listed pursuant to section 1533 of this title and by taking such action necessary to insure that actions authorized, funded, or carried out by them do not jeopardize the continued existence of such endangered species which is determined by the Secretary, after consultation as appropriate with the affected States, to be critical.[36]

Section 7 is more powerful than originally intended. The section was very short—the quote above encompasses virtually the entire section—and drew little attention during the congressional debate preceding enactment of the ESA. The strength of the provision became evident only after the U.S. Supreme Court's snail darter decision. Listing of the snail darter, a small minnow found in a Tennessee river, led to an injunction prohibiting further construction of a dam that would have caused the species' only known habitat to be inundated. At the time, the dam—a project being constructed by the federal Tennessee Valley Authority at a cost of well over $100 million—was nearly complete. The U.S. Supreme Court, in *Tennessee Valley Authority v. Hill,* ruled that the provisions of ESA Section 7 were mandatory, not discretionary, and upheld the injunction.[37]

The Supreme Court decided that Section 7 absolutely prohibited a federal agency from undertaking any activity that would directly or indirectly jeopardize the continued existence of an endangered species or its listed habitat. Citing the ESA's legislative history, the court emphasized that

> this subsection *requires* the Secretary and the heads of all other Federal departments and agencies to use their authorities in order to carry out programs for the protection of endangered species, and it further *requires* that those agencies take *the necessary action* that will *not jeopardize* the continuing existence of endangered species or result in the destruction of critical habitat of those species.[38]

The Supreme Court concluded that Congress intended protection of endangered species to take precedence over even the primary missions of the federal agencies. The import of this decision was enormous. For example, in a 1982 Nevada case a federal court ruled, following *TVA v. Hill,* that the secretary of the interior has the responsibility, under the ESA, to operate an existing reservoir for the purpose of protecting endangered and threatened species, even if such operation leaves no water for the other purposes for which the reservoir was built. The court stated that the secretary had a duty to defer all other uses until the listed species were no longer classified as threatened or endangered.[39]

The impact of Section 7 on water policy in the western states has been enormous. Most water development projects in the West not undertaken directly by the federal government receive at least some federal funding, or require at least a license or permit from a federal agency. Congress chose not to alter the language of Section 7 when it passed amendments to the ESA in 1978, and thereby implicitly ratified the Supreme Court's interpretation.[40] Congress did, however, create a process for agencies to seek exemption from the requirements of Section 7. An exemption was first sought—and rejected—in the snail darter case. Few agencies have used the process, and, at least to date, none has done so successfully.[41]

A key provision of Section 7 is the requirement that a federal agency consult with the secretary of the interior to determine whether actions proposed by the agency will result in jeopardy to a listed species.[42] Until consultations have occurred, no "irreversible" or "irretrievable" commitment of agency resources to a project may be made. For all formal consultations, the Fish and Wildlife Service or Marine Fisheries Service must provide a written biological opinion detailing the finding on which it has based its decision.

Most consultations under the ESA are quick and relatively easy. Others take months, even years. In a formal consultation the agency proposing the activity, license, or permit first requests information from the Fish and Wildlife Service or Marine Fisheries Service as to whether an endangered species might be present in the affected area. If a listed species is present, a "biological assessment" is prepared to identify possible negative impacts. If the assessment indicates potential adverse effects, the Fish and Wildlife Service or Fisheries Service prepares a "biological opinion" to determine whether the proposed project would in fact put the listed species in jeopardy. The ESA requires that all biological assessments and opinions be based on the best scientific and commercial data available. A variety of conclusions are possible, ranging from no jeopardy to total jeopardy. The ESA requires the secretary to "suggest reasonable and prudent alternatives" to the proposed course of action if a listed species would be jeopardized by the proposed action. Biological opinions therefore often recommend modifications to projects or suggest ways in which negative impacts could be mitigated. The consultation process may be ongoing, because of the inherent difficulties in assessing impacts for large projects all at once.

The formal consultation process can be long and cumbersome. Agencies avoid the formal process by engaging in an informal consultation with the wildlife agencies. The informal process usually involves some verbal communication between the development agency and the relevant wildlife agency, during which the wildlife agency may quickly determine that the proposed action is not likely to jeopardize a listed species. Or the wildlife agency may suggest changes that would accommodate the species. Once both agencies are satisfied, the communication is followed by the submission of appropriate documentation, essentially a letter. This process is often employed for small projects likely to have minimal or nonexistent impacts on listed species and helps to conserve the administrative resources of both agencies.

Implementation

The language of the ESA, particularly that of Section 7, is clear and forceful. Its impact on wildlife depends, however, on adequate implementation and enforcement. The degree to which this level of implementation and enforcement occurs is subject to some debate.[43] There are many reasons for incomplete implementation. Among the most frequently mentioned are a shortage of agency resources, political infeasibility, and political interference in the listing, consultation, and enforcement processes. Budgets of the Fish and Wildlife Service and Marine Fisheries Service are not adequate to comprehensively identify and list all species that might be in danger of extinction.[44] Species spend months or years on the waiting list before a determination of their status is made; other possibly endangered species do not even make it on to the waiting list; and the listing of critical habitats lags far behind the listing of species. In the face of a rising number of requests for consultation, the agencies have increasingly relied on informal, rather than formal, consultation processes. One critic noted that of the 28,000 informal consultations conducted by the Fish and Wildlife Service in 1990, only about 600 to 700 went to formal consultation. Less than 1 percent resulted in jeopardy opinions, and only two halted projects.[45]

Politics often interfere in the implementation and enforcement of the ESA. For example, the rate at which new species were listed as threatened or endangered dropped dramatically in the 1980s, during the tenure of an administration with less sympathy for environmental regulation. As recently witnessed with the controversy over listing of the northern spotted owl in the Pacific Northwest, the agencies assigned to implement and enforce the ESA are often faced with substantial pressure to modify their actions. A former director of the Fish and Wildlife Service noted, "The administrator of the act is always conscious of the possibility of powerful forces effecting a change in the act that may have shattering future consequences. One is always aware of the need to strike a balance, to proceed with caution, to accommodate today in order to prevail tomorrow."[46]

The reality of less than total implementation and enforcement should not obscure the fact that the ESA is very powerful and can be successful in meeting its stated objectives. One supporter of the ESA stated, "The strength of the Endangered Species Act has always been in modifying developmental activities, not in stopping them. And to the extent polls show that moderating negative development is precisely what the public supports, the law has been highly successful."[47] Development agencies have recognized the delay and controversy likely to result from their failure to adequately consider the impact of their actions on listed species and have been forced to modify their actions.[48] Or, as amusingly summarized by another commentator, developers have realized that "the project must be shaped to suit the requirements of the listed wildlife or the project, not the wildlife, will be dead in its tracts."[49]

The result can be an accommodation that meets the needs of all parties. Developers recognize the potential for delay; the administrative agencies fear political controversy; and environmentalists use provisions of the ESA sparingly out of fear of congressional backlash. "As a result, all three groups have the incentive to seek creative solutions that accommodate preservation and development concerns."[50] Even the direct intrusion of political considerations into the ESA listing and implementation processes should not necessarily be considered a complete loss:

> Political considerations clearly are present in the consultation process. Indeed, it would be surprising if they were not. . . . Rather than bemoaning

> the fact that politics enters into endangered species decision making, we should recognize the realities of the situation and work to exploit the benefits of political inputs—as sources of information about collective values and how intensely they are held—and minimize the negative effects of such forces on species preservation.[51]

Impact of the Endangered Species Act on Instream Flows

The ESA contains other provisions designed to prevent the extinction of species,[52] but those described above have had the largest impact on instream flow. Though much of the attention and controversy has focused on land-based species such as the grizzly bear, gray wolf, and northern spotted owl, the ESA has had a substantial impact on water-based species. Listed species such as the Snake River sockeye and chinook salmon, humpback chub, Colorado squawfish, whooping crane, least tern, and piping plover are dependent on characteristics of their aquatic habitat formed by particular regimes of river flow quantity and timing—in short, by instream flows. Protection and enhancement activities under the ESA have had a substantial impact on instream flow policies in the West, as evidenced by the following three examples.

Snake River Sockeye and Chinook. ESA management activities for several listed salmon species in the Columbia and Snake river watersheds are at the heart of a controversy surrounding the timing and quantity of flows released into the rivers of the northwest United States. The region has experienced a precipitous decline in the number of salmon using the watershed. Several runs of salmon have become extinct. There are numerous causes of these population declines: overfishing in rivers and oceans, degradation of spawning habitat as a result of a variety of land uses in upstream watersheds, the incidental diversion of salmon along with water into canals and other irrigation works, and the construction and operation of dozens of dams on the mainstem rivers and tributaries.

Some of the dams, such as the Grand Coulee Dam on the mainstem Columbia River in Washington, and the Hell's Canyon, Oxbow, and Brownlee dams on the Snake River along the border between Idaho and Oregon, were constructed without any fish passage facilities whatsoever, thereby eradicating all salmon runs that formerly used rivers and streams above the dams. Other dams were built with fish passage facilities but are still hazardous to fish. The facilities pose more of an obstacle to adult fish migrating upstream than do natural channels, and juvenile fish migrating downstream experience higher mortality when navigating the immense slackwater pools behind the dams or when run through the turbines of hydroelectric generating plants.

It is interesting to speculate whether these dams could have been built had environmental regulations like those currently in place been in effect at the time the dams were authorized. Now that the dams are in place, the issue is how to mitigate adverse impacts. At stake is the survival of at least three salmon runs. The Snake River sockeye salmon was listed by the National Marine Fisheries Service as an endangered species in April 1991, and the Snake River spring-summer chinook and fall chinook were listed as threatened in March 1992. The Fisheries Service is required to issue a biological opinion on the operations of eight large dams on the mainstem Columbia and lower Snake rivers, four dams on each river. Even the operators of the dams—the Army Corps of Engineers and the Bonneville Power Administration, both federal agencies—have acknowledged that current operations of the dams would kill 88 percent of migrating juvenile salmon and 66 percent of adults returning to spawn.[53] Nonetheless, the Fisheries Service originally issued a "no jeopardy" opinion for the eight dams. This caused an uproar among those wishing to protect the fish, followed by "lawsuits from all quarters."[54]

The courts usually grant broad latitude to the wildlife agencies in writing their biological assessments and opinions, deferring to the expertise of the agencies.[55] But a federal district court ruled in March 1994 that the Marine Fisheries Service's "no jeopardy" opinion for the eight dams was "arbitrary and capricious" and hence unacceptable. As a result, the Fisheries Service was forced to take action to protect the listed species. The Fisheries Service ordered water (and salmon) to be spilled over the dams starting May 10 and continuing through June 20 in an effort to aid migrating smolts.[56] In addition to bypassing the hydroelectric turbines, this pattern of increased spills—essentially a temporarily enhanced instream flow—was intended to create a stronger current that would help migrating smolts find their way through the reservoirs behind the dams. The enhanced instream flow was designed as an interim measure while the agency continued to study this and other, longer-term options for protecting fish. Such potential measures include a requirement for additional "flushing flows" drawn from reservoirs on the upper Snake River during the spring and the modification of dam operations and facilities on the mainstem dams downriver.[57]

The idea of using water stored behind dams in Idaho to help flush smolts to the sea is unpopular with many groups in Idaho, including the state legislature. Members of both houses of the Idaho legislature passed a resolution in 1996 stating that it is not the state's policy "to recognize as valid the concept of flow augmentation and the drawing down of Idaho's waterways and reservoirs for the purpose of Snake River Basin species recovery programs."[58] Opinions are divided, however. Just a month earlier Idaho's governor had announced a species recovery plan that would involve just such reservoir releases to aid the salmon.[59]

Challenges are still being made to the adequacy of federal efforts to protect Idaho's salmon. Eight conservation groups filed suit in U.S. District Court in Portland, Oregon, contending that the National Marine Fisheries Service, Army Corps of Engineers, and Bureau of Reclamation illegally failed to meet flow targets on the Snake and Columbia rivers, causing migrating juvenile salmon and steelhead to die.[60] The groups also alleged that the agencies had not yet implemented court-ordered long-term structural changes to federal dams on the two rivers.

Endangered Fish in the Colorado River. The ESA is also very much at the heart of an instream flow controversy on the Colorado River. At issue on the Colorado is the continued existence of four fish species, the razorback sucker, Colorado squawfish, humpback chub, and bonytail chub. The four species are faced by a variety of threats, including changes to their habitat resulting from the construction of several large dams, diminishment of flows due to numerous diversions, and the introduction of exotic sport-fish species such as trout.

For several years the Fish and Wildlife Service maintained that any water diversion from the upper Colorado would likely jeopardize the endangered species. Under great political pressure from development interests, the Service eventually shifted to a policy by which it would issue "no jeopardy" decisions in exchange for contributions to a fund that would be used to finance continuing impact studies and a handful of conservation measures. This shift in policy allowed development of Windy Gap, a project that transports Colorado River basin water to cities along Colorado's Front Range and an additional 30 water development projects. The Windy Gap policy drew criticism from wildlife interests.[61] In 1987, after three years of negotiation involving the Fish and Wildlife Service, Bureau of Reclamation, developers, environmentalists, and officials from the states of Colorado, Utah, and Wyoming, a consensus on a new plan was reached. The plan included a variety of measures designed to protect the listed species, including acquisition of instream flows, additional water releases from federal reservoirs, habitat improvements, and construction of fish passage facilities. The plan also called for the creation of a fund financed by one-time contributions from developers to be used to support continued fish recovery efforts.

Other actions to protect the four species throughout the watershed continue. In April 1994, the Fish and Wildlife Service issued its Final Rule on Critical Habitat Designation for the four fish. The agency designated 1,980 miles of critical habitat for the four species, including significant portions of the mainstem Colorado River and its tributaries in the states of Colorado, Utah, New Mexico, Arizona, Nevada, and California.[62] This designation will force modifications or denial of all development projects that have the potential to adversely affect the listed habitat or otherwise endanger the four resident fish species, which means that this measure too is likely to have a major impact on instream flows within the Colorado River watershed.

The Platte River. Another prominent example of the influence of the ESA on instream flows is the Platte River. Prior to its development, the Platte in Nebraska was said to flow "a mile wide and an inch deep," which was only a slight exaggeration. Most flow in the Platte occurred in the spring and early summer as a result of snowmelt in the river's high mountain tributaries in Wyoming (the North Platte) and Colorado (the South Platte and the uppermost reach of the North Platte). These wide, shallow, flushing flows created long stretches of sandbars and wetlands and, due to the widely shifting pattern of currents and channels from year to year, kept them largely clear of permanent vegetation. These conditions proved ideal for several wildlife species, primarily birds, which used the sandbars for nesting and as critical stopover points during migration.

Water development along the Platte and its tributaries drastically altered this environment. Over 70 percent of the river's natural flows are diverted from the river, and it is now unusual to find the river more than a couple of hundred yards wide at any given point.[63] Most dramatically, the river channel is now, in many places, lined and/or choked with vegetation, including large trees such as cottonwoods and willows. The absence of clear sandbars surrounded by water and unobstructed by vegetation has made the wildlife species using those sandbars vulnerable to predators. Two of these species, the whooping crane and least tern, have been listed as endangered.[64] A third species, the piping plover, has been listed as threatened. A fish species, the pallid sturgeon, which lives farther downstream in the Platte near the confluence with the Missouri, is listed as endangered.

Listing of these species has the potential to affect water users, primarily irrigators and municipalities, throughout the Platte River watershed. The Fish and Wildlife Service has stated its intention to take measures necessary to ensure that the existing habitat remains suitable for these species and is well aware of the historical importance of large springtime flows in the removal of vegetation and creation of sandbars. Options discussed so far include the possibility of using reservoir releases to simulate the large springtime water flows and limiting existing diversions to increase the total volume of water flowing in the river.[65] Other options, already being used to some extent, include the purchase and protection of remaining habitat along the river and habitat improvement projects utilizing people and machinery to remove vegetation.

Federal and state officials are still trying to put together a basin-wide recovery program. The federal government has made it clear that the program is likely to involve increased flows, particularly in the spring. These recommended flows would come primarily from existing reservoirs operating under federal permits in the three basin states. As an interim measure, the

Fish and Wildlife Service is using payments from project operators to fund habitat improvement projects and purchase additional water to maintain target flows for a 58-mile segment of the Platte River in Nebraska.

Concluding Comments

The ESA is a potent piece of legislation containing strong language regarding the protection of individual species. The act is often used in combination with other environmental laws, including the National Environmental Policy Act and Clean Water Act. Though implementation of the ESA has been tempered by political considerations and a shortage of agency resources, the ESA has proven to be a powerful tool in protecting the environment, including the provision of instream flows.

Wild and Scenic Rivers Program

The sheer size of the national reclamation and hydropower programs provides ample evidence of the value placed on river development by the federal government. The government has also taken steps to protect rivers from development. The most conspicuous of these efforts, and one specifically designed to balance the dominant policy of development with a countervailing policy of preservation, is the designation of rivers as components of the national Wild and Scenic Rivers Program.

Policy Statement

The Wild and Scenic Rivers Act of 1968 reads in part:

> It is hereby declared to be the policy of the United States that certain selected rivers of the Nation which, with their immediate environments, possess outstandingly remarkable scenic, recreational, geologic, fish and wildlife, historic, cultural, or other similar values, shall be preserved in free-flowing condition, and that they and their immediate environments shall be protected for the benefit and enjoyment of present and future generations. The Congress declares that the established national policy of dams and other construction at appropriate sections of the rivers of the United States needs to be complemented by a policy that would preserve other selected rivers or sections thereof in their free-flowing condition to protect the water quality of such rivers and to fulfill other vital national conservation purposes.[66]

Enactment of the Wild and Scenic Rivers Act (WSRA) marked the culmination of years of effort by river conservationists to provide federal protection for several carefully chosen rivers of national significance.[67] The Act designated eight rivers for immediate inclusion in the program and another

27 rivers for further study. The number of rivers included in the system has since grown substantially. As of November 1996, there had been 165 designations to the Wild and Scenic Rivers System, totaling approximately 10,815 river miles.[68] Most of these rivers are in the West. Over two-thirds of the designated rivers, and over 85 percent of the river miles within the system, are found in the eighteen western states (figure 10.1).[69]

General Characteristics of the Wild and Scenic Rivers

River designations under the WSRA do not necessarily encompass entire river systems. In fact, despite the common reference to designated "rivers," an entire river, from headwaters to mouth, is rarely designated a part of the Wild and Scenic Rivers system. Much more typical is the designation of one or more explicitly defined *segments* of a river. Tributaries are not included unless specifically designated and are designated individually rather than as a

FIGURE 10.1
Wild and Scenic Rivers in the Western States

group or system. Like mainstem rivers, tributaries are often designated in segments rather than as a whole.

River and tributary segments are classified at the time of designation as either "wild," "scenic," or "recreational," largely depending on river and shoreline characteristics such as the degree of development and availability of access.[70] These characteristics do not have to be consistent along the entire river. Designated river segments may include stretches of river that are given different classifications, so that a single segment may be designated as "wild" over some portion, "scenic" over another, and "recreational" over yet another. The Wild and Scenic Rivers Act states that rivers are to be managed to achieve the purposes for which they were designated, but with the exception of placing some restrictions on mining activity occurring on the shores of "wild" rivers, Congress did not explicitly require the use of different management methods corresponding to the different classifications.

River Designation

Rivers are designated as components of the Wild and Scenic Rivers system in either of two ways. The first method is for the governor of the state in which a river is located to petition the secretary of the interior to designate the river as a component of the national system. The secretary is authorized to make the designation if the river meets certain basic requirements, primarily that the characteristics of the river meet the same federal standards as other rivers in the system, and that the state possesses adequate mechanisms and resources to protect the river once it is designated. This method has been used for only about 10 percent of the designations, though its popularity may be increasing.[71] Of the western rivers in the Wild and Scenic Rivers system only seven—five of which are in California—entered the system in this way.

By far the most common method of designation is an act of Congress that defines and classifies the specific river segment to be included and states the purposes for which the river is being designated. Rivers considered for designation through this process are brought to the attention of Congress in several ways. Often the designations are preceded by the completion of intensive studies—the studies themselves having been previously authorized by Congress, as happened with the 27 rivers authorized for further study in the original 1968 Act—but this "dual-authorization" route has been followed less often in recent years. Private citizens have sometimes been successful in petitioning Congress directly for designation of particular rivers, and this method continues to be an important source of new river designations. The primary source of most recent designations, however, has been recommendations from federal government agencies. Federal agencies are required by the Wild and Scenic Rivers Act to consider and evaluate potential additions to the Wild and Scenic Rivers system, and this has resulted in the recommendation of hundreds of rivers for designation under the Act, as well as the identification of hundreds more for further evaluation.[72] In recent years many rivers have been added through "omnibus" legislation, in which several rivers in a single state are simultaneously designated by Congress as components of the system.[73]

Administration

The responsibility for administering a river once it has been designated as a component of the Wild and Scenic Rivers system is usually assigned to the government agency or other authority that has jurisdiction over the shoreline. Wild and scenic rivers are therefore administered by federal agencies, state agencies, Indian tribes, and other authorities. Most designations to date have been of rivers on federal land. Four federal land management agencies—the Forest Service, Park Service, Bureau of Land Management, and Fish and Wildlife Service—manage the vast majority of river miles in the system, managing 41, 23, 19, and 10 percent of all river miles, respectively. States manage 7 percent of the river mileage, and Indian tribes and all other authorities cumulatively manage less than 1 percent. Jurisdiction over a river is often split between different agencies, depending on land ownership patterns along the river, so that it is not uncommon for several agencies to be managing a single wild and scenic river. Managing agencies are required to develop management plans for the designated rivers, though the WSRA contains very little direction as to what must go into those plans.

General Features

The Wild and Scenic Rivers Act includes several substantive provisions designed to ensure that the purposes of the Act are achieved. Among the most important are provisions that authorize the federal agencies administering wild and scenic rivers to acquire land along the river corridor, to regulate uses of both land and water along the rivers, and to take other actions necessary to protect the natural, historic, and recreational values of the rivers. Though in many ways forming the very heart of the Act—and also usually attracting the most intense management efforts and greatest controversy—many of these provisions have little, if any, impact on matters related to instream flow. Our intent in the following sections is to focus only on those provisions of the Wild and Scenic Rivers Program that affect instream flows.[74]

Instream flows are affected by several different provisions of the Wild and Scenic Rivers Act. Two of the most significant are designed to prohibit federal agencies from doing anything that would undermine the purposes of the Act. First, the Act prohibits the Federal Energy Regulatory Commission (FERC) from licensing "the construction of any dam, water conduit, reservoir, powerhouse, transmission line, or other project works . . . on or directly affecting any river which is designated . . . as a component of the national wild and scenic rivers system."[75] Several rivers have been designated specifically to preclude FERC from licensing new hydropower developments on the designated reaches.[76] Second, the Act specifies that no other federal agencies "shall assist by loan, grant, license or otherwise in the construction of any water resources project that would have a direct and adverse effect on the values for which such river was established."[77] The protection is not absolute, for neither of the two provisions was intended to prevent the development of projects—whether upstream, downstream, or on tributaries—if they would not "invade the area or unreasonably diminish the scenic, recreational, and fish and wildlife values present in the area. . . ."[78] Nonetheless, because almost all major water development projects require either a federal license or federal assistance or both during at least some stage of development or operation, these provisions have great potential to protect the "free-flowing" nature of the wild and scenic rivers.

Federal Water Rights under the Wild and Scenic Rivers Act

The Wild and Scenic Rivers Act also provides a more direct means of protecting instream flows. As described in chapter 8, Congress has the authority to reserve flows of water if it chooses to do so. In most legislation having the potential to affect western water allocation, Congress has not chosen to address the topic explicitly, so there has been a great deal of controversy surrounding the issue of whether or not Congress did in fact intend to reserve water for federal purposes. But in the Wild and Scenic Rivers Act Congress specifically addressed the issue of water needs—even if in a roundabout fashion:

> Designation of any stream or portion thereof as a national wild, scenic or recreational river area shall not be construed as a reservation of the waters of such streams for purposes other than those specified in this chapter, or in quantities greater than necessary to accomplish these purposes.[79]

If stated more directly, this passage might have been written: "Sufficient water is reserved to accomplish the purposes for which rivers are designated under this Act, but only for those purposes specified in this chapter, and only in quantities necessary to accomplish those purposes." This, in fact, has been the interpretation reached by those who have analyzed the Act.[80] This means that managing agencies have the potential to claim federal water rights for the wild and scenic rivers. Given that the purpose of the Act is to protect "scenic, recreational, geologic, fish and wildlife, historic, cultural, or other similar values," many of which will require water within the stream channel, and that rivers "shall be preserved in free-flowing condition," there is no doubt that these rights were intended to be applied to instream purposes.

So how much water can actually be claimed for instream flows under the Wild and Scenic Rivers Act? Congress has never addressed this issue beyond the language quoted above, which stated that the quantity shall be limited to the amount necessary to fulfill the purposes of the Act. One could argue—and many have—that 100 percent of a river's flow is necessary to preserve a river in its "free-flowing" condition, which is the general purpose of the Act. However, Congress stated that the quantity of the federal water right for a river was to be limited to that quantity of water necessary to fulfill the "specified purposes." This language implies that some quantity less than the entire flow may be adequate to meet federal purposes. It is also important to realize that the purposes listed in individual pieces of legislation designating rivers for inclusion in the system usually encompass only a subset of the different purposes listed in the 1968 Act.

There are other questions about the quantity of water that would be associated with this federal water right. At what level should the specifically protected values be maintained? Optimal? Minimal? At some intermediate level? The Supreme Court's *New Mexico* decision, described earlier, said that the *implicit* nature of federal reserved rights required that the rights be interpreted as narrowly as possible. However, in the Wild and Scenic Rivers Act, Congress addressed the topic of federal water rights *explicitly,* and therefore the *New Mexico* standard should not apply. At least one scholar has argued that because federal water rights in WSRA were explicitly addressed, they should be interpreted broadly.[81] One factor complicating interpretation of the Act is that federal water rights arising under WSRA are not technically federal *reserved* rights—which were created to account for the implicit reservation of water accompanying federal *land* reservations—and therefore it is not known whether the courts will apply the same criteria that have been used in other federal water rights cases.

It has been widely assumed that the amount of *unappropriated* water in the river at the time of designation would serve as an upper limit on the amount of water that could be claimed, because the priority date assigned to a federal water right arising under WSRA would be the date that a specific river is actually designated for inclusion in the system. The legislative history of the Wild and Scenic Rivers Act gives credence to this assumption.[82] However, the Act itself states, "Under the provisions of this chapter, any taking by the United States of a water right which is vested under either State or Federal law at the time such river is included in the national wild and scenic rivers system shall entitle the owner thereof to just compensation."[83] This language clearly implies that the federal government may condemn existing water rights if condemnation is necessary to protect designated values. To date this has been a moot issue, however, and it is likely to continue in that status. The federal agencies managing the wild and scenic rivers have yet to condemn a private water right under WSRA and have expressed no interest in doing so.[84]

Other characteristics of the water rights arising under the WSRA remain unclear, both because Congress has not chosen to further clarify the issues and because the Act has not been extensively tested in the courts.[85] As is true with the "details" of other federal programs, matters that were not explicitly addressed by Congress have been left for the implementing agencies to resolve. The result is that the exact quantity of water needed for the wild and scenic rivers will be determined on a river-by-river basis, first by the managing agencies and then, perhaps, by the courts. Agencies will make a determination as to the precise purposes that must be protected, and then determine the amount of water necessary to achieve those purposes. If taken to court, the courts will then have to determine whether the agency's interpretation of designated purposes is reasonable and its methods of evaluating needs credible.

Protection of Instream Flows on the Wild and Scenic Rivers

The establishment of a water right for instream flow purposes constitutes one of the most effective methods of protecting instream values. There is little doubt either that the Wild and Scenic Rivers Act was intended to protect instream values or that it provided managing agencies with the authority to claim federal water rights for this purpose. To date, however, very few water rights have been granted for the wild and scenic rivers. There are several reasons for this disparity between potential and actual rights.

First, and perhaps most important, protection of the values for which individual rivers have been designated as components of the Wild and Scenic Rivers system does not necessarily require the creation of new water rights. Some of the wild and scenic rivers are already protected through other means. For instance, many of the wild and scenic rivers run through national parks, where the Park Service has previously been given both the mandate and the authority to preserve park resources.[86] In some cases rivers are protected by other rights, even if those rights were not specifically asserted for the purposes of protecting a wild and scenic river.[87] Instream flows for some rivers have been obtained through agreements with upstream dam operators.[88] Physical inaccessibility or extreme distance from sites of potential water use protect many rivers from development. Other rivers may carry enough water to supply both instream needs and reasonably foreseeable upstream consumptive needs. Protection of some of the purposes listed in individual pieces of river designation legislation—such as protection of historical or cultural values—may not require instream flows. In others, development may already have occurred, so that the very junior priority date of a wild and scenic rivers water right would have little, if any, effect in protecting flows.[89] Likewise, the threats facing some wild and scenic rivers aren't primarily of a type that can be handled within the context of the water rights system. For example, several wild and scenic rivers in Alaska are facing threats from state-authorized mining of riverbeds and shorelines. Although such mining has the potential to severely impair a variety of instream values—and already has, on rivers such as the Fortymile, which is one of Alaska's 26 wild and scenic rivers—the existence of instream water rights would not be sufficient to eradicate this threat.[90]

A second major reason for the scarcity of decreed water rights on the wild and scenic rivers is the substantial expense of the water rights application process. Much of the expense is a result of efforts to quantify the amount of water needed to protect instream values. Instream-flow quantification studies can be expensive, even under relatively ideal circumstances, because of the large investment in time and technical expertise required to make such an assessment.[91] The measures required to ensure that the quantification methods and procedures employed in the study will withstand intense scrutiny can involve even greater expense when claims are controversial or likely to be contested in court. The desire to avoid this expense accounts in part for the popularity of the idea that instream flow needs for the wild and scenic rivers should be quantified at the level of 100 percent of the flow legally unappropriated at the time of designation. This simple approach would sidestep the need for expensive and time-consuming studies.

The controversy surrounding federal water rights provides a third major reason for the lack of decreed rights for the wild and scenic rivers. The assertion of rights arising outside the context of state water allocation systems generates significant opposition. As discussed earlier, states resent the intrusion of the federal government in matters concerning "state" water;[92] holders of existing water rights become nervous at the possibility—even if remote—of condemnation of their rights;[93] and potential developers dislike the elimination of development sites and the addition of competing claims for scarce water. The federal agencies like to pursue "good neighbor" policies with states and surrounding landowners and to pay attention to the expressed needs and desires of local residents. Though authorized to assert federal water rights, the agencies may feel either that the benefits that would accrue to such rights are not worth the loss of good will that might result, or that the opposition stimulated by such assertions would frustrate achievement of other federal goals.

Congress has sometimes acted to alleviate potential conflict with the states by directing federal agencies to work with the states in protecting the necessary flows. For the Cache la Poudre in Colorado, Congress directed that "the reservation of water established by the inclusion of portions of the Cache la Poudre River in the Wild and Scenic Rivers system . . . shall be adjudicated in Colorado Water Court, and shall have a priority date as of the date of enactment of this title."[94] Though asserting that water would be reserved for the Poudre, Congress took pains to assure the state that designation of the Poudre "shall not interfere with the exercise of existing decreed water rights" nor "be utilized in any Federal proceeding, whether concerning a license, permit, right-of-way, or other Federal action, as a reason or basis to prohibit the development or operation of any water impoundments, diversion facilities, and hydroelectric power and transmission facilities" in a reach of the river downstream from the designated segments.[95] The Colorado Water Court with jurisdiction over the waters of the Poudre subsequently ruled that the amount of water reserved by Congress for the Poudre was "all of the native water arising upon or flowing through the designated segments of the Cache la Poudre River, subject to valid prior [existing] appropriations under Colorado law."[96]

Legislation designating Wyoming's Clark's Fork as a wild river contained similar language but went a step further by directing the secretary of agriculture to

> apply for the quantification of the water right reserved by the inclusion of the Clark's Fork in the Wild and Scenic Rivers System in accordance with the procedural requirements of the laws of the State of Wyoming: *Provided,* That, notwithstanding any provision of the laws of the State of Wyoming otherwise applicable to the granting and exercise of water rights, the purposes for which the Clarks Fork is designated . . . are declared to be beneficial uses and the priority date of such right shall be the date of enactment of this paragraph.[97]

Congress explicitly required the secretary of agriculture to work with the state in determining the amount of water necessary to achieve the purposes for which the Clark's Fork was designated, and directed that the secretary should apply for water rights in accordance with state procedural requirements. But Congress also took steps to ensure that state requirements would not be used to frustrate the intent of Congress. This direction resulted in the signing of a Memorandum of Understanding between the U.S. Forest Service and the State of Wyoming to jointly pursue a determination of water needs and the granting of a water right for the Clark's Fork.[98]

The Wild and Scenic Rivers Act itself does not require that river managers take steps to claim the federal water rights created under the Act, and river managers have been quick to cite the reasons above—lack of need, great expense, and the almost inevitable controversy—in support of this flexibility.[99] The application for water rights on the wild and scenic rivers is not a high priority when budgets are tight and resources scarce, unless there is an immediate and overwhelming threat to the river. A wide variety of measures not uniquely available to the federal government, or to the WSRA, have been used to protect the wild and scenic rivers.[100] For the most part, river managers have acted quickly and decisively in the face of immediate threats to the rivers, employing whatever means are available.

Prospects of Using the WSRA to Protect Instream Flows

The Wild and Scenic Rivers Program continues to be the most direct federal program for protecting selected national rivers and streams from development. The Wild and Scenic Rivers Act contains several strong provisions designed to protect rivers, and there has been rapid growth in the system in recent years. The identification and evaluation of potential wild and scenic rivers have now been integrated into agency resource management planning processes. Several hundred rivers are in the pipeline for further evaluation and possible recommendation to Congress, and two major omnibus river bills, in California and Arizona, are likely to be considered by the Congress in the next few years. Federal agency directors and conservation agencies seem uniformly optimistic about the potential for greatly expanding the number of designated rivers over the course of the next several years.[101]

Despite these obvious strengths, no one should jump to the conclusion that the Wild and Scenic Rivers Program by itself will provide adequate instream flow protection for America's rivers. Despite the 165 designations so far to the system, and the several hundred more likely to be considered, the Wild and Scenic Rivers Act has been designed to protect only those few select rivers that contain "outstandingly remarkable" characteristics of national importance. Less than 0.3 percent of the nation's total stream mileage—or less than 1 percent of the mileage in streams at least five miles in length—has been designated as part of the Wild and Scenic Rivers system.[102] Even dramatic growth in the system will not alter the basic fact that only a very small proportion of this nation's rivers will receive protection under the Wild and Scenic Rivers Act.

Even many rivers that have outstanding characteristics are not likely to be included in the system. Conservationists have pointed out that few of the nation's longest or biologically richest rivers have been designated for inclusion in the system. River segments are often dropped from consideration if they have significant development potential or run primarily through private land.[103]

There has been a tremendous amount of regional variation in the number of designations made under the Act. Numerous rivers have received protection in some states, but few, if any, rivers have received protection in other states. For instance, over three-quarters of the wild and scenic rivers in the western states are in Alaska, California, and Oregon. Idaho, Montana, and Washington contain most of the rest. Outside of the Pacific Northwest there have been very few rivers designated as components of the system. In the Southwest, New Mexico has five designations, but they cumulatively account for just 121 river miles. Arizona and Texas each have only one river in the system (the Verde and the Rio Grande, respectively). The Niobrara (Nebraska) and two different segments of the Missouri along the border between Nebraska and South Dakota are the only wild and scenic rivers in the Plains states. Neither of the Great Basin states of Utah or Nevada has a wild and scenic river. The Rocky Mountain states of Colorado and Wyoming each have only one wild and scenic river (the Cache la Poudre and Clark's Fork of the Yellowstone, respectively), despite the presence of hundreds of high-quality mountain rivers within their borders. Figure 10.1 makes it clear that to date, the Wild and Scenic Rivers Act has been of importance only in some parts of the West.

Other problems reducing the potential of the Wild and Scenic Rivers Program include the substantial investment of time, money, and political expertise necessary to shepherd river designations through Congress; the ability of local residents and state congressional delegations to derail potential designations that they oppose; substantial variation in the degree to which the different federal agencies have sought new river designations and the protection of existing rivers; the lack of clear guidance and requirements from Congress on what steps must be taken to manage the wild and scenic rivers once they are designated; the limited quantity of resources that has been made available for management activities; and split jurisdictions that fragment and often frustrate management efforts on individual rivers.

National political trends will affect the future assertion of federal water rights, the resources made available for management, and the number and identity of rivers subsequently added to the system. Much of what happens with the Wild and Scenic Rivers system will be unrelated to issues of instream flow. But policies and activities related to instream flows on individual wild and scenic rivers will continue to depend on a variety of circumstances peculiar to those rivers—the immediacy and magnitude of potential threats, management agency attitudes and resources directed toward protection, the presence or absence of political and/or legal opposition from local residents, and the availability of instream flow protection mechanisms under state law. The future of instream flow activities prompted by the WSRA is difficult to predict, but the Wild and Scenic Rivers Program is already an important component of this nation's efforts to protect rivers and streams, including instream flow.

National Wilderness Preservation System

All of the federal land management agencies have managed at least some of their holdings to preserve natural characteristics and values. Many of these special lands have been formally designated as "wilderness areas," official parts of the national wilderness preservation system. Because free-flowing rivers and streams are an integral part of the natural environment, the formal designation of lands as wilderness areas has substantial potential to affect the status of instream flows on federal land.

History of Wilderness on the Federal Lands

The first attempt to manage federal lands specifically to preserve wilderness characteristics probably occurred in 1924. In that year Aldo Leopold, a district ranger in the Gila National Forest in New Mexico, began efforts to administratively designate 574,000 acres of the forest as the Gila Wilderness Reserve. Leopold, a leading naturalist who wrote eloquently about the value of wilderness, viewed wilderness as "a continuous stretch of country preserved in its natural state, open to lawful hunting and fishing, big enough to absorb a two-week pack trip, and kept devoid of roads, artificial trails, cottages, or other works of man."[104] His efforts to preserve part of the Gila National Forest as wilderness succeeded in 1928, when the secretary of agriculture authorized the Forest Service to designate certain areas where no development would be permitted, subject only to exceptions authorized by the chief of the Forest Service or the secretary of agriculture.

The protection of wilderness values through administrative designation and management soon spread to other national forests. Protection was guided in large part by a set of rules developed by Bob Marshall—another leading wilderness advocate—known as the "U Regulations." A total of 63 wilderness or primitive areas, covering 14 million acres, were established in national forests during the 1930s. However, administrative protection of wilderness did not prove to be as permanent as wilderness proponents would have wished. Throughout the 1940s and 1950s the Forest Service found numerous reasons to relax protective regulations in circumstances where wilderness areas were thought to have strong development potential. The amount of national forest land designated for management as wilderness began to shrink, alarming wilderness proponents. Congress entered the controversy in 1956, when Hubert H. Humphrey introduced the first wilderness bill in the U.S. Senate. Introduction of the bill initiated the start of a long, highly visible series of debates that culminated eight years later in passage of the Wilderness Act.

The Wilderness Act of 1964 declared Congress's policy "to secure for the American people of present and future generations the benefits of an enduring resource of wilderness."[105] The Act was designed to "assure that an increasing population does not occupy and modify all areas within the United States and its possessions, leaving no lands designated for preservation and protection in their natural condition." The Act established a national wilderness preservation system created from existing federal lands and reserved for Congress the authority to designate new wilderness areas or revoke existing ones. The act stated that wilderness areas were to be managed by the department and agency having jurisdiction on the lands prior to the wilderness designation. Though the original act did not apply to BLM lands, BLM lands were subsequently included by provisions of the Federal Lands Policy and Management Act of 1976. All four of the major federal land management agencies—the Forest Service, Park Service, Fish and Wildlife Service, and BLM—now manage wilderness areas.

The Wilderness Act states that wilderness areas are to be "devoted to the public purposes of recreational, scenic, scientific, educational, conservation, and historical use." The Act defines wilderness areas as areas "where the earth and its community of life are untrammeled by man, where man himself is a visitor who does not remain" and declares that wilderness areas are to be "protected and managed so as to preserve [their] natural conditions" and left "unimpaired for future use and enjoyment as wilderness." Agencies managing wilderness areas are responsible for preserving the wilderness character of the designated lands. Virtually all forms of development are prohibited in wilderness areas, with certain exemptions for activities such as mineral prospecting and grazing. There is also an exemption that applies specifically to water:

> Within wilderness areas in the National Forests designated by this Act . . . the President may, within a specific area and in accordance with such regulations as he may deem desirable, authorize prospecting for water resources, the establishment and maintenance of reservoirs, water-conservation works, power projects, transmission lines, and other facilities needed in the public interest . . . upon his determination that such use or uses in the specific area will better serve the interests of the United States and the people thereof than will its denial.[106]

The National Wilderness Preservation System

The Wilderness Act declared all national forest lands then being managed as wilderness or primitive areas—approximately 9.1 million acres—to be the first components of the system. The federal land management agencies were required to study and recommend potential additions to the system. The system has grown substantially since that time. The Endangered American Wilderness Act created 17 new wilderness areas on 1.3 million acres of federal land in nine western states in 1978. The Alaska National Interest Lands Conservation Act added over 55 million acres in Alaska to the system in 1980. And Congress added an additional 7.2 million acres in 18 states in 1984. Since then, additional wilderness designations have been made on a state-by-state basis, usually on recommendations made by the federal land management agencies.

By November 1994, the federal land management agencies were managing approximately 600 wilderness areas covering 100 million acres of federal land.[107] Over 95 percent of this acreage is in the 18 western states, including Alaska, which by itself contains over half of the acreage in the national wilderness preservation system. California, Washington, Idaho, and Montana contain well over half of the remaining wilderness acreage. There is very little wilderness acreage in the Plains states, but all of the other western states each contain at least 1.6 million acres of designated wilderness, with the exception of Nevada and Utah, which each contain less than half that amount.[108]

Instream Flows in the Wilderness Areas

The flows of rivers and streams are an integral part of the natural environment, so any effort to preserve large areas of land in a natural condition would almost certainly have to include the protection of water resources. However, aside from the exception for development authorized by the president, as described above, Congress did not specifically address the question of water flows or water rights when it passed the Wilderness Act. The subject of water rights for wilderness areas has since become a source of bitter controversy. There have been several conflicting legal opinions on the subject, and the issue of water rights for wilderness areas has become a standard and divisive part of every congressional debate surrounding the designation of new wilderness areas.

Proponents of federal water rights for wilderness areas have argued that designation of lands as wilderness implicitly authorizes the creation of a federal reserved water right to enough water to accomplish wilderness purposes. Many proponents have argued for a reservation of 100 percent of natural unappropriated streamflow to preserve streams in their "natural" condition. Some commentators have pointed out that the accomplishment of wilderness objectives may be possible with flows substantially less than 100 percent.[109] Opponents of federal water rights for wilderness areas have argued that the designation of lands as wilderness does not constitute a new reservation and that water rights in wilderness areas should be governed by the status of the underlying reservation—national forests, national parks, national wildlife refuges, or BLM lands. Not coincidentally, the primary purposes of most of these lands are not usually sufficient to justify the creation of a federal reserved right for full natural flows.

There have been numerous milestones in the controversy over water rights in wilderness areas. In 1979 the Interior Department solicitor issued an opinion in which he concluded that the Wilderness Act of 1964 implicitly authorized the creation of federal reserved water rights for wilderness areas.[110] While solicitors' opinions do not constitute law, they are binding on the affected agencies, in this case the Park Service, Fish and Wildlife Service, and BLM. Commonly referred to as the Krulitz Opinion after its author, Solicitor Leo Krulitz, this conclusion was immediately and hotly contested on a number of fronts. It was also overturned in 1988, when Ralph

Tarr, the solicitor of a new administration, issued a new memorandum. Tarr reasoned that Congress had not intended to create federal reserved rights for wilderness areas when it passed the Wilderness Act, that Wilderness Act purposes were secondary to the purposes of the underlying reservation, and therefore—in accordance with the standards issued in the Supreme Court's *New Mexico* opinion—could not be the basis for asserting federal reserved rights.[111] The Tarr Opinion was approved by Attorney General Edwin Meese,[112] which prevented Department of Justice attorneys from making federal reserved water rights claims for wilderness areas.

Solicitor Tarr's opinion was not the final word on the matter. The new Interior Department solicitor in the Clinton administration, John Leshy, announced his intention to reconsider the Tarr Opinion soon after taking office. The announcement, issued in conjunction with the Department of Justice, was accompanied by the opening of a public comment period. Many comments were received, including a response from 14 western attorneys general that urged the solicitor not to create federal reserved water rights for wilderness areas.[113] After the comment period was over, a new analysis was written by the solicitor's office and submitted to the attorney general's office. No opinion has yet been released. It is widely expected, however, that the new solicitor—who participated in the creation of the Krulitz Opinion during an earlier stint in the Office of the Solicitor—will essentially reinstate the terms of the Krulitz Opinion and authorize Interior Department agencies to create policies allowing them to once again assert federal water rights claims for wilderness areas. In the meantime the Tarr Opinion has been suspended, allowing federal agencies to once again make claims for federal reserved water rights in wilderness areas, but only on a case-by-case basis. The suspension has not created a new policy regarding federal reserved water rights in wilderness areas.[114]

The existence of federal reserved water rights for wilderness areas has also been argued in court. The first, and only, major case addressing the topic, *Sierra Club v. Block,* was decided in 1985.[115] The Sierra Club had filed suit in federal district court in Colorado against the Departments of Agriculture and Interior in an effort to force them to claim federal reserved water rights for wilderness areas during state water rights adjudications.[116] The court concluded that the Wilderness Act did include implicit federal reserved water rights, that the designation of wilderness areas constituted a new reservation, and that the purposes of the Wilderness Act were primary:

> Moreover, contrary to defendant-intervenors' assertions, Congress intended to reserve previously unappropriated waters in the wilderness areas to the extent necessary to accomplish these purposes. It is beyond cavil that water is the lifeblood of the wilderness areas. Without water, the wilderness would become deserted wastelands. In other words, without access to the requisite water, the very purposes for which the Wilderness Act was established would be entirely defeated.[117]

While this decision provided a rallying point for proponents of wilderness water rights, it was eventually overturned by an appeals court on the grounds that the issue was not yet "ripe" for adjudication.[118] The Tenth Circuit Court reasoned that because the plaintiffs had not demonstrated that wilderness water values were facing an imminent or actual threat—especially given the Forest Service's assertion that sufficient administrative and/or geographic controls were in place to maintain wilderness characteristics—the case was not yet justiciable. This decision, and the fact that the U.S. Supreme Court has never addressed the issue, has left the status of water rights in wilderness areas very much up in the air.

The bitter fight over wilderness water rights in the administrative and legal arenas should not obscure the fact that Congress could have resolved the issue before it began by explicitly stating its intent with respect to water rights for wilderness. Congress has lately begun to do just that. For example, when Congress passed the Arizona Wilderness Act of 1990, it specifically reserved enough water to accomplish the purposes of the act:

> (1) With respect to each wilderness area designated by this title, Congress hereby reserves a quantity of water sufficient to fulfill the purposes of this title. The priority date of such reserved rights shall be the date of enactment of this Act.
>
> (2) The Secretary and all other officers of the United States shall take steps necessary to protect the rights reserved by paragraph (1), including the filing by the Secretary of a claim for the quantification of such rights in any present or future appropriate stream adjudication in the courts of the State of Arizona in which the United States is or may be joined and which is conducted in accordance with the McCarran Amendment.[119]

Congress did not specify exactly how much water would be needed, leaving this issue to be decided by the federal land management agencies and state courts. Congress acted similarly in recent wilderness bills for Nevada, New Mexico, and Washington, in which it reserved "sufficient" flows, "minimum" flows, and "necessary" flows, respectively.[120]

Congress can also resolve the water rights issue by specifying that water will not be reserved for wilderness purposes. A prime example of this approach occurred with the 1993 wilderness bill for Colorado. The Colorado wilderness bill had been held up for several years, largely because of controversy surrounding the issue of water rights. Wilderness proponents eventually agreed to language in the wilderness bill that specifically denied federal reserved rights for the new wilderness areas, reasoning that geography, together with other provisions of the bill, would still result in the protection of rivers and streams:

> (1) Congress finds that—
>
> (A) the lands designated as wilderness by this Act are located at the

> headwaters of the streams and rivers on those lands, with few, if any, actual or proposed water resource facilities located upstream from such lands and few, if any, opportunities for diversion, storage, or other uses of water occurring outside such lands that would adversely affect the wilderness values of such lands; and
>
> (B) the lands designated as wilderness by this Act are not suitable for use for development of new water resource facilities. . . .
>
> (C) therefore, it is possible to provide for proper management and protection of the wilderness value of such lands in ways different from those utilized in other legislation. . . .
>
> (2) The purpose of this section is to protect the wilderness values of the lands designated as wilderness by this Act by means other than those based on a Federal reserved water right.[121]

To make the point clear, Congress further stated that:

> Neither the Secretary of Agriculture nor the Secretary of the Interior, nor any other officer, employee, representative, or agent of the United States, nor any other person, shall assert in any court or agency, nor shall any court or agency consider, any claim to or for water or water rights in the State of Colorado, which is based on any construction of any portion of this Act.[122]

To ensure adequate protection without rights, the bill specifically forbade any development within the new wilderness areas, forbidding the possibility of even development through presidential authorization.

Wilderness proponents recognized that protecting wilderness areas without using federal reserved water rights would not necessarily be feasible in other states. Proponents succeeded in placing language into the Colorado Wilderness Act of 1993, stating that "Nothing in this section shall be construed as establishing a precedent with regard to any future wilderness designations."[123] Nevertheless, pending wilderness bills for Idaho and Montana have incorporated much the same terms, thus establishing a trend in which Congress has been specifically denying federal reserved water rights for new wilderness areas. Congress is not in a position to resolve the controversy surrounding water rights in existing wilderness areas, because that issue revolves around the intent of Congress at the time of designation.

The Colorado wilderness bill contained one other provision of note. Three of the areas initially recommended to Congress for wilderness status, the Piedra, Roubideau, and Tabeguache, were not located in headwater areas and were thus potentially vulnerable to upstream water development. Congress had previously, in 1991, directed the Forest Service to participate jointly with the State of Colorado in an effort to protect the Piedra's water resources through state law.[124] This directive resulted in execution of a memorandum of understanding between the United States and the state of Colorado and the completion of a water needs quantification study by staff drawn from the Forest Service, Colorado Water Conservation Board, and Colorado Division of Wildlife.[125] The federal-state study participants then sought to determine the minimum amount of water needed to maintain water-related resources and values "in approximately their current condition" and to "maintain the interaction of processes which have shaped the Piedra River and its ecosystems and would adequately protect the water related wilderness resources and values."[126]

The recommendation made by the study participants to the Colorado Water Conservation Board was for a water right that would specify a range of flows that would mimic existing patterns of streamflow. The proposed right would have been for relatively more water in high-flow years and relatively less water in low-flow years. The recommended right resembled, in fact, the dynamic claim made by the Forest Service for channel maintenance purposes in other national forests in Colorado, as described in chapter 8. This similarity may well have been the downfall of the joint effort. Numerous complaints and objections were made to the proposed right once it had been submitted to the Colorado Water Conservation Board. Many of the complaints argued that the state shouldn't give away to the Forest Service the kinds of water rights that it had fought so hard to oppose in the Colorado Water Division 1 channel maintenance proceedings.[127] The Colorado Water Conservation Board eventually rejected the recommendation, leaving the Piedra protection effort at a standstill.

Perhaps taking their cue from the stalemated efforts of the Forest Service and State of Colorado, Congress simply ignored water protection issues for the three downstream areas in the Colorado Wilderness Act of 1993. The bill directed that the Piedra, Roubideau, and Tabeguache areas be managed "so as to maintain the areas' presently existing wilderness character and potential for inclusion in the National Wilderness Preservation System" but did not actually designate the areas as parts of that system. Congress thereby sidestepped the issue of whether water rights were needed to protect downstream wilderness areas.[128] Rather than specifying any direct method of water resources protection in the three areas, Congress stated only that "the Secretary of Agriculture and the Secretary of the Interior, in consultation with the Colorado Water Conservation Board, shall compile data concerning the water resources of the areas described in subsection (a) [the Piedra, Roubideau, and Tabeguache] and existing and proposed water resource facilities affecting such values."[129] The three areas are therefore to be managed as wilderness but without any specified means of protecting wilderness water values from the threat of further development of water upstream.

Are Water Rights Needed in Wilderness Areas?

The heated controversy surrounding the issue of water rights in wilderness areas has all but obscured a basic question: Are water rights needed to protect instream flows in wilderness areas?

A quick answer, for almost all wilderness areas designated so far, is not really. Most of the wilderness areas in the contiguous 48 states are at high-elevation headwater sites, the so-called "rock and ice" wilderness areas, and there is relatively little pressure for water development on wilderness lands in Alaska. Imminent threats to wilderness area water resources seem few and far between. Wilderness areas, by definition, exist only in places that have not heretofore been developed, which means that they are not likely to constitute prime sites for development in the future. Moreover, Wilderness Act prohibitions on development are likely to protect the vast majority of wilderness rivers and streams from being dewatered. For some lands, such as those already managed for purposes of preservation by the National Park Service, any protection arising out of the designation of land as wilderness is likely to be redundant. Even the Wilderness Act's exception for water developments authorized by the president in the public interest is not likely to leave wilderness water sources exposed to development, given that the exception has not been invoked even once in the 30 years since passage of the Act. The presidential exemption is also subject to elimination by Congress in future wilderness bills, as was done in Colorado.

By the same token, the "threat" to existing water rights posed by granting federal water rights for wilderness areas is quite small. As noted by many commentators, water rights for wilderness areas would involve no diversions and very little consumptive use (only that amount of water that is consumed naturally, through evapotranspiration) and would have priority dates that are quite junior relative to other longstanding water uses in the West. Political and equitable considerations prevent federal agencies from asserting—or courts from granting—federal water rights that would have a severe impact on existing water rights. As summarized by one commentator, "In short, the wilderness water right would seem to be the paradigm illustration of what [longtime water lawyer and scholar] Frank Trelease was talking about—a theoretical potential for major interference with existing water uses that, upon closer examination, proves to be practically no threat at all."[130] Strenuous objections to water rights for wilderness areas are more likely a result of opposition to the concept of federal water rights or wilderness areas in general, or to simple misunderstanding.[131]

The quick answer that water rights for wilderness areas are not needed for the protection of wilderness water resources does, however, ignore a few situations in which the controversy may be more relevant. Not all wilderness areas are inaccessible or in headwaters, and many areas being studied for future designation—primarily on BLM lands—are at lower elevations where there may be significant development potential upstream, or contain large private land holdings within the wilderness. Water rights may be needed to protect against future diversions, or, more likely in the already fully allocated water systems prevalent in the West, to give standing to protest transfers of existing rights if such transfers would be injurious to the preservation of wilderness values.

For example, the existence of wilderness water rights could frustrate plans by downstream water rights holders to sell rights to purchasers who wish to divert water from higher locations, as frequently happens when farmers sell water rights to municipalities or industries. And, of course, the existence of a valid water right for wilderness areas would frustrate those who wished to make use of the water from an upstream location in the future.[132] Some of these disadvantages might be offset, however, by the advantages a wilderness water right might give to downstream water users. Because a wilderness right would be nonconsumptive and would not entail any diversions, it could eliminate potential competition for water supplies and/or other water uses that might adversely affect water quality. In situations in which wilderness water rights—or even the very designation of land as wilderness—cause such harm as to be contrary to the public interest, there is also a remedy of last resort. As is true of any other federal land reservation, Congress is always free to revoke the reservation—and associated federal reserved water rights—and return the land to its former status.

The Current Status of Water Rights in Wilderness Areas

Water rights continue to be a major factor in the consideration of new wilderness proposals, generating an ongoing series of contentious debates. In the absence of clear direction from Congress, the courts will have to decide the status of wilderness water rights for new wilderness areas just as they currently have to do for existing areas. So far it is not possible to know for sure whether federal water rights for wilderness areas exist, or whether federal agencies must assert such rights. There have been numerous but contrary indications from legal offices and the courts. The issue is not likely to be conclusively decided until the U.S. Supreme Court hears a wilderness water rights case. Much of what happens with rivers and streams in wilderness areas, and with the designation of new areas, will depend on shifting political trends. The current political climate is such that agencies are not likely to assert federal water rights unless such rights are absolutely necessary to meet mandated objectives. So far wilderness water resources are largely intact, protected from impairment by an array of geographical characteristics, statutory protections, and administrative actions.

CHAPTER ELEVEN

Reaching a Balance in Water Allocation

The 20th century has been an era of astonishing change, and no region typifies that change better than the American West. The population of the 18 western states has increased sevenfold, from 11.2 million in 1900 to 77.2 million in 1990.[1] The lifestyle of the "typical" westerner has been dramatically transformed. Only 6 percent of the West's population lived in large cities when the century began, but large cities are now home to 82 percent of westerners.[2] In 1900, 45 percent of the West's work force was engaged in agricultural pursuits. By 1990, this figure had fallen to just 3 percent.[3] Rather than working directly with the West's lands and waters for their livelihood, most westerners now live and work in cities.

The status of the West's water resources has also changed. The West's rivers today are highly regulated and diverted compared to the largely natural flows found by the settlers in the 19th century. The West had few reservoirs, fewer still of significant size, when the century began. Western reservoirs with a capacity of 100 acre-feet or more now number almost 10,000; there are an additional 20,000 reservoirs of smaller size.[4] The West's surface water is now being diverted to offstream uses at the rate of almost 119 million acre-feet per year.[5]

Water used offstream has caused the desert to bloom and cities and towns to prosper—but these developments have not been without cost. Only in some of the uppermost river stretches and smaller tributaries, and in a few of the West's most isolated areas, does streamflow remain unaltered by diversion or storage projects. Streamflow volumes throughout the rest of the West have been diminished. With the loss of instream flows has come a significant loss in aquatic and riparian habitat, recreational opportunities, and water quality.

Society values instream flows more highly now than at the turn of the century. The primary concern of the great majority of westerners at the dawn of the 20th century was to secure the basic elements of life, such as food, shelter, and safety. If these could be obtained by using water offstream, then the water was quickly diverted. As urban areas have grown to dominate the population and economy of the West, and average real incomes have risen substantially, residents have become more concerned about the quality of the surrounding environment and about the existence of recreational opportunities and other amenities. The beauty and proximity of the western landscape, including the West's rivers and streams, have much to do with the desirability of the West as a place to live and work. Today, people are moving West not to be farmers, ranchers, lumberjacks, or miners but to find work in the cities, to retire, and to seek a better quality of life, including access to the West's scenic beauty and outstanding recreational opportunities. Although it may dismay those westerners who still make their living directly off the West's land and water, the character of the West has changed dramatically. With that change has come an increase in the demand for the integrity of the natural environment, including the existence of instream flows.[6]

Changes in the social and economic character of the West have been accompanied by changes in western water law. Law adapts to the circumstances and values of the times, as clearly exemplified in the adoption of the prior appropriation doctrine in the mid-1800s by emigrants who had experienced only a riparian-based water law. More recently, western water law has evolved to recognize at least some instream water uses as beneficial and to establish procedures for protecting instream flows.

Where rivers still flow, instream flow protection measures can help assure that future appropriations will not deplete streams beyond the amounts deemed necessary to protect at least the most valuable instream water uses. Table 11.1 lists methods of protecting instream flow;[7] column 2 of the table identifies the methods that can be used to protect unappropriated flow from future diversion. Unfortunately, the increased value of instream flows and accompanying changes in western water law occurred largely after much of the flow in many western rivers had already been allocated to other uses. New instream-flow water rights, instream reservations, or other protective methods are typically so junior as to be ineffectual when most needed, in dry years when flows are most scarce. If instream flow is to be restored to a level commensurate with its new value, at least some water now allocated to offstream uses must be reallocated to instream uses.

Column 3 of table 11.1 identifies several instream flow protection methods that can be used to increase streamflow volumes beyond existing levels. Among the four methods available under state law, the most direct is a trans-

TABLE 11.1
State and Federal Methods of Protecting Instream Flow

Methods of protecting instream flow	Can method be used to protect existing instream flows?	Can method be used to increase instream flows?	Can method be imposed on unwilling parties?
Methods available under state law			
Specification of minimum flows or flow reservations by responsible state agencies	yes	no	no
Case by case evaluation of appropriative permit applications for consistency with public interest	yes	no[a]	no[a]
Instream flow rights	yes	no	no
Transfer of offstream right (permanently or temporarily) to instream purposes	no	yes	no
Movement of senior offstream right downstream	no	yes	no
Restrict movement upstream of a senior offstream right[b]	yes	no	no
Public trust doctrine	yes	yes	yes
State rivers programs	yes	no	no
Private water management agreements	yes	yes	no
Methods available under federal law			
Federal reserved water rights	yes	yes	yes
Federal land management agency administration of special use permits	yes	yes	yes[c]

(*Table Continues*)

TABLE 11.1 (CONTINUED)

Methods of protecting instream flow	Can method be used to protect existing instream flows?	Can method be used to increase instream flows?	Can method be imposed on unwilling parties?
Management of existing federal dams	yes	yes	yes[d]
FERC licensing and relicensing of nonfederal hydroelectric dams	yes	yes	yes[e]
Clean Water Act restrictions	yes	yes[e]	yes[c,e]
Endangered Species Act restrictions	yes	yes	yes
Wild and Scenic Rivers Act designation	yes	no	no

[a]California courts are an exception, as they have allowed review of existing permits. To our knowledge, other states have not explicitly allowed retroactive evaluation of permits based on the public interest.

[b]Restricting an offstream right from being transferred upstream avoids a potential future depletion of instream flow between the existing and the potential upstream point of diversion. Such restrictions might be purchased and are similar to conservation easements which restrict the owner's development options.

[c]Restrictions can only be imposed on the party applying for the permit, license, or relicense.

[d]The ability to impose constraints on unwilling parties depends on the specific contractual agreements governing management of the dams.

[e]This "yes" is tentative. The U.S. Supreme Court's 1994 decision in *PUD No. 1 of Jefferson County v. Washington Department of Ecology* affirmed the role of the Clean Water Act in protecting instream flow from water projects whose diversions would lower flow below the level deemed necessary by a state to meet its water quality standards for a stream segment's designated use. This case dealt with a 401 certification for a new project. Although the same authorization is likely to apply to any federal license, including a relicensing under FERC or a reissue of a special use permit for a water project on federal land, we know of no cases where the Clean Water Act has actually been used to reverse an existing lack of instream flows.

fer of water rights from offstream use to instream use. Transfers involve either permanent transfer via sale or donation or temporary transfer via leases or contingent agreements such as dry-year options. A related but less common method is the transfer of an offstream right to a downstream location, which improves instream flow in the stretch between the old and new points of diversion. The other methods available under state law include private water-management agreements, such as the alteration of reservoir operating rules, and the seldom used public trust doctrine.

The six methods of increasing instream flows under federal law are diverse and complex. Four of the methods—administration of special-use permits, management of federal dams, FERC licensing of private hydroelectric dams, and Clean Water Act restrictions—can be used only to enhance instream flows at facilities managed by federal agencies, or at facilities that require some kind of authorization under federal law. These methods are limited in effectiveness because they affect flow only at the point where a water management facility exists. Thus the methods cannot, in and of themselves, keep downstream water users from diverting the water made available by the change in facility operations. The other two methods, federal reserved water rights and Endangered Species Act restrictions, are very powerful and can be applied to significant lengths of rivers and streams. However, use of these two methods to date has drawn substantial opposition, and the methods have been used relatively infrequently to protect instream flows.

All of these methods of protecting instream flow are controversial to some extent, at least in some quarters.[8] However, the methods that have stimulated the most conflict are those that have the potential to increase instream flows by imposing restrictions on unwilling parties. Existing water users typically argue that the reallocation of water from offstream use to instream use—if allowed to take place at all—should take place only on a voluntary basis, with full compensation. Of the instream flow protection methods available under state law, only the public trust doctrine can be imposed on unwilling water users (see table 11.1, column 4). Several of the methods available under federal law can be used, and occasionally have been used, to force instream flow protection on unwilling water users. Use, and attempted use, of the federal methods in this manner are virtually always accompanied by vociferous objections from affected offstream water users and a good many others.

The rise in the popularity and use of water markets, and the unpopularity of federal involvement in state water matters, make it tempting to argue that we should rely solely on private parties acting through markets to identify and solve instream flow problems. Water purchases by groups such as The Nature Conservancy demonstrate the ability of private conservation groups to use the market to protect instream flows. However, many of the western states do not allow private parties to apply for, purchase, or own water rights for instream purposes. Some states do not allow any entity, public or private, to acquire water rights for instream purposes. In addition, the "public good" characteristics of instream flows, as described in chapter 4, indicate that even if private parties were allowed to participate fully in the transfer of offstream water rights to instream use, instream flows would still be underprotected without at least some intervention by the public sector. Thus we must look to public action—to use of our federal, state, and local governments—to at least some degree if we are going to achieve a balance between offstream and instream water uses that reflects contemporary values.

It is easy to understand why attempts to retroactively protect instream flows—i.e., to put water that has already been allocated to offstream use back into streams—raise serious objections. Some people argue that water uses that have lowered instream flows below the minimum levels necessary to, say, maintain healthy fish populations, should never have been allowed in the first place, so that retroactive reallocation is merely correcting a past wrong. But the fact is that these uses were allowed, and even encouraged, to occur at a time when values and conditions were different. Retroactive protection without compensation essentially asks just a portion of the population, current water users, to bear the entire cost of the larger public's benefit. This issue cannot be resolved solely by technical measurements of benefits and costs, for it is essentially an issue of fairness. The issue of who will pay for instream flow improvements in streams that have previously been allocated to offstream use will ultimately have to be settled in the legislatures, and perhaps in the courts.

It is less obvious why attempts to protect flows currently in streams from future appropriation should raise objections, particularly when such attempts do not go beyond protecting the most valuable instream flows. Part of the explanation lies in the general lack of information about instream flow protection among water users, water agency officials, and the general public. It is easy to fear, reject, or ignore that which is not known, particularly when long-established property rights are at issue. Many, probably even most, water users have little understanding of the effects that instream flow protection measures would have on their own current and future water uses. But objections based on lack of knowledge should gradually subside as information about instream flow protection becomes more widespread and people gain familiarity with instream flow protection measures that have already been implemented.

Some people object to instream flow protection on the grounds that it limits future diversions and therefore constrains future economic activities, such as agricultural expansion or urban development, for which diversions are necessary. It is true that instream flow protection measures constrain future diversions; that is, in fact, the purpose of the measures. But objections based on this fact should be tempered by at least four major considerations.

First, if there is not enough water to meet all possible needs—the usual situation virtually everywhere—then water should be allocated in such manner as to produce the greatest value. If a certain portion of the streamflow on a given reach of stream can produce more value to society if left instream than if diverted for offstream use, then it is entirely appropriate for that streamflow to be left instream.

Second, objectors often overlook the fact that more efficient use of diverted water can provide for new and expanded offstream uses of water without the need to increase diversions. Water conservation is often complicated by issues concerning ownership of water that has not previously been used efficiently, but conservation has nevertheless been growing in importance and effectiveness in many locations throughout the West.

Third, objectors frequently fail to realize that choosing to leave water instream does not categorically exclude offstream uses of water. Leaving some water instream, for some period of time, over some length of the stream, does not make water totally unavailable for offstream use. It is a rare situation indeed in which 100 percent of the natural flow of a watercourse is allocated to instream use. Instream flow protection generally involves the allocation of only a part of a stream's flow to instream purposes. And because instream uses are essentially nonconsumptive, the water is still available for offstream use after it has flowed through the protected reach. The decision to fight over every drop of water as if it were the only drop can be an effective political and legal tactic—often used by both sides of the instream flow controversy—but in most situations it does not square with reality.[9]

Fourth, objectors either don't realize or fail to mention that instream flow protection is not irreversible. Relative to existing allocations and values, instream flows currently have too little protection. But values and conditions can change. If at some point society determines that water being used instream would have greater value if used offstream, a new allocation can be made.

Fortunately, instream flow protection measures often do not have any adverse impact on other water uses. The fact that instream water uses are nonconsumptive greatly reduces the potential for conflict with other water uses. Rather than prohibiting the consumptive use of water, instream flow protection measures merely constrain the location at which diversions can occur. The impact of an instream flow requirement on other water uses decreases, all else being equal, as the location of the protected reach moves upstream. Though certainly not all river and stream reaches with valuable instream uses needing flow protection are located in headwater areas, the availability of sites where there is little potential for conflict—such as many sites in national forests and parks—provides instream flow protection advocates with opportunities to demonstrate to other water users that protection measures do not necessarily have to disrupt existing patterns of water use.

We do not mean to imply, however, that instream flow protection does not ever require hard choices and substantial costs. There are still many unprotected instream flows, particularly at lower elevations, that would provide greater benefits if left instream than if diverted for offstream use. In fact, the relative ease with which higher-elevation streams can be protected has caused much of the existing protection activity to take place there, leaving stream segments at lower elevations with very little protection. Many of the most substantial instream flow protection controversies are taking place, and will continue to take place, over river segments flowing through lower elevations. It is in these areas that the most care will have to be taken to ensure, first, that the water truly would be more valuable if used for instream purposes than if diverted for offstream use, and second, that the concerns of existing offstream water users are adequately addressed.

Because the protection of instream flows will require that at least some water currently allocated to offstream use be reallocated to instream use, we should not close without briefly discussing the sensitive topic of where that additional flow is to come from. Clearly, the answer in any given case will depend on the physical, legal, and economic circumstances of the particular river or stream at issue. And all current offstream uses are potential sources of increments in instream flow. Cities, because of their rapid growth, are generally viewed as a source of expanding need for offstream water use. However, the extensive water infrastructure built by municipalities—dams, reservoirs, canals, pipelines, and treatment facilities—provides many opportunities for enhanced instream flows through more efficient, better coordinated water flow management. And, as the Mono Lake decision made clear (see chapter 6), even relatively high-valued urban water uses are not immune to the sacrifices necessary to accommodate the new water allocations demanded by changing societal values.

The agricultural industry, however, is likely to provide most of the water that will eventually be reallocated to instream use, primarily because it is by far the largest offstream user of water. Irrigated agriculture accounts for 77 percent of all freshwater withdrawn from rivers, streams, and lakes and 90 percent of all water consumption in the West.[10] Furthermore, most irrigation water in the West is used to grow feed grains and forage for livestock, or to grow cotton, rather than to directly feed the urban masses.[11] The value added by water consumption in irrigated agriculture tends to be lower—especially when the water is used to grow feed grains, forage, and other low-value crops—than in other consumptive water uses.

It is important to keep in mind that the possibility of transferring water from agricultural use to instream use is likely to be fiercely resisted by many—though definitely not all—agricultural water users.[12] Much of this resistance is due to fears that water will be taken from farmers involuntarily. Much is also due to the fact that instream flow protection is often perceived as a threat to a rural way of life that is increasingly challenged for other reasons as well. Instream flow advocates are not the only ones looking at the huge proportion of the West's water being used for agriculture. Municipalities, in particular, have been actively seeking to obtain a greater share of that water, and the adverse impacts of agricultural practices on water quality have drawn increased scrutiny now that so much headway has been made in controlling point sources of pollution. Other threats to the rural way of life such as expanding residential development, economic difficulties, and changing values and uses of our public lands have also contributed to a defensive mind-set on the part of many longtime rural residents.

There are several factors that may temper the resistance of agricultural water users to the transfer of water to instream use. Transfers can often be accomplished through voluntary measures, whether through sales or through water conservation. Particularly for farmers facing economic pressures from other sources, the sale of some water may provide monetary resources that enable farmers to stay in business in the long run. Water conservation—the use of more efficient irrigation methods—is a particularly promising potential source of water for instream use, as irrigation efficiencies tend to be quite low in many areas of the West.[13] Conservation raises the interesting possibility that more water can be made available for instream use without having any adverse effect whatsoever on agricultural production.[14] Even in situations in which less land is irrigated as a result of the transfer of water to instream flows, the sheer enormity of the amount of water currently used in agriculture means that even a relatively small reduction in irrigated agriculture might result in a relatively large increase in instream flow. Adverse effects will be further reduced by efforts to target water used on the lowest-value crops, and on the least productive lands, for transfer to instream flow.

It is also worth noting that many agricultural water users are likely to be sympathetic to instream protection goals. Though fears that water will be taken from them involuntarily often reduce their public expressions of support for instream water uses, many residents of the rural West have more frequent contact with, and appreciation for, flowing rivers than do most urban residents. So it is certainly possible, if approached with patience and concern for the legitimate fears and issues raised by agricultural water users, that significant quantities of water currently being used for agricultural purposes can eventually be returned to rivers and streams to support instream uses of water.

Instream flow protection is a difficult and complex issue because of (1) the large mix of often competing instream and offstream water uses; (2) the mix of rather complicated state and federal laws and policies and agencies that control water allocation; (3) inadequate knowledge about the relation of flow to some water uses; (4) the stochastic nature of stream flow; and (5) the critical importance of water to so many human endeavors. Because of this complexity, it is easier to elucidate general principles about a balance among competing water uses than it is to determine the best instream flow level for a given reach of stream. Physical and biological scientists and engineers have not learned enough about the effects of water availability on different water uses, and social scientists are not good enough at measuring the public's values, to achieve precise instream flow recommendations. Any proposed balance, even the best intentioned, will be an educated guess.

Indeed, we as a society can be certain of little, except that there will be change that we cannot adequately anticipate. The West has changed greatly and will continue to change. One day, perhaps under severe population pressure, instream flow protection may recede into the background as a desirable but impractical concern of forebears. But for now we must respond to today's circumstances and values, which indicate that more protection for instream flows is needed. We can, however, rest assured that in protecting instream flow today, we are not making irreconcilable mistakes that will haunt us tomorrow, because instream flow protection does not require irreversible commitments of resources. To the contrary, instream flow protection preserves our future options.

There is no single instream flow protection policy in the West. The status of instream flows is, for the most part, a result of myriad choices made to accomplish a variety of ends. Most decisions that affect streamflows almost assuredly were not made with instream flow in mind. Just as there has been no single policy or source of our instream flow situation, solutions to the instream flow problems will not come from any one source. Solutions will come from water users of all sorts, from land and waterrights owners, from natural resource managers, from state and federal agencies, from legislators, and from the public at large.

To date, few people have been actively involved in making the choices that affect the status of instream flows. Though the choices that are made affect everyone who lives in or visits the West, most people have been involved only tangentially, if at all, with instream flow protection issues. Better, more creative, and more effective solutions to water resource problems will be possible only if more people—policymakers, natural resource managers, and citizens alike—have access to information about instream flow uses, issues, and values and about instream flow protection methods and options.

There is no question that farmers, urban developers, and others relying on water diversions enjoy seeing a beautiful stream, just as environmentalists or river recreationists do, and that concern about aquatic species and their environments is shared, albeit in varying degrees, by all. The difficult questions arise when one must decide how a given stream segment should be managed. Hard choices between instream and offstream water uses will continue to be faced in thousands of locations throughout the West as pressures intensify for various kinds of water uses. Decisions will, and should, vary at different locations, to accommodate differing circumstances and relative values at each site. If the decisions are made with due consideration for the best available information about effects, values, rights, and methods, we can hope to achieve a worthy balance in the use of our water resources.

Notes

Chapter One

1. Doerksen related the anecdote in a 1991 article in the journal *Rivers* (Doerksen, 1991).

2. Trelease (1976), pp. 8–9.

3. Ibid., p. 18.

4. Ibid., where he presciently noted that "if a new law is needed, you will find that law reform is no sport for the short winded."

5. Doerksen (1991), p. 99.

6. Scientists and engineers, particularly in other countries, sometimes employ metric units when working with water, but for the remainder of the text we will use non-metric units, as these are much more commonly used by water users and administrators in the United States.

Chapter Two

1. Haury (1967).

2. Haury (1945), Meyer (1984), and Reisner (1986). Meyer noted that the loss of freedom implied by the social control necessary to organize multitudes of people to build and maintain extensive irrigation systems may have dissuaded other tribes from following a similar path. Reisner contended that the difference between the world's great early civilizations and the more modest ones was that the former used irrigation extensively, while the latter did not (see generally at pp. 474–76).

3. Haury (1967). See also Reisner (1986), pp. 474–76.

4. Meyer (1984) addresses this topic with respect to water ownership patterns among early Indian cultures of the American Southwest, p. 18.

5. Though it is not possible to determine the population of the Hohokam civilization with a high degree of certainty, Reisner (1986) offered the insight that it was probably not until the years following World War II, when Arizona's population reached 400,000, that the state's population once again was as high as it was during the peak of the Hohokam civilization some 750 years earlier.

6. Hollon (1961), Reisner (1986). Reisner reports the possibility of this explanation for the demise of the Hohokam at pp. 264–66 and the problems of irrigation, drainage, and salt more generally at pp. 477–88.

7. The San Juan River arises in the San Juan Mountains of southwestern Colorado and flows west to join the Colorado River in southern Utah. The San Juan River watershed encompasses much of what is now known as the Four Corners area, where the states of Colorado, Utah, New Mexico, and Arizona come together. The area contains many remnants of prehistoric dwellings and other artifacts, indicating that it has been almost continually inhabited for many centuries. It currently is home to several Indian tribes, including the Navaho, Ute, and Jicarilla Apache, as well as to several Pueblo tribes of northern New Mexico.

8. Hollon (1961). Climatic conditions of prehistoric times can be ascertained through the study of tree rings. Trees grow much more quickly during wet years than dry years, so tree rings are larger during wet periods. By matching up ring patterns from a series of increasingly older trees, researchers have been able to establish a continuous chronology of wet and dry years extending back several centuries from the present.

9. Much of the following discussion of water uses and institutions in the Spanish West is derived from Meyer (1984), unless otherwise noted.

10. Hollon (1961).

11. Meyer (1984). See generally p. 50 and pp. 84–88 for a discussion of the amalgamation process and the critical nature of water in determining mineral output from the mines.

12. According to Meyer (1984), regulations to this effect were issued by King Philip II in 1573. The general principles and concepts to be used in making water distributions were later brought together in the "Plan de Pitic," the founding ordinance for the town of Hermosillo, Sonora, in 1789. Among other things, the Plan states that water is to be shared by all (including both Spaniards and Indians), and that all shall cut back their use during times of shortage. See generally pp. 29–37.

13. Meyer (1984), p. 133, and Pisani (1992), p. 39.

14. Book IV, Title 17, Law 5 of the *Recopilacion de leyes de los reynos de las Indias*, a compilation and codification of Spanish New World law completed in 1681, cited in Meyer (1984), p. 135.

15. Meyer (1984), pp. 156–57.

16. Common law consists of the decisions accruing from case-by-case adjudication of disputes by the courts; the resolution of new disputes is guided by precedents set in older disputes. Common law is thereby distinguished from civil law, which

consists of ordinances and statutes issued by the executive and legislative branches of government.

17. Two influential law review articles written in 1918 and 1919 by water rights scholar Samuel Wiel attributed the origins of American riparian law to the civil law of France, the Code Napoleon (Wiel, Samuel C., "Origin and Comparative Development of the Law of Watercourses in the Common Law and in the Civil Law," *California Law Review* 6[1918]: 245, 342; "Waters: American Law and French Authority," *Harvard Law Review* 33[1919]: 133, 147). His argument was that two influential American judges had relied heavily on the Code Napoleon in deciding American water law cases, and that the English in turn had been influenced by developments in America to move their own water law system from an appropriative rights scheme to a riparian structure. According to Hutchins (1971, p. 181), this thesis was widely accepted by water rights scholars for several decades. Doubts were raised about the accuracy of Wiel's thesis in the 1950s, however, and a review of Wiel's articles by Arthur Maass and Hiller Zobel in 1960 strongly contradicted the original conclusions (Maass, Arthur, and Hiller B. Zobel, "Anglo-American Water Law: Who Appropriated the Riparian Doctrine?" Graduate School of Public Administration, Harvard, *Public Policy* 10:[1960]: 109–156). Since publication of the Maass and Zobel article it has been widely agreed that English common law, and the water law of the eastern American states, have always been riparian. See generally pp. 181–83 of Hutchins (1971). Excerpts of Wiel's 1919 *Harvard Law Review* article and of the 1960 article of Maass and Zobel are found in Beuscher (1967) at pp. 87–95.

18. Ausness (1986), citing W. Buckland, *A Textbook of Roman Law from Augustus to Justinian* (3rd ed., 1963), and R. Lee, *Elements of Roman Law* (3rd ed., 1952).

19. Johnson and DuMars (1989). "Riparian" is derived from the Latin *riparius* and refers to things that are "of, pertaining to, or situated or dwelling on the bank of a natural watercourse or body of water" (*Random House Dictionary of the English Language,* unabridged edition, 1979).

20. Leshy (1990), p. 5.

21. 4 Mason 397 (1827). Excerpts of the case can be found in Beuscher (1967), pp. 81–86.

22. An important exception to the general rules stated above is the development of water rights under the riparian doctrine through direct appropriation. Appropriation of water for out-of-stream uses is legal under the English-American common law system if the new water user is able to obtain the consent of all affected riparian landowners. Consent may be explicit but may also be assumed if the new water use negatively impacts riparian owners but is nonetheless allowed to continue without interruption or objection for a specified period of time, usually twenty years. Rights developed through implied consent (or through "adverse possession") are often referred to as "prescriptive" rights. Rights developed through grant or prescription are as valid as traditional riparian rights and receive the same legal protection.

23. Goldfarb (1988), pp. 21-24.

24. This limitation of the riparian right was not likely to have caused significant

problems in the eastern states, because water sources were plentiful and out-of-stream uses were few (Leshy, 1990).

25. Ausness (1986), p. 417.

26. Pisani (1992), p. 15.

27. Ibid., pp. 15–17. Pisani states that hydraulic mining was preceded by "ground sluicing" methods starting as early as 1850, and that the initiation of "modern" hydraulic mining is generally attributed to Edward E. Mattison, who first used the method at American Hill, near Nevada City. Despite the massive works required to divert and transport water to the mining site, hydraulic mining quickly became established as the fastest and cheapest way to mine large quantities of gold. Hydraulic mining was also extremely hard on the environment, causing nearby rivers and even the remote San Francisco Bay to become choked with sediment. For these reasons the State of California abolished hydraulic mining in the 1880s.

28. Ibid., pp. 18–19.

29. Hutchins (1971), pp. 164–67. The precise form of the rules depended on local conditions and the preferences of individual miners, but the basic principles were often quite uniform. Hutchins notes that two earlier scholars had recognized striking similarities between the rules adopted by the California miners and those used in much earlier mining enterprises in the Old World. One, William E. Colby, compared the right of free mining and free use of water in California with the customs used by Germanic miners in the Middle Ages, and claimed that the doctrine of prior appropriation was in wide use throughout the world's mining regions (Colby, William E., *The Freedom of the Miner and Its Influence on Water Law,"* published in *"Legal Essays, in Tribute to Orrin Kipp McMurray,* pp. 67–84 [1935], cited by Hutchins, p. 165).

30. Cal. Stat. 1851, ch. 5, sec. 621, cited by Hutchins (1971), p. 164, note 24.

31. Tarlock (1989), sec. 5.02[2], pp. 5–7, posed the difficulties facing the state courts in two questions: "First, once the territory became part of the United States, either as a state or territory, why did the common law of riparian rights not apply? All states received the common law. Second, how could state law be the source of a water right on federal land? Federal law governed federal patents so it should follow with them that all federal patents either carried with them riparian rights or other federal water rights."

32. *Eddy v. Simpson,* 3 Cal. 249 (1853).

33. *Irwin v. Phillips,* 5 Cal. 140 (1855).

34. The court's struggle to resolve conflicts arising from different sets of expectations was not yet over, however. The *Irwin* decision was one of several of the Supreme Court's rulings on water issues in the 1850s, and not all of them were consistent with this ruling. For example, two years later the Supreme Court used common law riparian principles rather than prior appropriation principles to resolve a dispute between two groups of miners that were both working claims located next to a stream (*Crandall v. Woods,* 8 Cal. 136 [1857]). The court indicated in its discussion of *Crandall* that the decision in *Irwin* also would have been decided on common law rather than prior appropriation principles if the riparian miners had been first to establish their claims.

35. *United States v. Gratiot,* 39 U.S. (14 Pet.) 526 (1840), and *United States v. Gear,* 44 U.S. (3 How.) 120 (1845), respectively. Cases cited by Tarlock (1989), sec. 5.02[2], pp. 5–6, and Coggins and Wilkinson (1987), p. 100.

36. Tarlock (1989, pp. 5–7) cited Wiel, *Water Rights in the Western States* (1911, p. 88), to the effect that the miners "took possession of the public lands, mines, water and timber wherever they located, following out as between themselves the customs and rules of prior appropriation of all things prevailing in California, and not hearing from Congress one way or the other."

37. 14 Stat. 251 (1866), as amended 43 U.S.C. sec. 661 (1964). The actual title of the legislation was an Act granting the Right of Way to Ditch and Canal Owners over the Public Lands; the deceptive title was part of a strategy to bypass Senate committees dominated by unsympathetic eastern senators. See Wilkinson (1992), pp. 41–43. The 1866 Mining Act is discussed in more detail by Meyers (1971), Hutchins (1971), Dunbar (1983), and Coggins and Wilkinson (1987).

38. Cited by Meyers (1971), p. 27.

39. Tarlock (1989, pp. 5–7) characterized the recognition of prior appropriation, first by the state courts and later by Congress, as "a classic example of a post-hoc rationalization of a fait accompli."

40. 16 Stat. 217 (1870), as amended 43 U.S.C. sec. 661 (1964).

41. 19 Stat. 377 (1877), as amended 43 U.S.C. sec. 321 (1964).

42. The Desert Land Act applied specifically to Arizona, California, Idaho, Montana, Nevada, New Mexico, North Dakota, Oregon, South Dakota, Utah, Washington, and Wyoming. It was extended to Colorado by amendment in 1891 (Hutchins, 1971, p. 173).

43. Cited in Meyers (1971), p. 27. Although the Act recognized water needs for irrigation, mining, and manufacturing, it was silent on the question of stock watering on public land.

44. In the years following its passage, the supreme courts of some of the western states had disagreed about whether or not the water policies expressed in the Desert Lands Act were limited to desert portions of the public domain. The Oregon Supreme Court had ruled in *Hough v. Porter,* 51 Oreg. 318, 95 Pac. 732 (1908), that the policies applied to all of the public domain, not just to the desert lands. The South Dakota Supreme Court, in *Cook v. Evans,* 45 S. Dak. 31, 185 N. W. 262 (1921), agreed. The Washington Supreme Court came to a different conclusion, ruling in *Still v. Palouse Irr. & Power Co.,* 64 Wash. 606, 117 Pac. 466 (1911), that the Act applied only to the reclamation of desert lands, and that common law riparian rights still attached to nondesert portions of the public domain. The California Supreme Court came to a similar conclusion in *San Joaquin & Kings River Canal & Irr. Co. v. Worswick,* 187 Cal. 674, 203 Pac. 999 (1922).

On appeal from a case arising in Oregon, the U.S. Supreme Court ruled in *California Oregon Power Co. v. Beaver Portland Cement Co.,* 295 U.S. 142 (1935), that the interpretation of the Oregon Supreme Court was correct—the Desert Lands Act, together with the other two acts, "applied to all the public domain in the States and Territories named, and that it severed the water from the public lands and left the unappropriated water of nonnavigable sources open to appropriation." However, the court also ruled that the states were not necessarily obliged to apply the prior appropriation doctrine; it said that the Congress had left it to the individual states to decide what doctrine they would use, and thus it was permissible for Washington and California, and any other state, to choose to apply common law riparian principles in one set of circumstances and prior appropriation principles in another set of circumstances.

45. Brough (1898), p. 52. Where they could, the Mormons utilized existing canals; Mormons that settled Mesa, Arizona, were actually able to use an old Hohokam canal running from the Salt River (Dunbar, 1983).

46. Brough (1898).

47. For example, George Hansen founded a colony in Southern California (Anaheim) in 1857 that used a pattern of settlement similar to that of the Mormons and was probably influenced by them. In 1870 Nathan Meeker established a colony in Colorado, later named Greeley in honor of Horace Greeley, who had visited Salt Lake City in 1859 and been favorably impressed by Mormon irrigation institutions (Dunbar, 1983).

48. Brough (1898), p. 41.

49. *Lux v. Haggin,* 69 Cal. 255, 10 P. 674 (1886).

50. Texas later adopted appropriative principles for the entire state, and, like most other mixed-doctrine states, took strong steps to limit the application of riparian rights. The state did, however, continue to recognize preexisting rights developed under the riparian doctrine.

51. Colorado Constitution, Article XVI, Section 6.

52. *Yunker v. Nichols,* 1 Colo. 551 (1872).

53. *Schilling v. Rominger,* 4 Colo. 100 (1878).

54. *Coffin v. Left Hand Ditch Co.,* 6 Colo. 443 (1882).

55. *Hutchinson v. Watson Slough Ditch Co.,* 16 Idaho 484, 101 Pac. 1059 (1909).

56. *Hough v. Porter,* 51 Oreg. 318, 95 Pac. 732 (1908). The United States Supreme Court later agreed with this reasoning in the famous case of *California Oregon Power Co. v. Beaver Portland Cement Co.,* 295 U.S. 142 (1935).

57. 1909 Water Code, Oreg. Laws 1909, ch. 216, now referred to as the Water Rights Act, Oreg. Rev. Stat. sec. 537.010 (Supp. 1973).

58. Session Laws 1861, p. 67, sec. 1, as cited by the court in *Coffin v. Left Hand Ditch Co.,* 6 Colo. 443 (1882).

59. Session Laws 1862, p. 48, sec. 13, as cited by the court in *Coffin v. Left Hand Ditch Co.,* 6 Colo. 443 (1882).

60. As cited by Dunbar (1983), p. 75.

61. *Thorp v. Freed,* 1 Mont. 651 (1872), cited by Pisani (1992), at p. 351, note 65.

62. Dunbar (1983), pp. 71–72.

63. As quoted by Pisani (1992), p. 36.

64. John Norton Pomeroy, *A Treatise on the Law of Water Rights,* St. Paul, Minnesota, 1893, as cited by Pisani (1992), pp. 46-47.

65. Cited by Pisani (1992), p. 36.

66. The Board of Control referred to by Brown was an administrative feature that was ultimately adopted by the convention and became a part of the state of Wyoming's constitution; the rise of administrative systems is discussed in more detail below.

67. Cited by Pisani (1992), p. 60. According to Pisani, Brown was not alone in his dislike of the prior appropriation system.

68. Dunbar (1983, p. 16) states that the changes became necessary because a growing population led to more conflicts, which placed great strains on the older system. An alternative explanation, offered by Thomas (1920, pp. 55–56), was that the territorial legislature privatized rights because they feared takeover of local government positions by federal officials, with whom they had been engaged in a series of conflicts.

69. See generally Brough (1898) and Thomas (1920).

70. Thomas (1920).

71. This information came from a publication by Traub, published in 1988. Unfortunately, we have misplaced that reference and have been unable to find it in the library. We would greatly appreciate it if a reader who is familiar with the Traub reference would contact us with the details.

72. Reisner (1986).

73. P.L. 102-575, 43 U.S.C.S. sec. 390h (1995).

74. The Reclamation program originally encompassed the 16 westernmost states, not including Alaska or Texas. Texas was added in 1905.

75. U.S. Bureau of Reclamation (1993). Irrigated acreage in Utah in 1894 totaled 417,455 acres (Brough, 1898, table 7, p. 81). The size of the federal reclamation program greatly exceeds that of previous reclamation efforts: the total extent of canals built by the Hohokam measured in the hundreds of miles, while the Mormons were able to build just over one thousand miles of canals during the very active development decade 1860–70.

76. 34 Stat. 386. Traub (1988; see note 71). Dams were not allowed to impair navigation.

77. Traub (1988), p. 18. See note 71 to this chapter.

78. Pisani (1992), p. 31.

79. See generally Dunbar (1983).

80. See generally Hutchins (1971), Dunbar (1983), and Grant (1987).

81. Dunbar (1983), p. 109.

82 Wyoming Constitution, Article 8, Section 3 (1995).

83. See generally Hutchins (1971), Robie (1977), Dunbar (1983), and Grant (1987).

84. Alaska Stat. sec. 46.15.080(b) (1984), as cited by Grant (1987), p. 689.

85. See Hutchins (1971) for a discussion of water-use preferences in each of the western states.

86. Colorado Constitution, Article XVI, Sec. 5.

87. Solley, Pierce, and Perlman (1993).

88. Ibid., table 2, p. 11.

89. Ibid., table 6, p. 15.

90. See Palmer (1991) for an extended discussion of the loss of streamflows in the Snake River.

91. The map and figures were developed from the National Inventory of Dams database produced in 1992 by the U.S. Army Corps of Engineers and the Federal Emergency Management Agency (FEMA). The database is available on compact disc from FEMA's National Dam Safety Program, 500 C Street, S.W., Washington, D.C. 20472. The precise figures are 9,981 dams (excluding stock ponds) of at least 100 acre-feet and 30,091 dams in total, in the western states.

92. Figure 2.2 includes dam and stock ponds of all sizes. The source is given in the preceding endnote.

93. Robie (1977).

Chapter Three

1. In many instances the instream flow requirements of fish have served as a surrogate for other instream flow needs, since flows considered adequate for maintaining healthy fish populations are often assumed to be adequate for meeting a number of other instream needs as well.

2. Marcus et al. (1990).

3. California State Lands Commission (1993).

4. Depending on the particular species, region, and riverine conditions involved, anadromous fish may spend anywhere from 2 months to a few years in freshwater before migrating to the ocean.

5. The most common anadromous trout is the steelhead, which is a variety of rainbow trout. Some cutthroat trout species also spend most of their lives in the ocean, though most do not.

6. Lewis (1991) provides a succinct and readable description of salmon types and life cycles.

7. Salmon in the Columbia and Snake river systems formerly were able to reach parts of Nevada, Wyoming, and Montana, but since extensive development of the river basins occurring this century, they are now largely restricted to the states of Oregon, Washington, and Idaho.

8. Quoted and cited by Marcus et al. (1990).

9. Marcus et al. (1990); California State Lands Commission (1993).

10. The shading effect of riparian vegetation is most significant on smaller streams, where riparian vegetation has the potential to shade the entire water surface. On very wide rivers, where relatively little of the river's surface area is shaded by veg-

etation, the effect is negligible. The angle at which the sun strikes the surface of the water depends on latitude, season, and time of day.

11. Marcus et al. (1990).

12. For example, the culture and livelihood of the Paiutes of Pyramid Lake, Nevada, were closely tied to the life cycle of the now extinct Lahontan cutthroat. Their culture continues to be tied to the life cycle of the endangered cui-ui fish, which lives in Pyramid Lake and spawns in tributary rivers.

13. For example, Wolfe and Walker (1987) document the reliance of Alaska natives on subsistence fisheries.

14. California State Lands Commission (1993).

15. "Riparian" is frequently used in conjunction with the words "zone" or "area," i.e. "riparian zone" and "riparian area." These two terms are generally interchangeable. In this context, the riparian zone is the area of transition between aquatic and terrestrial habitats—areas that are influenced by their proximity to a water source but are not actually aquatic. The boundaries of this transitional zone are often roughly similar to the limits of the floodplain and, like floodplains, vary enormously in areal extent between different rivers and between different reaches of the same river. Some riparian areas—particularly those associated with very small streams at the bottom of narrow, steep valleys—may be just a few feet wide, while others—usually associated with larger streams along wide valley bottoms—may extend for over a mile in width. The Arizona Riparian Council states that the term *riparian* is intended to include "vegetation, habitats, or ecosystems that are associated with bodies of water (streams or lakes) or are dependent on the existence of perennial, intermittent, or ephemeral surface or subsurface water drainage" (*Sonorensis,* 1988). See also note 19, chapter 2.

16. Cottonwood trees are a familiar sight along many irrigation canals as well as along the banks of natural rivers and streams.

17. For example, a single mature Fremont cottonwood tree may transpire 100 gallons of water per day (*Sonorensis,* 1988).

18. For a detailed description of the relationship between streamflows and riparian vegetation on a desert stream in Arizona, see Stromberg, Pattern, and Richter (1991).

19. For example, Stromberg et al. (1991) determined that the average recruitment interval for cottonwoods living along the banks of the Hassayampa River in Arizona was 12 years.

20. For example, the Arizona-Sonora Desert Museum (*Sonorensis,* 1988) reported these facts about riparian areas and birds: (1) riparian areas can contain up to 10.6 times as many spring migrants per hectare as found on adjacent nonriparian habitats; (2) of 134 breeding bird species in the Sonoran Desert, 37 percent are totally dependent on riparian areas; (3) the 1,059 pairs of breeding birds found on less than 100 acres of cottonwood gallery forest along the Verde River in central Arizona is the highest breeding-bird density found anywhere in the continental U.S.; and (4) in many regions, riparian areas may be the only suitable sites for birds to feed or rest.

21. In the Southwest, where riparian areas are especially important as wildlife

habitat because of the prevailing aridity, estimates are that just 5–10 percent of the region's riparian areas remain intact.

22. In earlier times the Soil Conservation Service of the U.S. Department of Agriculture encouraged ranchers to use herbicides to remove woody riparian vegetation such as willows in order to improve livestock forage production.

23. Brown, Taylor, and Shelby (1991).

24. The general categories and characteristics of streamflow-dependent boating—and other forms of recreation—as well as methods and concepts for evaluating their relationship to instream flows, are described at greater length by Doug Whittaker, Bo Shelby, William Jackson, and Robert Beschta in *Instream Flows for Recreation: A Handbook on Concepts and Research Methods.* The document was published by the National Park Service in January 1993. Copies may be obtained through the Alaska Region of the National Park Service, 2525 Gambell Street, Anchorage, AK 99503.

25. Whittaker et al. (1993).

26. Whittaker et al. (1993). As an example of the importance of streamflows in creating good camping sites, one need only look at the enormous effort and substantial sums of money spent in recent years in the Grand Canyon, where an ongoing study has attempted to determine sediment loads associated with variable flow levels emanating from the Glen Canyon Dam. The study is designed to assess the impact of flows in creating and destroying beaches and river bars in the Canyon, in large part because of the perceived shortage of high-quality campsites along the popular river.

27. James E. Fletcher and Michael King, "Attitudes and Preferences of Inland Anglers," report to California Department of Fish and Game, Survey Research Center, University Foundation, California State University, Chico, 1988, cited in California State Lands Commission (1993).

28. A river is also a metaphor for change, as in the saying from an early Greek philosopher "A man never steps twice in the same river."

29. Litton (1984).

30. Kaplan (1977).

31. Litton (1977).

32. Kaplan (1977).

33. Bradshaw (1993).

34. For example, Litton (1984) and Whittaker et al. (1993). This scenic preference for moderate flows is not based only on the opinions of landscape architects; Brown and Daniel (1991) obtained public judgments for a whole series of riparian scenes, each at a range of flow levels, and found using multiple regression models that scenic beauty rose with flow to a point and then fell for further increases in flow.

35. Litton (1984).

36. Ibid.

37. The expenses that must be subtracted from total willingness to pay to estimate economic value include all costs for actual goods and services, such as costs for equipment, travel, and lodging, but do not include simple transfer payments, which

do not represent a commitment of resources. Thus, for example, the cost of entry to a site or of a hunting permit would not be subtracted if it were, like a tax, only a transfer payment but would be subtracted if the collected money were spent to facilitate the activity at issue.

38. U.S. Department of the Interior, Fish and Wildlife Service, and U.S. Department of Commerce, Bureau of the Census, *1991 National Survey of Fishing, Hunting, and Wildlife-Associated Recreation,* U.S. Government Printing Office, Washington D.C., 1993.

39. Regions defined in the survey don't correspond exactly to those presented here, but it is clear that fishing is an extremely popular activity for westerners. Approximately 21 percent of all residents in the Mountain region fished in 1991, as did 15 percent of the residents of the Pacific states, 27 percent in the West North Central region, and 23 percent in the West South Central region. The survey groups Alaska, Washington, Oregon, California, and Nevada in the Pacific region; Idaho, Montana, Wyoming, Utah, Colorado, Arizona, and New Mexico in the Mountain region; North Dakota, South Dakota, Nebraska, and Kansas, in the West North Central region (which also includes the states of Minnesota, Iowa, and Missouri); and Oklahoma and Texas in the West South Central region (which also includes Arkansas and Louisiana).

40. See, for example, Cordell et al. (1990) and Crandall, Colby, and Rait (1992).

41. Crandall, Colby, and Rait (1992).

42. Hansen and Hallam (1990).

43. The authors obtained the value of an acre-foot of water used in irrigation from NARIM, the National Agricultural Resources Interregional Model. The value of the same amount of water passing downstream for recreational fishing was based on the change in the number of days fished multiplied by the value of a day of fishing to the recreationist. Estimates of the change in the number of days fished were based on changes in annual streamflow that would occur as a result of the transfer, while the authors assigned a value of $10 to represent the value of a day fishing to the recreationist, a figure they believed to be conservative. The authors found that the marginal value of water used for recreational fisheries varied significantly across regions, ranging from a few cents per acre-foot to over $100 per acre-foot. The marginal value of water for recreation was highest in the arid southwestern states.

When interpreting these results it is necessary to keep in mind the consequences arising from the economic concepts of marginal value and diminishing marginal returns. The study indicated that, given current allocations between irrigation and fishing, the transfer of an acre-foot of water from irrigation to fishing would yield an increase in total value. But if additional quantities of water were to be transferred, it is expected that this net increase in total value would diminish with each additional acre-foot of water transferred and—at some point—could even become negative, so it is not necessarily true that net benefits would continue to be positive if additional increments of water were shifted from irrigation to fishing.

44. An important reason for this result is that an acre-foot of water left in the stream will pass through—and hence add value to—many fishing sites but could be used only once for a consumptive purpose.

45. National Water Commission (1973), p. 5.

46. Krutilla (1967) was the first to describe such concepts in economic terms and suggest that they be measured.

47. Brown (1993) reviewed 31 studies published since 1980 that estimated existence and bequest values and concluded that, for most individuals, including recreational users, existence and bequest values combined were over twice as large as use values. Other studies that have focused on river flow values include Duffield, Brown, and Allen (1994), Loomis (1987a), and Sanders, Walsh, and Loomis (1990).

48. Some prime examples include Glen Canyon Dam on the Colorado River in Arizona, Navajo Dam on the San Juan River in New Mexico, Flaming Gorge Dam on the Green River in Utah, Yellowtail Dam on the Bighorn River in Montana, Seminole Dam above the "miracle mile" stretch of the North Platte River in Wyoming, Cheesman Dam on the South Platte River in Colorado, and Rudi Dam on the Fryingpan River, also in Colorado. This phenomenon attracts less attention below dams along rivers with anadromous fish, where the primary effect of dams on sport fisheries tends to be negative.

49. Recreational rafting is an example. The rafting season may be lengthened into the summer or fall by dams that store high spring flows for later release; the Cache La Poudre River upstream of Fort Collins, Colorado, and the Colorado River through the Grand Canyon are examples of this. (In addition, Cache La Poudre boating benefits from upstream importation of water from a neighboring watershed.) Dams may also allow increased-use rates early in the boating season by lowering the high spring flows that pose a safety hazard for many users. Note, however, that upstream dams tend also to reduce the variety of flows over the season, and therefore to lower the overall value of the river for those recreationists who especially enjoy the more extreme (high or low) flows.

50. Strictly speaking, *power* refers to generation capacity, and *energy* refers to the electricity that is generated. However, *hydropower* is a widely used shorthand term that may refer to the generation capacity of a hydroelectric plant, the energy produced at such a plant, or even (as in the first sentence of this section) to the production of electricity using the force of falling water.

51. For purposes of broadly categorizing types of water use, many researchers, including Solley, Pierce, and Perlman (1993), do not distinguish between instream and out-of-stream hydropower water uses. For instance, Solley et al. considered all hydropower uses to be instream. Nonetheless, distinction between instream and out-of-stream hydropower uses is important, given the very different impacts on riverine uses and environments resulting from the two.

52. According to Solley, Pierce, and Perlman (1993), thermoelectric power generation throughout the United States accounted for the withdrawal of 195,000 million gallons of water per day (Mgal/day) in 1990. Two-thirds of this was drawn from freshwater surface sources, and one-third was drawn from surface water saline sources. Groundwater accounted for less than 1 percent of all water withdrawn for use in thermoelectric power plants. The amount used by thermoelectric generators was 1.4 times as large as the amount withdrawn for the nation's second largest use of water: irrigation (137,000 Mgal/day).

Hydroelectric power generation (not considered by Solley et al. to be a water withdrawal) in 1990 made use of 3,290,000 million gallons of water per day, an amount several times larger than that used by thermoelectric power generators and irrigators combined. There is, of course, a difference between the amount of water that is withdrawn and the amount of water that is consumed, and it is interesting to note that the rankings of water use among these three uses are completely reversed when looking at water consumption rather than water withdrawals: only a negligible fraction of the water used in hydropower generation is consumptively used, and thermoelectric generation consumes just 3,980 Mgal/day, while irrigation consumptively uses 76,200 Mgal/day.

53. The two most common generating units in thermoelectric power plants are the steam turbine and the gas turbine. The steam turbine unit uses heat produced by the combustion of fossil fuels or the reaction of nuclear materials to create steam, which is then used to turn the turbine blades that turn the shaft of the generator that produces the electricity. In a gas turbine unit, hot gases produced by the combustion of natural gas and oil are used to turn the turbine blades.

54. Solley, Pierce, and Perlman (1993).

55. It does not matter whether such flows are natural or are regulated by some other facility upstream; the important feature of run-of-the-river plants is that they do not themselves have the capacity to alter the timing or magnitude of water flows.

56. Because of inherent inefficiencies in the conversion of one type of energy (the kinetic energy of falling water) to another type of energy (electrical energy, by way of the mechanical energy of spinning turbines), there is a net loss of energy associated with pumped storage operations. Nonetheless, the ability to use excess capacity to pump water when demand for electricity is low and then use the stored water to generate additional electricity when demand is high enables a utility to meet demand more cheaply than would be possible if the utility were forced to build all plants with capacity sufficient to meet peak demand.

57. Figures presented here were derived from data presented in U.S. Energy Information Administration (1994), tables 12 and 13.

58. Western Area Power Administration (1984).

59. Western Area Power Administration (1991).

60. This figure was determined by multiplying total revenues from electric utility sales in the 18 western states in 1992—$56.7 billion (Solley, Pierce, and Perlman 1993, table 28)—by 18.0 percent, the proportion of all electricity produced in those states using hydropower. The result was $10.2 billion. Because there was no attempt to measure the actual revenues accruing directly to electricity sales based on hydropower, this figure should be used only as an approximation.

61. Solley, Pierce, and Pearlman (1993), table 28.

62. Solley, Pierce, and Pearlman (1993, p. 56) noted that the estimated 3,290,000 Mgal/day used in hydroelectric power generation nationwide was 2.6 times the average annual runoff in the conterminous United States, a result made possible only by the repeated use of the same water by hydropower facilities downstream. They also noted that although the water losses directly attributable to hydropower production are quite small, substantial losses do occur as a result of evaporation from the reservoirs backed up behind the dams of large storage plants.

63. The general equation, as given by the Western Area Power Administration (1984), is

> $P = Q H e / 11.8$, where P equals power in units of kilowatts, Q represents water flow in units of cubic feet per second, H is head in units of feet, and e is the efficiency of the turbine-generator unit, which is usually in the range of 80–90 percent (0.80–0.90). 11.8 is a constant used to convert units to kilowatts.

Energy production is usually measured in kilowatt hours (kWh), which is the instantaneous power output (P above) multiplied by the time during which the unit is in operation. As a general rule of thumb, a hydropower facility generates approximately 0.87 kWh of electricity for each acre-foot of water falling through one foot of head (Butcher et al., 1986).

64. This assumes that there is sufficient generating capacity in place to make use of the flow.

65. It is also possible to improve electrical production by increasing the efficiency of the turbine generator, but efficiencies for hydropower plants are already quite high and in most cases are not susceptible to much improvement. Electrical production may of course be significantly improved by the addition of new capacity in situations where the capacity of existing turbine generators is not sufficient to take advantage of existing flows and elevation drops.

66. They may, however, still be in conflict with other instream uses. For example, a run-of-the-river hydropower facility may constitute a significant river channel obstacle and hazard to migrating fish and to boaters.

67. Based on historic flow levels and legal availability.

68. Western Area Power Administration (1984).

69. Ibid.

70. The primary alternative to the use of hydroelectric facilities for generating peaking power is the use of gas turbine plants. Coal-fired plants cost much less to operate, per kilowatt of energy produced, than gas turbine plants, but coal-fired plants cannot be brought on- and off-line as quickly as gas turbine plants. Thus, the marginal cost of producing peaking power is determined largely by the cost of producing energy at the more expensive gas turbine plants. This means that when a hydropower plant that is capable of producing peaking power is constrained from doing so, the cost of making up for that lost peaking power is relatively high.

71. Run-of-the-river plants do not have the capacity to alter streamflows.

72. Hamilton, Whittlesey, and Halverson (1989).

73. See Butcher, Wandschneider, and Whittlesey (1986), Hamilton Whittlesey, and Halverson (1989), and Miller (1990) for examples of situations in the Pacific Northwest in which water may be more valuable if left in the river channel to generate electricity rather than being diverted and used for irrigation.

74. Storage plants would have the capacity to store extra flows for later release, while run-of-the-river facilities would have to pass higher flows through the plant in an essentially unaltered form.

75. Data on the quantity of goods transported on the Columbia and Snake rivers,

and other waterways discussed below in the text, are from the U.S. Army Corps of Engineers (1988).

76. A depth of eight feet allows for vessel draft of 7.5 feet. "Full service"—water depth of nine feet, allowing 8.5 feet of draft—requires a flow of 31,000 cfs. "The level of navigation service to be provided is determined by the amount of water in storage [in the six federal dams upstream] on March 15 and July 1 of each year. On March 15, if there is more than 54.5 MAF [million acre-feet] in total mainstem storage, full service is provided. If there is 46 MAF or less, then only minimum service is provided. On July 1, full service is provided if there is 59 MAF or more water in storage, and minimum service is provided if there is less than 50.5 MAF. When the quantity of water in storage is in between, navigation service flows are reduced in proportion" (U.S. Army Corps of Engineers, 1994, pp. 3:97–98).

77. Martin, Hamilton, and Casavant (1992).

78. Gibbons (1986).

79. Young and Gray (1972), as cited by Gibbons (1986). Young and Gray found that in the short run (counting only operation and maintenance costs), the economic values for the two rivers were positive, although low. However, in the long run (adding in capital costs), the values were negative.

80. U.S. Army Corps of Engineers (1982).

81. For a comprehensive but readable introduction to the topic of sediment transport and stream dynamics, see Heede (1992).

82. There are a variety of factors that influence river discharge, sediment loads, and channel slope, shape, and length. Because changes to any one of these factors can lead to a new set of "equilibrium" conditions, rivers are frequently in a state of flux. For this reason, water resource scientists generally refer to the concept of "dynamic equilibrium" when describing rivers. In a dynamic equilibrium, riverine processes continually adjust in order to move toward the equilibrium consistent with the new set of prevailing conditions and may, therefore, never actually be in a true or static equilibrium. Even in the absence of a change in long-term conditions, rivers may temporarily be thrown out of equilibrium, as in the case of infrequent but very large floods. Rivers may take years, centuries, or even longer to recover from such events but do tend to move back toward an equilibrium consistent with long-term "average" conditions.

83. The relationship between storage dams and sediment transport is quite complex, because, in addition to altering the timing of streamflows, dams trap sediment being carried by rivers, thereby reducing the volume of sediments that need to be transported. The ultimate effect on river channels downstream from storage dams will depend on a number of factors, including the exact relationship between reduced sediment loads and altered streamflow and the input of sediments below the dam from tributaries.

84. The flows needed to maintain stream channels, and the methods of identifying those flows are very complex. The flows needed to maintain the channel of any given river will need to be identified on an individual basis, probably with the use of multiple methods and substantial professional judgment. Some of the methods used

to quantify channel maintenance flows are described in more detail in a later section of this chapter.

85. The influence of specific processes on sediment transport and channel shape and form varies depending on the conditions surrounding individual rivers and streams. For example, researchers have determined that large and infrequent storm events dominate the channel-forming process for many high-desert streams in the Southwest, where rainfall and streamflow are less frequent.

86. The term *fluvial geomorphology* is not as enigmatic as it might first sound. *Fluvial* refers to anything pertaining to a river, *geo* refers to the earth, and *morphology* is a term used to connote the study of form and structure. "Geomorphology"—the study of the characteristics, origin, and development of earth forms—is a standard component of the study of geology. "Fluvial geomorphology" is the study of the characteristics, origin, and development of earth forms that are associated with rivers.

87. Gordon, McMahon, and Finlayson (1992).

88. El-Ashry (1980).

89. Howe and Ahrens (1988).

90. Trelease (1976). The Federal Water Pollution Control Act of 1972 made it clear, however, that storage and release of water was not an acceptable substitute for adequate treatment or other methods of controlling waste at the source.

91. Wahl (1990).

92. U.S. Bureau of Reclamation (1992).

93. See generally Dunning (1993b).

94. Karim et al. (1995).

95. Among the most useful sources of basic information about instream flow quantification methods, and those relied on most heavily by the authors in writing this section, are Stalnaker et al. (1995), Shelby, Brown, and Taylor (1992), and Whittaker et al. (1993).

96. The wetted perimeter concept is illustrated by a simple example. Imagine a stream channel that is perfectly rectangular and measures ten feet in width and five feet in height. Flows sufficient to fill this channel to a depth of one foot would have a wetted perimeter of twelve feet—ten feet of river bottom plus one foot of bank on each side. Flows sufficient to fill the channel to a depth of four feet would have a wetted perimeter of eighteen feet—ten feet of river bottom plus four feet of bank on each side. Natural stream channels are not perfectly rectangular, only occasionally symmetrical, and may in fact assume all manner of shapes, making the calculation of wetted perimeter at different levels of flow more difficult.

97. "PHABSIM" is pronounced with three syllables: P-Hab-Sim.

98. Substrate is the material on the bottom of the stream channel, such as rocks and vegetation.

99. See Stalnaker et al. (1995).

100. The description of these techniques is based on Lamb (1993).

101. Brown, Taylor, and Shelby (1991) and Whittaker et al. (1993) review meth-

ods for and past studies of understanding the relation of recreation quality to flow quantity.

102. Corbett (1990) found that canoeing zero flow increased at a decreasing rate relative to mean annual flow—for every doubling of mean annual flow, canoeing zero flow increased by about 50 percent.

103. Hyra (1978).

104. Whittaker et al. (1993, p. 46) stated, "As a category of methods for evaluating flows or flow-dependent conditions, survey methods are generally the best. . . . No other set of methods so directly allows the potential 'client' of the river [to] help determine the 'product' that will be provided." Comparing user surveys to incremental methods used to quantify the flow needs of fish, Whittaker et al. stated, "Fish biologists developed a model as complicated as IFIM because they are unable to talk to fish and find out directly which flows are best" (p. 50).

105. Shelby, Brown, and Taylor (1992).

106. See Bleed (1987) and Gordon, McMahon, and Finlayson (1992) for more complete reviews of the methods used to identify channel maintenance flows.

107. See Gordon, McMahon, and Finlayson (1992), pp. 305–6 for examples of the range of values researchers have established for different rivers.

108. Schmidt (1994).

109. The Wild and Scenic Rivers Program is discussed in more detail in chapter 10.

110. U.S. Bureau of Land Management (December 1992).

111. Based on table 19 in U.S. Bureau of Land Management (December 1992), p. 65.

112. The following discussion of the Rio Chama instream flow needs assessment study is taken from U.S. Bureau of Land Management (December 1992).

113. Though it appears from the table that flows necessary to support whitewater boating are in conflict with eagles, that is not actually the case; eagles are not present during the summer, when whitewater recreation occurs.

Chapter Four

1. Brown (1991) lists 14 transactions, which occurred from 1987 to 1990, of water rights and one-time transfers of water quantities gleaned from issues of the *Water Intelligence Monthly* and its predecessor, the *Water Market Update*. For detailed descriptions of western water markets, see Saliba et al. (September 1987 and Summer 1987) and Colby (1990).

2. Annear (1993), p. 40.

3. The establishment of water rights and the transfer of senior water rights to a downstream location as methods of protecting instream flows are discussed in chapters 5–7.

4. Municipalities often need to have water delivered to sites well removed from the stream channel, so diversion points high in the watershed may become more common as senior irrigation water rights are purchased by municipalities.

5. See Fradkin (1984) for a history of the conversion of the Colorado River from a natural river into what has become essentially a highly engineered plumbing network.

6. See Kahrl (1982) for a detailed description of Southern California's development of the state's water.

7. The irrelevance to efficiency of the initial allocation of resources, as long as transaction costs are zero, is known as the Coase theorem.

8. An allocation of resources is said to be Pareto Optimal when it produces the maximum possible return.

9. The common problem with return flows (as explained more thoroughly in chapter 5) is that downstream appropriators, even if junior, who rely on the return flows have a right to maintain their use. If a transfer would affect downstream appropriators, only the consumptive-use portion of a right may be transferred. Quantification of that portion may be difficult and easily questioned by others. Of course, transfers from an offstream use to an instream use technically avoid return flow problems, because they actually increase, rather than decrease, the amount of water flowing to downstream diverters. However, administrative simplicity requires that all transfers be subjected to the same rule, and the prevailing rule is that only the consumptive portion of an existing water right can be transferred.

10. Public interest criteria are also applied to the transfer process in many states. Public interest criteria that protect instream flows are considered to be a special protection mechanism and are described in more detail below.

11. This assumes that the offstream use that has been retired was at least partially consumptive, as is normally the case with offstream water uses. The amount of "new" water available for use downstream will be the amount of water that was previously consumed, not the amount that was previously diverted.

12. For a good discussion of the concepts of exclusion and rivalry, see Randall (1989).

13. Some of these studies were cited in chapter 3. See also Colby (1993), Loomis (1987b), and Brown, Taylor, and Shelby (1991) for reviews of economic studies related to the value of instream flows.

14. Duffield, Neher, and Brown (1992).

15. Marginal benefits diminished to zero because, though higher flows continued to produce positive value to downstream hydropower users, higher flows eventually caused negative benefits to recreationists.

16. The fact that the marginal benefit of diversions to agriculture was estimated to be constant over the range of flows depicted in figure 4.1 should not suggest that the marginal benefit is constant for all levels of diversion. At lower levels of diversion (i.e., greater levels of instream flow), marginal benefits of irrigation could rise, depending on the nature of production and crop prices.

17. Just (1990), p. 311. See also Brown, Taylor, Shelby (1991).

18. "Typical" is discussed in chapter 3.

19. The diminishing marginal value of water in any one consumptive use gives the total cost curve its positive slope as instream flows increase.

20. For this reason, total cost is positive even at very low flow levels. This secondary cost component would extend into future time periods, but is attributable to an action in the current time period. Technically, this cost would be represented in the graph by the present value of current and future costs attributable to the action.

21. The choice of flow level to define as "minimum" is of course arbitrary. One could just as easily define zero flow, or any other low-flow condition, as minimal, but it is unlikely that anyone would act to preserve flows at a level lower than Q_{min} in figure 4.2b, because that would result in negative benefits. It makes sense to us to define minimum flow as the flow level at which net instream benefits first start to become positive.

Chapter Five

1. Traub (1988). See note 71 of chapter 2.

2. Ibid.

3. Doerksen (1991).

4. Ibid.

5. Danielson (1989), Sims (1993), Fassett (1993).

6. A more systematic instream flow protection program was recommended by a blue-ribbon committee in 1982, but the recommendation was not followed by the California state legislature (Dunning, 1990).

7. This history of Arizona's instream water rights program is derived from Dishlip (1993) and Kulakowski and Tellman (1994).

8. *McClellan v. Jantzen*, 547 P.2d 454, Arizona Court of Appeals, 1976.

9. The certificated right was for The Nature Conservancy's Ramsey Creek Preserve. The other permit, for the Conservancy's Canelo Hills Preserve, was resubmitted to DWR for approval of a higher flow after the Conservancy determined, as a result of the monitoring process, that there was more water available in the stream than it had earlier realized.

10. Kulakowski and Tellman (1994).

11. Davidson (1995).

12. Olson (1995). There is some question as to the value, and validity, of one of the rights. The other, for a state park in the eastern part of the state, has been more effective.

13. Information about this transfer comes from Ahadi (1995) and Olson (1995).

14. Information about the LaCreek case is taken from the South Dakota Supreme Court's holding *In Re Water Right Claim No. 1927-2* and *DeKay v. United States Fish and Wildlife Service*, 524 N.W.2d 855 (S.D. 1994).

15. Some were disappointed that the South Dakota Supreme Court did not use the *DeKay* decision to explicitly endorse the concept of instream water rights, especially given that federal attorneys had submitted extensive briefs to support such an endorsement (Kunz, 1995).

16. Beneficial uses are listed in sections 11.023 (a) and (b) of the Texas water code (Kaiser, 1994).

17. Underwood (1994), citing Skillern (1993), p. 69.

18. Kaiser and Kelly (1987), p. 1141, note 132, citing Tex. Water Code Ann. sec. 11.002(5).

19. Templer (1981). It is not known why Parks and Wildlife decided to withdraw the claim.

20. Moss (1995).

21. Slade (1995).

22. Staff (1995).

23. *Nevada v. Morros,* 104 Nev. 709, 766 P.2d 263 (Supreme Court of Nevada, 1988). See Potter (1990) for a more detailed description of the case.

24. McLelland (1995).

25. Duerr (1995), McLelland (1995), Randall (1995).

26. *State ex. rel. Reynolds v. Miranda,* 83 N.M. 443, 493 P.2d 409 (1972). See Kury (1973).

27. Hatch (1994, 1995). One reason that the ruling may not apply in other situations is that the water right claim at issue in *Miranda* preceded the passage of New Mexico's statutory appropriation law in 1907. Prior to the creation of an administrative system, diversions were necessary to provide tangible notice and evidence of the intent to appropriate prior to the creation of an administrative system. After 1907 this function was accomplished by requiring water users to apply to the state engineer's office for a permit.

28. Hatch (1994).

29. Hatch (1995).

30. Bolton (1995).

31. Ibid.

32. Ibid.

33. Dyke (1995).

34. 594 P.2d 570 (Colo. 1979), often referred to as the Crystal River decision.

35. 96 Idaho 440, 530 P.2d 924 (1974), often referred to as the Malad Canyon decision.

36. 463 N.W.2d 591 (Neb. 1990), often referred to as the Long Pine decision.

37. 830 P.2d 915 (Colo. 1992).

38. White (1993).

39. *In the Matter of the Adjudication of the Existing Rights to the Use of All the Water, Both Surface and Underground, Within the Kootenai Tributaries of the Kootenai River in Flathead and Lincoln Counties, Montana* (In the Water Courts of the State of Montana, Clark Fork Division—Kootenai Rover Basin, Case No. 76D-48 and Case No. 76D-49, July 23, 1986), as described and cited by McKinney (1993), pp. 15–19, note 58.

40. White (1993).

41. McKinney (1993).

42. Williams (1996).

43. *Bonham v. Utah State Engineer,* 788 P.2d 497 (Utah, 1989), cited by Holden (1993).

44. Johnson (1980), p. 265. Emphasis in the original.

45. For example, fish and wildlife are listed last in the water-use preference list in North Dakota (Dyke, 1995), and recreation and pleasure are at the end of Texas' preference list (Kaiser and Kelly, 1987). Legislation that created the Idaho Water Resource Board in 1965 stated: "Subject to *the primary use of water for beneficial uses now or hereafter prescribed by law,* minimum stream flow for aquatic life and the minimization of pollution shall be fostered and encouraged. . . . ," which suggested that instream flow for fish and other aquatic life was not considered to be a primary water use (Higginson, 1989, p. 50, citing Section 42-1734(b)(4) of the Idaho Code; emphasis by Higginson). However, the Board's first state water plan, issued in 1976, stated that " . . . all new water uses: . . . such as irrigation, municipal, industrial, power, mining, fish and wildlife, recreation, aquatic life and water quality will be judged to have equal desirability as beneficial uses . . ." and that "water rights should be granted for instream flow purposes" (Higginson, 1989, p. 50).

46. Harbour (1995).

47. Harle and Estes (1993).

48. "Ironically, states have turned to 'command and control' structures with respect to instream flows; turned away from market-based principles which underlie the prior appropriation doctrine. Prior appropriation was meant to provide a system free of unnecessary government interference" (Meyer, 1993).

49. This is not to imply that the appropriation doctrine is everything that it has been claimed to be. For example, consider this comment of Wilkinson (1985, p. 344): "Western water law, often hailed as a laissez-faire, market-oriented system, in fact is riddled with federal subsidies, restrictions against alienation, and preferences for inefficient uses; to the extent that the market does operate, prior appropriation is a reminder that the market traditionally has been ineffectual in protecting minorities and the environment."

50. Meyer (1989, p. 135) wrote that the control of instream water rights by the states "sounds more like creeping socialism than it does the commitment to the principles of private enterprise upon which the prior appropriation doctrine is supposed to be grounded. A lot of folks parading around as staunch conservatives sound more like Bolsheviks when it comes to instream flows. Is our western water allocation system riddled with communist sympathizers? Perhaps the time has come for a little peristroika [sic] right here in our own water courts."

51. Shelby et al. (1992).

52. Trelease (1976), Wahl (1990).

53. McKinney and Taylor (1988).

54. Danielson (1989).

55. Fassett (1993). Jeff Fassett is currently the Wyoming state engineer.

56. Hutchinson (1995).

57. Purkey (1995).

58. Fritz (1989).

59. Mattick (1995).

60. Harbour (1995), Hutchinson (1995).

61. Anderson (1991) noted that as a practical matter, the time and cost necessary to obtain an appropriation give private individuals very little incentive to pursue instream water rights or reservations on a broad scale.

62. Meyer (1989, p. 135) argued: "The point is simple. If we want efficiency in the allocation of water for both consumptive and nonconsumptive uses, we must harness the engines of private enterprise. We mustn't fear the people." Writing at the time that communism was crumbling in Europe, he added: "Anyone who wonders why western instream flow programs administered by bureaucrats devoted to sluggish central planning have accomplished so little, must also wonder why East Germans are streaming west. Anyone who asks 'What's wrong with a state board made up of unelected political appointees making all the decisions with respect of the allocation of this resource?' must be bewildered by recent events in Poland, Hungary, and Romania" (p. 135).

63. Gray (1993), pp. 11–23. A similar argument was made by Wahl (1990).

64. Many state water agencies are finding it increasingly difficult to monitor and enforce water rights. For example, a shortage of resources for enforcement at Arizona's Department of Water Resources has meant that many water rights owners have had to enforce their own rights in the courts (Tellman, 1995). Washington's Department of Ecology has been barred by the state supreme court from enforcing rights that have not yet been adjudicated by the courts. Because very few rights have yet been adjudicated, this has effectively prevented Ecology from enforcing most water rights in the state (Nelson, 1995). At least one commentator has noted that weak enforcement in Montana has made it difficult to ensure that water flows are actually available to those who have reserved water for instream purposes (Fritz, 1989). Some believe that monitoring and enforcement is the single weakest link in creating viable instream flow protection programs (Purkey, 1995).

65. For example, this is one reason that some water rights owners have given for not transferring their rights to the state of Washington under the provisions of that state's trust water rights program (Nelson, 1995). A Fish and Game official in Wyoming noted that several water rights holders had inquired about converting their rights to instream purposes but lost interest after discovering that the right would have to be donated to the state (Annear, 1995).

66. Meyer (1993, pp. 2–6) provided several examples of the value of instream rights to private individuals and organizations: "For instance, instream flows also may be used to protect multi-million dollar investments in water treatment systems (and to ensure compliance with permit requirements) which depend on a particular flow regime to assimilate treated waste. They also may be of critical use to developers who are required to undertake mitigation measures to offset the environmental impacts of their projects. . . . Cities may make use of instream rights to protect their investment in parks and to make their communities more appealing to desirable new

industries. Some home builders and commercial developers are beginning to recognize that protecting instream flows can add significantly to the attractiveness (and value) of their projects. Surprisingly, perhaps, agricultural users may be among the biggest beneficiaries. By enabling them to enter into leases and other voluntary arrangements that commit unneeded water rights to instream uses, they may protect themselves from forfeiture and abandonment actions while gaining a critical financial advantage."

67. Harle and Estes (1993), Harbour (1995).

68. Just (1990).

69. Livingston and Miller (1986) noted that it is not likely that all beneficiaries of instream flows will find representation for their interests via environmental groups. Wahl (1990) maintained that sole reliance on private acquisition of rights would result in too few instream flow rights.

70. Mirati (1995), Mattick (1995).

71. Hutchinson (1995). The Game and Parks Commission's application was also protested by some irrigators along the Central Platte. A hearing on the application was scheduled for February 1996.

72. Sometimes it is the state water rights administrative agency that comes under attack for granting an instream flow right. For example, the Nevada state engineer was sued by the state Board of Agriculture after granting an instream right to the federal Bureau of Reclamation. The state supreme court upheld the state engineer's action. See Potter (1990).

73. Potter (1993) claimed that "new water users" increasingly turn to other strategies, including the use of federal statutes, when they are locked out of instream flow protection mechanisms at the state level. This may well apply to both individuals and the federal agencies. Meyer (1993) noted that if federal agencies find themselves barred from pursuing instream flow protection under state law, they'll use federal law, so an effective instream flow protection program at the state level may be one of the best ways to prevent conflict with federal agencies.

74. McKinney and Taylor (1988), Beecher (1990).

75. Slattery and Barwin (1993).

76. Harbour (1995). Of course, applicants must realize they are likely to face more protests the more water they request, and the Department of Water Resources is under no obligation to grant a permit for water in the amount requested.

77. Information about Oregon's instream flow protection programs is drawn from Mattick (1993, 1995).

78. Skinner (1995).

79. Ibid.

80. Annear (1995).

81. Fassett (1993). Annear (1995) noted that instream flow opponents probably thought that the "minimum" language in the state's instream flow protection law would be more restrictive than it has actually proven to be. Annear also noted, however, that people have grown much more accepting of Wyoming's instream flow

protection efforts as they have gained more experience with the program, and that the program enjoys substantial support.

82. Aiken (1993), citing *In Re Application A-16642,* 463 N.W.2d 591 (Neb. 1990), more commonly referred to as the Long Pine case.

83. Wigington (1990).

84. The following discussion is derived from Beecher (1990).

85. Beecher (1990).

86. Annear (1993, 1995).

Chapter Six

1. Grant (1987).

2. Protection for the Rogue was aimed primarily at preventing the construction of dams. The protection was incomplete, because many of the river's tributaries were left unprotected, and future appropriations for domestic, stock, irrigation, and municipal purposes were allowed (Trelease, 1976).

3. Beeman (1993).

4. McKinney and Taylor (1988).

5. Tarlock (1990). The National Water Commission noted in 1973 that "state laws creating and protecting rights to the use and enjoyment of water fail to give adequate recognition to social (that is noneconomic) values in water. This omission derives in the west from the law of appropriation, which embodies the social preference during the period of its formulation for economic development over protection of such social values as esthetics, recreation, and fish and wildlife propagation. . . ." (National Water Commission, *Water Policies for the Future,* 1973, p. 27, as cited by Johnson, 1980).

6. McKinney (1993).

7. See Slattery and Barwin (1993) for a more detailed description of Washington's instream flow protection programs.

8. See McKinney (1993) for a comprehensive description of instream flow protection in Montana.

9. Fritz (1989).

10. Ibid.

11. Mattick (1995). However, the state does still administer approximately 17 minimum flows that were in process at the time that the others were converted to instream rights.

12. Rolfs (1995).

13. California Fish and Game Code 5937 states: "The owner of any dam shall allow sufficient water at all times to pass through a fishway, or in the absence of a fishway, allow sufficient water to pass over, around or through the dam to keep in good condition any fish that may be planted or exist below the dam." See Roos-Collins (1993) for more information on the Fish and Game Code provisions.

14. Interference of federal programs with state river designations is most common with regard to federal hydropower programs. This situation is described in more detail in chapter 9.

15. McKinney (1993).

16. Fritz (1989).

17. Trelease (1976), p. 7.

18. As described in chapter 5, the most basic of these criteria is that unallocated water must be available for a new appropriation, and that neither a new appropriation nor a change to an existing appropriation will harm other water rights holders. In all of the western states but Colorado, these criteria are applied in an administrative process by a designated state agency. In Colorado, the criteria are applied by the courts in an adjudicative process.

19. Grant (1987).

20. States with an explicit list of instream flow values included in public interest statutes include Alaska and Oregon. Oregon has been a leader in public interest protection of instream flows; a 1929 law mandated consideration of the effects of new appropriations on public recreation and commercial and game fishing as part of the public interest review. Refer to chapter 2 for a more complete description of the history and content of public interest criteria in the western states.

21. Grant (1987).

22. Changes to public interest criteria that made them more amenable to the protection of instream flows became especially widespread in the 1960s (Grant, 1987).

23. Robie (1977), p. 96, quoting from Cal. Water Code section 1253 (West, 1971).

24. *Johnson Rancho County Water District v. State Water Resources Control Bd.*, 235 Cal. App. 2d 863, 874, 45 Cal. Rptr. 589, 596 (1965), as cited by Gray (1993), Ch. 11, p. 11.

25. Gray (1993).

26. Andrews and Buchsbaum (1984). Regional Water Quality Control Boards and the Department of Forestry and Board of Forestry may also comment on water rights applications.

27. Wahl (1990).

28. Grant (1987), Gray (1993).

29. Dunning (1990).

30. Smith (1995).

31. *California Trout, Inc. v. Superior Court,* 218 Cal.App.3d 187, 266 Cal Rptr. 788 (1990).

32. Smith (1995). Because the Department of Fish and Game does not have an established administrative hearing process, enforcement of Fish and Game Code section 5937 is essentially left to the State Water Resources Control Board.

33. Just (1990).

34. Utah Code Ann. sec. 73-3-8 (Supp., 1986), as cited by Grant (1987), p. 690.

35. Holden (1993).

36. Underwood (1994).

37. Slade (1995).

38. Underwood (1994).

39. Staff (1995).

40. Moss (1995).

41. Morgan (1989), Higginson (1989).

42. Matthews (1995), Ahadi (1995).

43. Staff (1995).

44. The influence of agency attitudes and experiences on the effectiveness of instream flow protection was discussed at greater length in chapter 5.

45. Gray (1993). The State Water Resources Control Board performs a balancing function, judging the relative merits of different needs. From the standpoint of Fish and Game officials, this balancing process can make it difficult to protect the full amount of water the agency believes is necessary to protect instream uses (Smith, 1995).

46. Gray (1993).

47. Ibid. The essential difference between administrative protection and minimum flow levels is that with minimum flow levels the desired level of protection is determined but once, and thereafter applied to all applications for new appropriations. Administrative protection occurs only by conditioning or denying new appropriations each time that someone submits an application.

48. Tarlock (1988).

49. Policies regarding the amounts of water that can be appropriated for instream use were described in chapter 5.

50. Instream-flow water rights have sometimes been created under federal law and subsequently recognized by the states. These rights are discussed in chapter 8.

51. Nelson (1995). Washington's trust water rights program is described in more detail later in this chapter.

52. Gray (1993) has argued that without rights, instream flows are directly or indirectly sacrificed to new uses more readily than are other, consumptive rights. Mattick (1995) noted that water rights are not as subject to political manipulation as are minimum flows, which are essentially just administrative rules.

53. For example, the very junior priority dates of instream flow rights in Oregon have often meant that the rights have very little direct effect in protecting flows, but the existence of instream rights has helped to slow down the rate at which new off-stream appropriations have been approved (Mirati, 1995).

54. Rights can be lost for nonuse, or limited to only the amount of water used beneficially, i.e., without waste. In practice, regulatory agencies and the courts have been extremely loathe to eliminate or reduce water rights on an involuntary basis, and it rarely happens, though it may become increasingly common in the future as pressure on existing water supplies grows ever greater.

55. Administrative procedures used to evaluate transfer applications are a major source of the transaction costs that occur in water markets.

56. Aiken (1993), Hutchinson (1995). The prohibition on transfers between uses was probably created for the purpose of protecting agricultural water from municipalities, but it affects all other potential transfers as well. Nebraska does allow the purchase of stored water to be used for instream purposes.

57. Olson (1995).

58. Harbour (1995), Cahoy (1995).

59. For example, the transfer of a water right to instream purposes has been authorized in Oregon since 1987, but very few have yet occurred (Mattick, 1995). Both Washington (trust water rights) and Wyoming have explicitly authorized the transfer of water rights to instream purposes, but no such transfers have yet occurred in either state (Nelson, 1995, Annear, 1995).

60. Rolfs (1995).

61. McKinney (1993).

62. Gillilan (1992).

63. Wigington (1990), U.S. Bureau of Reclamation (1992).

64. The instream advocate may be able to contract with the downstream water user to ensure that the senior right is not subsequently forfeited or transferred in a way that would damage the instream flow, but she or he would not be able to force the downstream user to protest new appropriations or other transfers that would reduce streamflow.

65. For an example, see the discussion by Butcher, Wandschneider, and Whittlesey (1986) about the value of hydropower and irrigation on the Snake and Columbia rivers. Water left instream to generate electricity may be especially valuable given that it can be used over and over again; hydropower generation, like most other instream uses of water, is nonconsumptive.

66. See the discussion of hydropower in chapters 3 and 9 for a more comprehensive description of the relationship between hydropower and instream flows.

67. Wahl (1990).

68. Ibid.

69. Gray (1993).

70. Ausness (1986).

71. Tenets of the public trust doctrine have also been applied to other resources, such as the public lands (Wilkinson, 1980). Sax (1970) is often cited as one of the seminal articles written on the public trust doctrine.

72. The public trust doctrine has also been recognized by the federal courts. The doctrine was first recognized by the United States Supreme Court in the case of *Martin v. Waddell,* 41 U.S. (16 Pet.) 367 (1842), and subsequently recognized in a number of other cases (Ausness, 1986, Gardner, 1989). But, so far at least, the trust has been viewed as a responsibility of the states, not of the federal government, and so the doctrine has been developed primarily in the state courts.

73. Dunning (1993a).

74. *Marks v. Whitney,* 6 Cal.3d 251, 98 Cal.Rptr. 790, 491 P.2d 374, Supreme Court of California, 1971; as excerpted by Sax, Abrams, and Thompson (1991), p. 569.

75. *State v. Superior Court* (Lyon), 29 Cal.3d 210, 172 Cal.Rptr. 696, 625 P.2d 239 (1981), cert. denied 454 U.S. 865, 102 S.Ct. 325, 70 L.Ed.2d 165 (1981), and *State v. Superior Court* (Fogerty), 29 Cal.3d 240, 172 Cal.Rptr. 713, 625 P.2d 256 (1981), cert. denied 454 U.S. 865, 102 S.Ct. 325, 70 L.Ed.2d 165 (1981), as cited by Sax, Abrams, and Thompson (1991), p. 571, note 3. Much earlier, the U.S. Supreme Court had held that the public trust doctrine was not limited to tidal areas; the court's 1876 ruling in *Barney v. City of Keokuk,* 94 U.S. 324, held that the doctrine extended to lands under navigable fresh waters (Ausness, 1986).

76. *National Audubon Society v. Superior Court of Alpine County,* 33 Cal.3d 419, 189 Cal.Rptr. 346, 658 P.2d 709, Supreme Court of California, 1983.

77. The Water Commission Act of 1913 gave the Division of Water Resources authority to reject applications that were not in the public interest, but 1921 amendments to the Water Commission Act declared that domestic purposes constituted the highest use of California's water. The amendments are what led the Division of Water Resources to conclude that it could not interfere with Los Angeles' diversions in the Mono Basin.

78. Quotes from the *National Audubon* case are excerpted by Sax, Abrams, and Thompson (1991), pp. 578–88.

79. State of California, Water Resources Control Board, "Decision and Order Amending Water Right Licenses to Establish Fishery Protection Flows in Streams Tributary to Mono Lake and to Protect Public Trust Resources at Mono Lake in the Mono Lake Basin," Water Right Decision 1631, Sacramento, California, 1994.

80. *California Trout, Inc. v. Superior Court,* 218 Cal.App.3d 187, 266 Cal.Rptr. 788 (1990). At issue in this case was a provision of the Fish and Game Code requiring releases from dams to maintain fisheries; this case was mentioned above in the section on administrative protection of instream flows.

81. Ausness (1986), p. 414, referring to the observations of a number of commentators.

82. A century ago the U.S. Supreme Court stated, "There is no universal and uniform law on [the public trust doctrine] . . . each State applies the doctrine to the lands under the tide waters within its borders according to its own views of justice and policy. . . . Great caution, therefore, is necessary in applying the precedents in one State to cases arising in another" (*Shively v. Bowlby,* 152 U.S. 1, 26, 14 S.Ct. 548, 38 L.Ed. 331 (1894), excerpted by Cook, 1993, p. 4). The difficulty of defining the public trust doctrine for all of the states continues. Cook (1993, p. 23) noted that "each state . . . is free to develop its own unique body of law regarding the obligations of the state as trustee," and Campbell (1994, p. 81) wrote that judicial formulations of the doctrine have remained "invitingly vague."

83. For example, in *J.J.N.P. Co. v. State,* 655 P.2d 1133 (Utah 1982), the Utah Supreme Court "established the principle of public ownership of water, thereby affirming that the state must act as a responsible trustee for the waters of Utah. The court held that if lake or stream water levels get so low as to harm the public's in-

terest, the public ownership doctrine could be invoked to protect water flows" (Mortensen, 1994, p. 126). Mortensen (p. 130, note 81) also quoted Jaqualin Friend ("Nephi City v. Hansen: The Utah Supreme Court Sidesteps Public Trust Principles in Allowing Forfeiture of Municipal Water Rights," *Journal of Energy Natural Resources & Environmental Law* 11:[1991] 369, 378): "This increasing recognition of public interest considerations suggests that Utah is moving toward a broad view of the public trust doctrine at least in spirit, if not name."

84. *United Plainsmen Ass'n v. North Dakota State Water Conservation Com'n*, 247 N.W.2d 457 (N.D. 1976).

85. *Montana Coalition for Stream Access, Inc. v. Hildreth*, 684 P.2d 1088 (Mont. 1984).

86. Quoted and cited by Cook (1993), p. 5 and note 15.

87. *Owsichek v. State, Guide Licensing and Control Board*, 763 P.2d 488, 493 (Alaska 1988).

88. *Kootenai Environmental Alliance v. Panhandle Yacht Club*, 105 Idaho 622, 671 P.2d 1085 at 1095-96 (1983), excerpted by Johnson (1988), p. 130.

89. At 671 P.2d 1094, cited by Reed (1992), p. 663 and note 78.

90. Reed (1992). The case was *Shokal v. Dunn*, 109 Idaho 330, 707 P.2d 441 (1985).

91. This dam, to be located at Auger Falls, already had a permit from the Federal Regulatory Energy Commission and would have produced 43.6 megawatts of electricity. The project would have dewatered several popular rapids and been the seventh dam along the Middle Snake River. The Land Board, which made its decision on February 13, is composed of the state's five highest elected officials.

92. *Water Intelligence Monthly*, April 1996.

93. Ibid, p.8.

94. *Aspen Wilderness Workshop, Inc. v. The Colorado Water Conservation Board*, Certiorari to the Colorado Court of Appeals Pursuant to C.A.R. 50, Supreme Court, State of Colorado, No. 93SC740, June 19, 1995, pp. 19–22.

95. Ibid.

96. See Dunning (1993a) for an explanation of three possible scenarios.

97. For example, Johnson (1980, p. 233) stated, "The public trust doctrine and the appropriative water rights system are headed on a collision course in the West."

98. For example, Johnson (1988, p. 127) wrote, "A basic premise of this essay is that the prior appropriation system is flawed; from inception it has failed to protect public rights to clean water, recreation, fish and wildlife, and environmental quality." See also quote at note 5, this chapter.

99. A decision to reallocate water based on public trust criteria is not expected—barring new developments in public policy related to government takings of private property—to be accompanied by any compensation. The reasoning is that the initial allocation of water was always conditional, that it was always subject to public trust obligations, so that no taking occurs when the trust is exercised.

100. The cost in this case was borne by millions of people—the residents of Los

Angeles—rather than by any single individual. This fact may have made it easier for the court and the SWRCB to support their actions in reconsidering the city's initial appropriation. In addition, the California state legislature subsequently offered up to $60 million to help Los Angeles fund replacement water supplies, largely through water reclamation and conservation (Environmental Water Act of 1989, California Water Code sec. 12929 et seq.). This amount was expected to fall well short of making up for the total financial loss to the city resulting from the decreased diversions.

101. Quoted by Sax, Abrams, and Thompson (1991), pp. 578–88.

102. "We believe that before state courts and agencies approve water diversions they should consider the effect of such diversions upon interests protected by the public trust, and attempt, so far as feasible, to avoid or minimize any harm to those interests" (quoted by Sax, Abrams, and Thompson 1991, pp. 578–88). The court explained that its decision in *National Audubon* was a result of the fact that "this is not a case in which the Legislature, the Water Board, or any judicial body has determined that the needs of Los Angeles outweigh the needs of the Mono Basin, that the benefit gained is worth the price."

103. The *Audubon* court made numerous references to the need to integrate the two doctrines. For example, the court stated that its conception of California water law was as "an integration including both the public trust doctrine and the board-administered appropriative rights system. . . ." The court also maintained that:

> In our opinion, both the public trust doctrine and the water rights system embody important precepts which make the law more responsive to the diverse needs and interests involved in the planning and allocation of water resources. To embrace one system of thought and reject the other would lead to an unbalanced structure, one which would either decry as a breach of trust appropriations essential to the economic development of this state, or deny any duty to protect or even consider the values promoted by the public trust.

(As excerpted by Sax, Abrams, and Thompson, 1991, pp. 580 and 583, respectively.)

104. The federal Wild and Scenic Rivers Program is described in more detail in chapter 10.

105. Mattick (1993, 1995).

106. Harle and Estes (1993).

107. Estes (1995).

108. Gray (1993).

109. Beeman (1993).

110. Lamb and Lord (1992).

111. Matthews (1995).

112. Bolton (1995). The state attorney general's office has determined that there probably is no legal barrier to the creation of an instream flow right under state law, but the issue probably will not be settled until it has been adjudicated in a court of law.

113. Material on Washington's Scenic Rivers Program was drawn from Starlund (1995).

114. Information on Montana's Recreational Waterway Program was drawn from McKinney (1993).

115. McKinney (1993).

116. Clark (1995).

117. This account of The Nature Conservancy's arrangement with the North Poudre Irrigation Company is derived in its entirety from Wigington (1993).

118. The Colorado Water Conservancy Board and Division of Wildlife were involved with the St. Vrain Corridor Committee but were there primarily to give advice and provide technical help. Neither agency played any official role in the provision of instream flows in North St. Vrain Creek.

119. In practice, administrative agencies and the courts have been loathe to enjoin wasteful uses of water except in the most egregious of circumstances.

120. Ore. Rev. Stat. 537.455–537.500, effective September 27, 1987.

121. Proportions vary depending on local conditions (Reed, 1990).

122. Codified at Chapter 90.42 RCW.

123. Washington State Department of Ecology (1992), p. 1.

124. Ibid., p. 20.

125. RCW 90.42.005 states that

> conservation and water use efficiency programs, including storage, should be the preferred methods of addressing water uses because they can relieve current critical water situations, provide for presently unmet needs, and assist in meeting future water needs. *Presently unmet needs or current needs includes the water required to increase the frequency or occurrence of base or minimum flow levels in streams of the state,* the water necessary to satisfy existing water rights, or the water necessary to provide full supplies to existing water systems with current supply deficiencies. (Emphasis added.)

126. Mattick (1995), Nelson (1995).

127. Reed (1990).

128. Ibid.

129. Mattick (1995).

130. Purkey (1995).

131. Nelson (1995).

Chapter Seven

1. The only exception to this statement would be a case in which someone would later wish to make a diversion within the protected reach. As is true in the general case of upstream junior water rights users, the situation of the person wishing to make the diversion within the protected reach would be a consequence of standard prior appropriation principles—the instream right was created first, and therefore has priority—rather than of the fact that the water is being used for instream purposes.

2. Protection of freshwater flows at the San Francisco Bay Delta, where the Sacramento and San Joaquin rivers join before flowing into San Francisco Bay, may be a prominent exception to the general rule that instream protective measures—not

actually rights in this situation—aren't usually implemented at a river's mouth to the sea.

3. See Palmer (1991).

4. Livingston and Miller (1986).

5. Hatch (1994).

6. "It seems entirely plausible that such a strategy could be expanded statewide and adapted to serve environmental welfare and aquatic-based recreation objectives" (Hatch, 1994, p. 5).

7. For example, water rights are often constrained by laws intended to protect basins of origin, or certain classes of water users. For a general discussion of water rights transfer policies and issues in the West, see National Research Council (1992).

8. This rule does not apply, of course, to hydrologic changes that may occur, since hydrologic variation is a natural and expected part of the environment in which water uses take place. Hydrologic variation means that in any given year there may or may not be enough streamflow to supply water in the amount that was in existence at the time that a right was created. The "existing conditions" that are protected refer to the legal status and pattern of uses being made of the watercourse at the time that a right is created.

9. Part of the difference may also be accounted for by the fact that water rights are usually specified for diversions at different times, so that not all water rights holders may be diverting water at the same time.

10. See chapter 2 for information about the proportion of water that is consumed by different kinds of uses.

11. Withdrawal, and subsequent offstream use, may mean that the water is also diminished in quality; return flows often contain more constituents than were present when the water was first diverted.

12. This rule is also applied to prevent the transfer of water in amounts that were never used at all. Many water rights, particularly many of the earliest water rights, were issued for quantities that greatly exceeded the amounts actually diverted, because few water users had the expertise necessary to measure their water use with any degree of precision and erred on the side of claiming too much rather than too little.

13. This result assumes that the return flow reenters the river above the site of diversions that use the return flow. The possibility that return flows would not reenter the river above the site of subsequent diversions complicates the situation and raises the possibility that a downstream water user could be injured. The need to account for this possibility, and others, is discussed immediately below in the text.

14. It may certainly be possible to transfer more than just the consumptive-use amount in some situations. For example, few are likely to object to a transfer of the full diversion amount in a situation in which it is quite clear that consumptive use is unlikely to change. Because administrators typically rely on other water users to object to potentially injurious transfers, the absence of objections to a proposed transfer is likely to make it easier for such a transfer to be approved.

15. Livingston and Miller (1986).

16. National Research Council (1992).
17. Danielson (1989).
18. Livingston and Miller (1986).
19. National Research Council (1992).
20. Tarlock (1990).
21. Livingston and Miller (1986).

Chapter Eight

1. U.S. Supreme Court in the case of *The Daniel Ball*, 77 U.S. (10 Wall.) 557, 563 (1871), as cited by Tarlock (1989), sec. 9.03[1][b-c].

2. See Tarlock (1989), sec. 9.03, for a more detailed description of the evolution of the navigability concept.

3. Tarlock (1989), sec. 9.03.

4. A prominent exception to this statement is the recent wilderness bills, in which the Congress has explicitly claimed or denied federal water rights for specific new wilderness areas. The Wild and Scenic Rivers Act also contains explicit language regarding the amount of water needed for wild and scenic rivers purposes. The status of water rights in wilderness areas and on the wild and scenic rivers is discussed in greater detail in chapter 10.

5. 207 U.S. 564, 28 S.Ct. 207, 52 L.Ed. 340 (1908). The possibility of water rights associated with federal lands had been hinted at in the 1899 case of *United States v. Rio Grande Dam & Irrigation Co.*, in which the Supreme Court ruled: "In the absence of specific authority from Congress a State cannot by its legislation destroy the right of the United States, as the owner of lands bordering on a stream, to the continued flow of its waters, so far at least as may be necessary for the beneficial uses of government property" (174 U.S. 690 (1899), p. 703).

6. 373 U.S. 546, 83 S.Ct. 1468 (1963).

7. Tarlock (1989), pp. 9–49.

8. There were prior hints as to the direction that the court was moving. An earlier case, *Federal Power Commission v. Oregon*, 349 U.S. 435 (1955), referred to as the Pelton Dam case, had hinted at the principle that reserved water rights might be associated with other types of federal land reservations, but the full impact of this line of reasoning was not felt until the *Arizona* ruling was handed down. In *Pelton Dam* the court essentially ruled that the Federal Power Commission had authority to grant a license for a dam in Oregon because of the property clause and the federal government's ownership of the land on which the dam was to be located.

9. 426 U.S. 128, 96 S.Ct. 2062, 48 L.Ed.2d 523 (1976).

10. As described in the following section, the federal government may well be required to assert and quantify its reserved water rights claims in state courts, but the state courts are required to apply federal law in drawing their conclusions.

11. 43 U.S.C.A. sec. 666 (1988).

12. As many predicted, the state courts have been "predictably grudging" in their

determinations of both the scope and the purposes of federal reserved rights. Blumm (1992), p. 451.

13. 438 U.S. 696, 98 S.Ct. 3012, 57 L.Ed.2d 1052 (1978).

14. The *New Mexico* case is discussed in more detail later in this chapter.

15. 438 U.S. at 715, cited by Gray (1988), p. 572.

16. As discussed in more detail below, the Forest Service has subsequently attempted to claim federal reserved water rights under authority of the Multiple-Use Sustained-Yield Act of 1960.

17. Some other common units of the national park system include national historic sites, national historical parks, national seashores, national trails, national battlefields, national memorials, and national lakeshores. The four units listed in the text—parks, monuments, recreation areas, and preserves—are of most importance with respect to instream water issues in the western states. Another designation—national rivers—is also relevant to this topic, but such rivers are essentially the same as wild and scenic rivers, discussed in more detail in chapter 10.

There are no rigid rules governing the titles given to units of the national park system, but the following general definitions from U.S. Department of the Interior, National Park Service (1993) may be useful in understanding differences between types of units: *National park*—contains a variety of resources and encompasses large land or water areas to help provide adequate protection of the resources. Examples include Yellowstone (WY-MT-ID), Denali (AK), Grand Canyon (AZ), and Big Bend (TX). *National monument*—intended to preserve at least one nationally significant resource. Usually smaller than a national park and lacks a park's diversity of attractions. Examples include Devils Tower (WY), Natural Bridges (UT), Craters of the Moon (ID), and White Sands (NM). *National recreation area*—originally used for lands surrounding reservoirs impounded by dams built by other federal agencies, but later expanded to include other lands and waters set aside for recreational use, especially in or near urban areas. Examples include Lake Mead (AZ-NV), Glen Canyon (AZ-UT), Ross Lake (WA), and Golden Gate (CA). *National preserve*—established primarily for the protection of specific resources. Activities such as hunting and fishing or the extraction of minerals and fuels may be permitted if they do not jeopardize the natural values. Examples include Big Thicket (TX), Mojave (CA), Gates of the Arctic (AK), and Glacier Bay (AK); most of the West's national preserves are in Alaska.

18. Moffett and Carson (1994).

19. National Park Service Organic Act, 16 U.S.C. sec. 1 et seq.

20. National Park Service Management Policies, revised 1988, as quoted and cited by Doppelt et al. (1993), pp. 297 –300.

21. Water resources have been central to efforts to designate lands as components of the national parks system. For example, the flow of the Colorado River through the Grand Canyon, waterfalls in Yosemite, and an array of water-related features such as hot springs, geysers, lakes, and waterfalls in Yellowstone were primary purposes for which those national parks were created. See Tarlock (1987), pp. 30–31.

22. As excerpted and cited by Sax, Abrams, and Thompson (1991), p. 818. The court was referring specifically to language in the National Park Service Organic Act of 1916.

23. *United States v. City and County of Denver*, 656 P.2d 1, Colorado Supreme Court 1982, often referred to as Denver I.

24. Memorandum of Decision and Order, Case No. W-8439-76 (W-8788-77), Concerning the Application for Water Rights of the United States of America for Reserved Rights in Rocky Mountain National Park, December 29, 1993, p. 9.

25. Judge Behrman of the district court went on to say that "in the national forests preservation of natural features such as stream channels to a reasonable degree was sufficient. In a national park, and particularly in Rocky Mountain National Park, those features were to be preserved 'unimpaired.'" Ibid., p. 8.

26. 16 U.S.C. sec 195, as cited at p. 7 of the Memorandum of Decision and Order, ibid.

27. NPS Management Policies at 4:17, as cited by Doppelt et al. (1993), p. 299 and note 785.

28. Williams (1994).

29. Federal reserved water rights, as well as appropriative rights, are claimed by the Park Service in most if not all general adjudications in the West. However, most of these claims are not for instream purposes.

30. Williams (1994). Tarlock (1987) noted that there is some question as to whether the Park Service is legally required to assert water rights for the parks. The same point was made by Gray (1988).

31. There are, of course, some situations in which park resources are threatened by the possibility of water development occurring outside park boundaries. For example, see National Parks and Conservation Association (1995) for a more detailed report on threats to the water resources of Zion National Park in southern Utah.

32. Tarlock (1987), p. 31, noted that "the fortunate geographic location of many parks and units makes it possible to share without conflict rivers and streams that arise in the park with the major consumptive claimants, downstream users." However, greatly increased rates of visitation to many parks has led to increased water consumption within the parks themselves. In such cases, conflicts with downstream water users are more likely to occur (Williams, 1996).

33. Some developments already in existence at the time of designation, including dams and reservoirs, have been allowed to remain in the parks. Development also occurred in parks before Congress clearly specified the purposes for which parks were to be managed. Probably the best known of these developments is Hetch Hetchy Reservoir in Yosemite National Park. The controversy surrounding construction of Hetch Hetchy by the City of San Francisco in 1913 is thought by many people to have been a major factor in motivating subsequent passage of the National Park Service Organic Act, in 1916.

Amendments to the Federal Power Act, in 1921, prohibited the issuance of licenses for hydropower-generating facilities in the national parks, unless such projects were specifically authorized by Congress. For a discussion of hydropower and the national parks, see Bodi (1988). Occasionally, Congress has placed language in legislation designating new parks that would allow the Federal Energy Regulatory Commission to issue licenses for the future construction of hydropower facilities within the boundaries of the new park; for example, Congress specifically left two sites open for such development in the Grand Canyon when it designated the area as a national park. The proposed projects in the Grand Canyon faced immediate and significant opposition from environmentalists and others, and in 1968 Congress specified that licenses for the projects could not be issued unless subsequently authorized by Congress.

34. Williams, Owen (1994).

35. 16 U.S.C. sec. 431–33.

36. As quoted by Sax, Abrams, and Thompson (1991), p. 810.

37. 656 P.2d 1, Colorado Supreme Court, 1982.

38. Wilkinson (1988b), pp. 266–67.

39. At least one legal scholar maintains that, based on the inclusion of the national monuments within the National Park Service Organic Act in 1916, federal water rights for the national monuments are likely to be more generously acknowledged in other settings than they have been in Colorado. See Wilkinson (1988b), pp. 266–67. Creation of the National Park Service Water Rights Branch in 1984, discussed in more detail later in this chapter, was in part a reaction to this decision.

40. The issue of whether reserved rights exist for administratively designated lands has not been conclusively decided. Reserved rights for Indian reservations created by executive order have been granted, so it is possible that the courts would grant reserved rights for other administratively designated lands. However, such lands would still have to meet the *New Mexico* primary purposes test.

41. McGlothlin (1994).

42. Data on current refuge numbers and acreage are drawn from U.S. Fish and Wildlife Service (1993).

43. More emphasis on protecting fish, another goal of the refuges, would probably lead to greater efforts to protect instream flows. However, despite the emphasis placed on fish by other units of the U.S. Fish and Wildlife Service, fish protection in the refuges is clearly a secondary goal. Though 281 refuges have fishery resources, only four refuges have been established for the specific purpose of conserving and enhancing fisheries (U.S. Fish and Wildlife Service, 1993, p. 1:18).

44. Williams (1996).

45. U.S. Bureau of Land Management (1993). The Plains states of North Dakota, South Dakota, Nebraska, Kansas, Oklahoma, and Texas have very little land managed by the BLM, ranging from less than 0.3 million acres in South Dakota to none in Texas.

46. Or the president, as authorized by Congress.

47. The BLM will also assert federal reserved rights if it is joined in an adjudication that meets the requirements of the McCarran amendment. As explained earlier in this chapter, the McCarran amendment allows the states to join the federal government in general adjudications to establish the status of water rights in a watercourse or watershed.

48. Subsequent to passage of the Federal Lands Policy Management Act in 1976, BLM must claim new water rights at public water holes under state rather than federal law.

49. Corrigall (December 1994).

50. The total land area of the national forests in the eighteen western states is 163,681,640 acres, as derived by the authors from data presented in U.S. Forest Service (February 1994). That figure includes only the national forests; it does not include purchase units, national grasslands, research and experimental areas, land utilization projects, water areas, and other acreage that is also managed by the Forest Service. All but the lattermost ("other acreage") are considered by the Forest Service to be part of the "national forest system," which totals 191,553,355 acres nationwide. Gross acreage administered by the Forest Service is 231,553,013 acres nationwide and 182,482,180 acres in the eighteen western states.

51. U.S. Public Land Law Review Commission, *One Third of the Nation's Land: A Report to the President and to Congress*, 1970, p. 140.

The national forest lands are by no means evenly distributed throughout the eighteen western states. Most of them are found in the Pacific Northwest and Intermountain states. Alaska, California, and Idaho each contain over twenty million acres of national forest, Montana and Oregon each contain over fifteen million acres, and Colorado and Arizona over ten million acres. New Mexico, Washington, Utah, Wyoming, and Nevada each have between five and ten million acres of national forest, while the Plains states of Kansas, Nebraska, North Dakota, Oklahoma, South Dakota, and Texas each have less than two million acres. There is no national forest land in either Kansas or North Dakota, though both states contain some other units (primarily national grasslands) that are administered by the Forest Service (U.S. Forest Service, February 1994).

52. 373 U.S. 546, 83 S.Ct. 1468 (1963), as described above.

53. There are several national forests in the lower Colorado River watershed, though the Gila National Forest in New Mexico was the only one specifically mentioned in the decision.

54. Tarlock (1987), p. 40, noted, "These inchoate federal rights were not sought by the federal government pursuant to a comprehensive scheme of federal water administration. Rather, in an ad hoc fashion the rights were thrust upon a somewhat surprised federal government, which had been forced to participate in the litigation by the Supreme Court."

55. For a comprehensive description and analysis of the case, see Fairfax and Tarlock (1979).

56. Fairfax and Tarlock (1979), p. 525.

57. 438 U.S. 696, 98 S.Ct. 3012, 57 L.Ed.2d 1052 (1978). This decision was discussed earlier in this chapter, in the context of the development of the federal reserved rights doctrine.

58. Quotations from footnote 14 of the decision, as cited by Sax, Abrams, and Thompson (1991), p. 817.

59. Mr. Justice Powell, joined by Mr. Justice Brennan, Mr. Justice White, and Mr. Justice Marshall, dissenting in part, as cited by Sax, Abrams, and Thompson (1991), p. 821.

60. Ibid. Powell was quoting in part from the decision in *Mimbres Valley Irrigation Co. v. Salopek*, 90 N.M. 410, 412, 564 P.2d 615, 617 (1977).

61. This was true even though many felt at the time that the decision would not

have much influence on future cases. For example, Fairfax and Tarlock (1979, p. 554) stated, "Advocates of state control can take little heart from *New Mexico* for the case is too flawed and hence unstable to have a long term influence. It is unlikely that the Supreme Court will reverse itself on the narrow issue of the effect of the 1891 and 1897 acts, but it is by no means certain that the broad dicta and attitudes which run through the opinion will prove a reliable guide to future reserved rights controversies."

62. The initial result in the state district and supreme courts took the Forest Service by surprise in large part because it was felt that "Water users in New Mexico were not particularly interested in the claims of the United States in this case. The users and the state were looking to other adjudications where the conflict between federal reserved right claims and subsequent state appropriations was much greater" (Fairfax and Tarlock, 1979, p. 524).

63. In doing so, the Forest Service seemed to be taking the words of Justice Powell to heart. As part of his dissent, Powell had strongly implied that the two primary purposes of the national forests—even as construed by the majority in the *New Mexico* decision—were broad enough to support a claim for federal reserved water rights for instream flow purposes.

64. The state of Colorado has been divided into seven water divisions based on hydrographic features for the purposes of water rights adjudication and administration. Each division is associated with a major river and its tributaries.

65. Emery (1984).

66. Denver based its claims on three arguments. "First, that the United States had abandoned any interest in water on the public domain by virtue of the Act of July 26, 1866 [the Mining Act discussed in chapter 2], the Desert Land Act [1877], and entry of Colorado into the Union upon ratification of its constitution by Congress; second, that the McCarran Amendment required the federal government to submit to state water law; and third, that the Supreme Court decisions creating and interpreting the reserved rights doctrine were either dicta or somehow inapposite. . . ." (Emery, 1984, p. 74; citations omitted).

67. 656 P.2d 1 (Colo. 1982).

68. As will be described below, the Colorado Supreme Court later held that the court had not ruled out the possibility that the federal government could pursue its "primary purposes," instream claims based on a stronger factual situation. See *United States v. Jesse*, 744 P.2d 491, Supreme Court of Colorado, 1987.

69. 656 P.2d 1, at 23, as cited by Emery (1984), p. 75.

70. In making the latter argument, the Colorado Supreme Court relied heavily on the U.S. Supreme Court's decision in the *New Mexico* case, perhaps erroneously so. The United States argued in the *Denver* case that the discussion of MUSYA claims in the *New Mexico* decision was dictum and erroneous: dictum (i.e., not part of the holding, because not actually at issue in the case) because the United States had asserted MUSYA in the *New Mexico* case simply as a confirmation of national forest purposes as found in the 1897 Act, rather than as the basis for additional claims with a 1960 priority date, and erroneous because the purposes of MUSYA were "secondary" only in the sense that they could not stand alone as purposes for which a na-

tional forest was created, not in the sense that they were less important than the "primary" purposes mentioned in the 1897 Act. There was solid evidence to support both of these claims: Justice Powell of the U.S. Supreme Court had noted in his dissent that the court's discussion of MUSYA was dictum, and the House Report on MUSYA "explained that the resources listed in the act were to be treated equally, no priority was to be given to any, and that Congress neither intended to upgrade nor downgrade any resource" (H.R. Rep. No. 1551, 86th Cong., 2d Sess. 3, as cited by Emery, 1984, p. 81, note 108).

71. Fluvial geomorphology was discussed in the channel maintenance section of chapter 3.

72. The story of the introduction of the field of fluvial geomorphology to the courts is, of course, much more involved than described here. One account, by a former attorney at the United States Department of Justice office in Denver who was involved in the process, is that he knew of the work that hydrologists in Forest Service Region 2, with headquarters near Denver, Colorado, were doing in the field and saw the potential relevance of this work to the instream water flow claims issues being argued in the courts. After consultation with these hydrologists and other experts he asked the Forest Service hydrologists to develop and propose a methodology for determining the amount of streamflow needed to maintain stream channels in conditions suitable for "securing favorable conditions of water flows," which they did. (Hill, 1994). The Forest Service methodology was developed with the support of numerous people both within and outside the Forest Service, but was primarily developed by Forest Service hydrologists David Rosgen and Hilton Silvey, based in large part on Leopold, Wolman, and Miller (1964), which was published about the same time as the *Arizona v. California* decision.

73. *In Re The General Adjudication of All Rights to Use Water in the Big Horn River System and All Other Sources,* Civil No. 4993, District Court of the Fifth Judicial District, State of Wyoming.

74. The adjudication of all water rights claims in the two watersheds was split into three parts, one covering the federal reserved rights claims made by the Shoshone and Arapaho tribes of the Wind River Reservation, one for non-Indian federal reserved rights claims, and one for individual appropriators. Claims asserted within the non-Indian federal portion of the proceedings for purposes other than the national forests were for the entire natural flow of Middle Creek within Yellowstone National Park and for reserved rights in public water reserves, stock driveways, water-producing gas wells, wildlife habitat, and the Big Horn National Recreation Area (Mead, 1986).

75. Shupe (1985), pp. 28–29.

76. As quoted by Mead (1986), p. 444. The quote illustrates the prominence of passion over logic that frequently prevails in debates over instream flow protection. Water left to flow instream through the national forests would still be available for subsequent offstream use after leaving the confines of the forest rather than being captured or "secreted away" within the forest.

77. Shupe (1985).

78. Partial Interlocutory Decree and Supporting Documents Regarding the United States' Non-Indian Claims, *In Re The General Adjudication of All Rights to Use Water in the Big Horn River System and All Other Sources,* State of Wyoming, No. 4993 (5th Dist. Wyo. Feb. 9, 1983). The United States abandoned close to 60 percent of its claims in the Bighorn and Shoshone national forests (Mead, 1986, p. 445).

79. Shupe (1985) noted that the subordination of the claims meant that both existing uses and large-scale future development would cause additional depletions to the National Forest streams. Mead (1986) referred to the United States' claims in the national forests as having been "gutted" and "dramatically reduced" and noted that diversions by senior rights holders during years of drought would make the rights granted in the Partial Decree "completely worthless." It is important to realize, however, that the Forest Service retained other methods—not involving water rights—of protecting streamflows in the national forests. These methods are discussed later in the text.

80. The rights to the natural levels of springs and seeps were limited to those that would not interfere with existing rights under state law but would be actionable under state law if threatened by subsequent withdrawals. The precedential value of this result may be reduced by a clause in the agreement that states that the agreement "shall not be used as precedent in any other adjudication of federal water rights" (Mead, 1986).

81. Continuation of the Forest Service's water rights claims described in the text was preceded by several other events. The United States first amended its pending claims in Colorado Water Divisions 1, 2, 3, and 7 in order to allow quantification of its claims for instream flows using the revised methodology. At that time the State of Colorado and other objectors then moved for partial summary judgment against the Forest Service in Water District 2, arguing that the Organic Act did not implicitly reserve water for the purposes of instream flow in the national forests, and that the Colorado Supreme Court's decision in *United States v. City and County of Denver* had precluded the United States from claiming federal reserved rights for those purposes. The water court for Division 2 agreed with these arguments and granted partial summary judgment in favor of the objectors, but the United States appealed. In *United States v. Jesse* (744 P.2d 491, Supreme Court of Colorado, 1987), the United States argued before the Colorado Supreme Court that recent advances in the science of fluvial geomorphology justified a hearing of its instream flow claims for the national forests based on the primary purposes of the Organic Act. The Colorado Supreme Court agreed, stating that the case presented "genuine issues of material fact" and remanded the case back to the water court with instructions that "if, after a full consideration of the legislative history and factual circumstances, the water court determines that the purpose of the Organic Act will be entirely defeated unless the United States is allowed to maintain minimum instream flows over the forest lands, the United States should be granted such reserved water rights under the Organic Act. Otherwise, the claims should be denied" (744 P.2d 491, Supreme Court of Colorado, 1987, as excerpted by Sax, Abrams, and Thompson (1991), p. 840–43).

82. Memorandum of Decision and Order, Case No. W-8439-76, District Court, Water Division No. 1, State of Colorado, February 12, 1993.

83. For a detailed summary of the technical testimony, see Gordon (1995).

84. Among the advantages accruing to the storage of water in the mountainous national forest regions, as noted by the court, were: 1) the presence of geologic formations that made the construction of reservoirs easier and less expensive, 2) cooler temperatures and greater depth of mountain sites, which reduced the amount of evaporation from reservoir surfaces, and 3) the fact that water could be delivered by gravity, which enables water users to avoid the additional money and energy costs of pumping. The court also emphasized the increased flexibility resulting from storage high up in the mountains, since it enabled the water to be used many times before passing out of the state and also increased the possibility of upstream water exchanges. District Court, Water Division No. 1, 1993, pp. 6–7.

85. District Court, Water Division No. 1, 1993, p. 20.

86. Bankfull flow was one of about the 1.5-year return period.

87. District Court, Water Division No. 1, 1993, p. 7.

88. After noting the many problems encountered by the Forest Service in its attempts to document its needs, the court added, that "this is not to denigrate the efforts of the Forest Service. It was confronted with a monumental problem, one that is perhaps insurmountable." District Court, Water Division No. 1, 1993, p. 26.

89. The Forest Service recognized the inadequacy of some of its methods during the course of the trial and attempted to amend its claims in light of new developments in theory and methods. This effort was, however, rejected by the court, primarily because the proposed amendments were made after many months of trial and involved a substantial change in the Forest Service's proposal, which the court felt would have triggered a new round of investigations by all parties and more months of trial, which would have been unfair to the objectors.

90. It is likely that most of the subject channels had been receiving the flows necessary to adequately maintain the stream channels, in part because diversions were usually operated to avoid high flows that could damage the diversion ditches, and in part because by the time really high flows occurred, the reservoirs were often already filled. Thus, existing flows had largely been sufficient to maintain the channels. The Forest Service case was aimed primarily at protecting flows from future additional diversions if such diversions would unduly interrupt channel maintenance.

91. The Indians have had a hard time securing aid necessary to develop their water on any significant scale:

> While the Federal Reclamation Act of 1902 successfully subsidized water development for non-Indian agriculture, there has never been equivalent legislation for the development of Indian agriculture in the West. The few irrigation projects initiated by the Bureau of Indian Affairs (BIA) are largely unfinished, primarily because of a lack of funds. (Shay, 1992, p. 557; citations omitted.)

In 1973 the National Water Commission stated that:

> Following *Winters*, more than 50 years elapsed before the Supreme Court again discussed significant aspects of Indian water rights. During most of this 50-year pe-

> riod, the United States was pursuing a policy of encouraging the settlement of the West and the creation of family-sized farms on its arid lands. In retrospect, it can be seen that this policy was pursued with little or no regard for Indian water rights and the *Winters* doctrine. With the encouragement, or at least the cooperation, of the Secretary of the Interior—the very office entrusted with the protection of all Indian rights—many large irrigation projects were constructed on streams that flowed through or bordered Indian Reservations, sometimes above and more often below the Reservations. With few exceptions, the projects were planned and built by the Federal Government without any attempt to define, let alone protect, prior rights that Indian tribes might have had in the waters used for the project. Before *Arizona v. California* . . . actions involving Indian water rights generally concerned then-existing uses by Indians and did not involve the full extent of rights under the *Winters* doctrine. In the history of the United States Government's treatment of Indian tribes, its failure to protect Indian water rights for use on the reservations it set aside for them is one of the sorrier chapters. (National Water Commission, *Water Policies for the Future: Final Report to the President and to the Congress of the United States*, pp. 474–75, 1973, as quoted and cited by Shay, 1992, at pp. 586–87 and note 291.)

In referring to this report, Shay (1992, p. 587) also noted that "the assertion that the Federal Government merely failed to protect Indian water rights for use on the reservation is a grave understatement."

92. Collins (1985).

93. Rusinek (1990), p. 359.

94. For example, Shay (1992, p. 586) cites the Big Horn case, described below in the text, as evidence that " . . . the purpose a treaty evinces depends strongly on who is scanning the treaty and what they would like it to mean."

95. The quantification of water rights using the PIA standard can be complex and involve substantial discretion. It has been the subject of much commentary by the courts and others but is beyond the scope of the issues presented here.

96. It may also be possible to protect instream uses with reserved water rights if the courts recognize the purpose of a reservation as the establishment of a "homeland" for the tribes. Maintenance of a viable homeland could conceivably encompass a number of instream water uses. However, though the term *homeland* appears frequently in treaties, the courts seldom recognize this general purpose as one of the specific purposes for which reserved water rights will be quantified. Recognition of homeland purposes would have several advantages for the tribes and for the rest of society (see Rusinek, 1990; also Kinney, 1993). However, Collins (1985, pp. 491–92) noted that the prospects of having the homeland argument recognized by the courts are not very good, stating that "the 'homeland' argument that water was reserved for all purposes that contribute to tribal self-sufficiency will probably fail."

97. *Colville Confederated Tribes v. Walton*, 647 F.2d 42, 48 (9th Cir. 1981), *cert. denied* 454 U.S. 1092 (1981); *United States v. Adair*, 723 F.2d 1394, 1412-14 (9th Cir. 1983), *cert. denied*, 104 S.Ct. 3536 (1984); *Muckleshoot Indian Tribe v. Trans-Canada Enters.*, 713 F.2d 455 (9th Cir. 1983), *cert. denied*, 104 S.Ct. 1324 (1984), as cited by Collins (1985) at p. 491, note 76.

98. Williams, Wes (1994).

99. As noted by Squillace (1993), p. 53: "Indeed, given the cost of new reclamation works, many Indian reserved rights will have little value if the tribes are not free to use their water for other beneficial purposes."

100. As quoted and cited by Kirk (1993), pp. 475–76, note 69.

101. Kirk (1993).

102. Non-Indians farm most of the irrigated land on the reservation "mainly due to the inability of the tribes to secure adequate capital to develop and upgrade their own water projects" (Rusinek, 1990, p. 381). Also, see Ambler (1987).

103. For more information on the *Big Horn* adjudication see Mead (1986); Ambler (1987); Abrams (1990); Rusinek (1990); Hannum (1992); Collins (1992); Kirk (1993); Kinney (1993); Squillace (1993); and Williams, Wes (1994).

104. Report Concerning Reserved Water Right Claims By and On Behalf of the Tribes of the Wind River Reservation, Wyoming, by Teno Roncalio, Special Master, Dec. 15, 1982, Civil No. 4993, in Appendix H of Wyoming's Petition for a Writ of Certiorari to the Supreme Court of Wyoming, Aug. 19, 1988, as cited by Kirk (1993), p. 468, note 12.

105. Decision Concerning Reserved Water Rights Claims By and On Behalf of the Tribes of the Wind River Reservation, Wyoming, Fifth Judicial District, State of Wyoming, May 10, 1983 (Civil No. 4993), amended May 24, 1985 (Docket No. 101-324), as cited by Kirk (1993), at p. 469, note 15.

106. Amended Judgment and Decree, First Judicial District, State of Wyoming, May 24, 1985 (Docket No. 101-234), as cited by Kirk (1993), p. 469, note 21.

107. *In Re The General Adjudication of All Rights to Use Water in the Big Horn River Sys.*, 753 P.2d 76 (Wyo. 1988).

108. Rusinek (1990).

109. *Wyoming v. United States*, 492 U.S. 406 (1989). The Wyoming Supreme Court's opinion was upheld by the U.S. Supreme Court on a 4–4 vote, with Justice O'Connor abstaining. Rusinek (1990) believed the tie vote signaled future problems for the PIA standard in subsequent proceedings before the Supreme Court.

110. Hannum (1992), p. 684, and note 11.

111. Kirk (1993).

112. Rusinek (1990, at p. 393) noted that "years of dewatering streams on the reservation to provide irrigation water for both non-Indians and Indians had resulted in the 'almost total devastation of game fish populations' and constituted another blow to the reservation's already shattered economy." The instream flows were also to be used to recharge the groundwater basin and benefit downstream irrigators (Williams, Wes, 1994).

113. Hannum (1992), p. 685, note 12.

114. The other two phases were to settle the non-Indian federal reserved rights claims and the claims of individual appropriators in the basin.

115. Letter from Gordon W. Fassett, state engineer, to Gary P. Hartman, district judge, Fifth Judicial District (March 12, 1991), as cited by Kirk (1993), p. 471, notes 36 and 37. There is a substantial amount of confusion about the reason or reasons given by the state engineer for not enforcing the tribes' permit (Kirk, 1993), so the reason offered in the text may not be definitive. Hannum (1992, p. 685 and note 13), citing a brief submitted to the court by the Shoshone and Northern Arapaho tribes, wrote that the state engineer refused to enforce the permit on the grounds that the permit attempted to authorize an impermissible use of water. A third explanation was offered by Kinney (1993, p. 852)

> The State Engineer's refusal to enforce the Wind River Tribes' instream flow permit was based on a two step interpretation of the federal Indian reserved right awarded to the Wind River Tribes in *Big Horn I.* The first step of the interpretation was that the future federal Indian reserved water right could not be changed to an instream flow use until it had been diverted and beneficially used to irrigate the future projects. The diversion requirement was tied to the nature of the federal Indian reserved water right as evidenced by the state supreme court's use of the phrase "right to divert, or to have water diverted" in *Big Horn I.*

The second step of the State Engineer's interpretation was that once "actual" diversion and irrigation had been accomplished, the Wind River Tribes could change the use of their federal Indian reserved water right to a secondary instream flow use pursuant to state law. The basis for this interpretation was the Supreme Court decision in *United States v. New Mexico.* [Citations omitted.]

116. "The only restrictions on this freedom of use are those imposed by federal law: that water use be confined to the boundaries of the reservation, and that consumptive use not be increased above the quantity reserved" (Hannum, 1992, p. 685, citing the Report and Recommendation of the Special Master, October 4, 1990, *In Re The General Adjudication of All Rights to Use Water in the Big Horn River System and All Other Sources*, State of Wyoming, Fifth District Court of Wyoming [No. 86-0012]).

117. Judgment and Decree, March 11, 1991, *In Re The General Adjudication of All Rights to Use Water in the Big Horn River System and All Other Sources,* State of Wyoming, Fifth District Court of Wyoming (No. 4993), p. 18, as cited by Hannum (1992) p. 696 and notes 20 and 96.

118. Kirk (1993).

119. *In Re The General Adjudication of All Rights to Use Water in the Big Horn River System and All Other Sources*, State of Wyoming, 835 P.2d 273 (Wyo. 1992), as quoted by Hannum (1992) p. 703.

120. The legal experts apparently believed that the composition of the U.S. Supreme Court in 1992 was not conducive to a decision that would favor the tribes, and the tribes did not want to risk giving the decision greater precedent by taking it to the top federal court. See Collins (1992).

121. The case may not even be good law; many commentators think the case was poorly decided. For example, see Williams, Wes (1994) and Kirk (1993). Kinney (1993) believed that the court's decision was correct, but that the justices reached

their conclusion through flawed analysis. Hannum (1992, p. 703) wrote that "The court's disposition of the case can be most charitably described as confusing." Squillace (1993, p. 52) wrote, "The lack of cogent legal analyses and the vastly different reasons offered by the justices who spoke the majority render the decision a doubtful source of precedent." University of Colorado law professor Charles Wilkinson called the decision "splintered" and a "step backward" (as quoted by Collins, 1992).

122. "Tribes have a legitimate concern that states may employ administrative jurisdiction as a means of imposing substantive state water laws on them" (Collins, 1985, p. 491).

123. The issue may not even be settled for the Wind River Reservation, as the tribes may be able to pursue other means to achieve essentially the same purposes. For example, Kinney (1993) listed possible scenarios in which the tribes could still achieve some of their goals through negotiation of remaining issues. Collins (1992) noted that the Wind River tribes were considering the protection of instream flows by using the water for other purposes—new irrigation projects—downstream. It is clear, however, that Shay (1992, p. 580) was premature in asserting, "It is settled that Indians do not have to irrigate reservation land with their water, even if it was acquired under the PIA standard."

124. Shurts (1984). The opinion was titled "Federal Water Rights of the National Park Service, Fish and Wildlife Service, Bureau of Reclamation, Bureau of Land Management" and is found at 86 Interior Dec. 553, 594–602, 1979, Solicitor Opinion No. M.-36914, June 25, 1979.

125. Sax, Abrams, and Thompson (1991), pp. 903–5.

126. Shurts (1984). The Martz Opinion is found at 88 Interior Dec. 253 (1981).

127. "Non-Reserved Water Rights—United States Compliance With State Law," 88 Interior Dec. 1055 (1981).

128. Office of Legal Counsel, U.S. Department of Justice, Legal Memorandum: "Federal 'Non-Reserved' Water Rights," 6 Op. Off. Legal Counsel 328 (1982).

129. Office of Legal Counsel, U.S. Department of Justice, Legal Memorandum: "Federal 'Non-Reserved' Water Rights (June 16, 1982)," as quoted and cited by Shurts (1984), pp. 119–20.

130. Shurts (1984).

131. As cited by Shurts (1984), p. 124.

132. As cited by Shurts (1984), p. 126.

133. U.S. Bureau of Land Management (1993), table 36, p. 55.

134. Shurts (1984), p. 145. Shurts also notes, p. 146 that "the mandate of FLPMA is far from clear."

135. 16 U.S.C. sections 1600–1614 (1982). Wilkinson and Anderson (1987) provide an extensive description and analysis of the National Forests Management Act and were the first to suggest the possibility of federal nonreserved rights in the national forests based upon the Act.

136. Wilkinson and Anderson (1987).

137. Sax, Abrams, Thompson (1991), p. 898.

138. 44 Cal. 3d 448, 243 Cal.Rptr. 887, 749 P.2d 324 (1988), cert. denied 488 U.S. 824, 109 S.Ct. 71, 102 L.Ed.2d 48 (1988).

139. Romm and Bartoloni (1985).

140. *In Re Waters of Long Valley Creek Stream System*, 25 Cal.3d. 339, 158 Cal.Rptr. 350, 599 P.2d 656, (1979).

141. 766 P.2d 263, 1988.

142. Wigington (1990).

143. Water Education Foundation (1993).

144. Romm and Bartoloni (1985).

145. Lee (1990).

146. 16 U.S.C. sec. 1761(a)(1). The secretary of the interior continued to administer right-of-way permits granted prior to 1976 while the secretary of agriculture took over responsibility for administering all subsequent permits and renewals. Passage of the "Colorado Ditch Bill" in 1986 amended FLPMA and transferred the administration of the older existing Department of Interior grants to the Forest Service as well. (Condon [Colorado], "Briefing Paper," undated.)

147. "The standard of review for the by-pass flow or other condition used by the courts is 'whether the decision was based on a consideration of relevant factors, whether there has been a clear error of judgment and whether there is a rational basis for the conclusions approved by the administrative body'" (Smith, 1984). These standards are derived from administrative law, which guides the regulatory activities of all other federal agencies as well as the Forest Service.

148. "Over the years the permit system has proven adequate to control development to an extent consistent with the purposes of the national forests" (Colorado District Court, 1993, p. 9).

149. The court in Colorado's Water Division No. 1 did not see the logic of this latter argument, noting, "It is likely that even if the application herein [for a federal reserved right] is granted, much of the monitoring would have to be done by the applicant" (Colorado District Court, 1993, p. 10).

150. Mayors of the cities of Greeley, Loveland, and Boulder stated in their March 19, 1992, letter to the congressmen: "We believe that there is no federal law which requires bypasses, and that these requirements violate federal law, injure existing water rights, and constitute a taking of property interests by the USFS." Eleven members of Congress, from several western states, then sent a letter to Secretary of Agriculture Madigan, on August 12, 1992, reiterating this point:

> We have been informed that the Forest Supervisor for the Arapaho/Roosevelt National Forest takes the position that the Forest Service has the authority to impose bypass flow requirements as a condition of the renewal of the special use permits or approvals of the maintenance and rehabilitation of municipal water diversion and storage facilities located on the Forest. If implemented, these requirements would result in the loss of the historic yields relied upon by these water providers. This position violates the law, injures vested property rights, destroys established man-

> agement practices, and would result in the implementation of environmentally damaging alternatives by the cities which would be forced to replace these supplies from other sources.

The congressmen's letter went on to say:

> We cannot overemphasize the importance of this issue. If the Forest Service is allowed to proceed in this manner, this administration will have taken private property rights, interfered with the development and use of state and interstate water allocations, and replaced state water administration systems with a federal permit system.

151. The letter stated in part: "The Forest Service will reissue permits for existing water supply facilities for 20 years with provisions to recognize and respect both the rights of the applicants and the multiple use objectives of the national forests. New bypass flow requirements will not be imposed on existing water supply facilities."

152. There is no doubt that the mere existence of the letter posed a difficult problem for the Forest Service, as it appeared to direct an action that would not be in compliance with existing statutes and regulations. In his Record of Decision the forest supervisor, Skip Underwood, chose to treat the letter this way: "I read this letter as one which encourages me to work within the confines of existing law to meet its objectives. . . . When, in my judgement, voluntary mitigation will not reduce significant resource damage to minimum acceptable levels, I will not be able to implement the Madigan letter consistent with applicable laws" (U.S. Forest Service, July 1994, p. 6).

153. The Joint Operations Plan was proposed by the cities of Fort Collins and Greeley and the Water Supply and Storage Company.

154. Comment of Senator Hank Brown, R-Colorado, as reported in High Country News, May 16, 1994, p. 7. Senator Brown was also quoted in a Boulder, Colorado, newspaper, the *Daily Camera,* on January 2, 1993, as saying that the Forest Service "is literally asking to extort water from the municipalities for renewing the permits. . . . This is clearly the greatest endangerment of Colorado water rights the state has ever faced, because once this precedent is established, then almost all of the water storage projects in the state are in danger. . . . I like minimum streamflow, and I think it's a worthy objective. . . . The proper way to go about it is to purchase the water, not steal it."

155. Authorities cited and explained in the Records included analysis of the property clause of the U.S. Constitution, the Organic Act of 1897, the Multiple-Use Sustained-Yield Act of 1960, the National Forest Management Act of 1976, the Federal Land Policy and Management Act of 1976, and the Land and Resource Management Plan for the Arapaho and Roosevelt National Forests and Pawnee National Grassland, among others.

156. U.S. Forest Service (July 1994), p. 7.

157. Ibid.

158. As of the summer of 1995, Forest Service Region 2 is proposing guidelines that are more sensitive to natural variability in hydrologic conditions. The proposed

guidelines offer two options for determining minimum flows for naturally perennial streams capable of supporting fish. The first requires that summer flows not drop below the natural median August flow level and that winter flows not drop below the median February flow level, except when natural flows would also fall below those levels. The other option uses three criteria, also employed by the Colorado Water Conservation Board, focusing on flow velocity, flow depth, and wetted perimeter.

159. U.S. Forest Service (July 1994), pp. 9–10.

160. The amount to be contributed by each entity is based on the relationship of the depletions made by each specific project relative to the total depletions occurring within the Platte River Basin.

161. Smith (1984), p. 3.

162. U.S. Forest Service (1994), p. 10.

163. Regional forester Elizabeth Estill, as reported by *High Country News* (May 16, 1994).

164. Senator Hank Brown and Representative Wayne Allard of Colorado sought, as an amendment to the 1996 Farm Bill (H.R. 2854), to legislatively preclude the Forest Service from requiring, as a condition of renewal of a right-of-way, any limitation on the operation of an existing water supply facility that would reduce the water yield of the facility or cause an increase in the cost of the water supply provided by the facility. However, during the time the Farm Bill was in conference, negotiations between the congressmen and Secretary of Agriculture Glickman resulted in a final Farm Bill provision that calls for careful study of the issue. Specifically, Title III, Subtitle H, Section 389, of the 1996 Agricultural Market Transition Act (the Farm Bill) calls for "an 18-month moratorium on any Forest Service decision to require bypass flows or any other relinquishment of the unimpaired use of a decreed water right as a condition of renewal or reissuance of a land use authorization permit" and the establishment of a Water Rights Task Force to study the following subjects: 1) whether federal water rights should be acquired for environmental protection on national forest land, 2) measures necessary to protect the free exercise of nonfederal water rights requiring easements and permits from the Forest Service, 3) the protection of minimum instream flows for environmental and watershed management purposes on national forest land through purchases or exchanges from willing sellers in accordance with state law, 4) the effects of any recommendations on existing state laws, regulations, and customs of water usage, and 5) measures that would be useful in avoiding or resolving conflicts between the Forest Service's responsibilities for natural resource and environmental protection, the public interest, and the property rights and interests of water holders with special-use permits for water facilities. See also Thomas (1996).

165. Though the branch takes responsibility for virtually all of the work leading to these determinations, the actual legal work is performed by staff in the Solicitor's Office of the Department of the Interior, and the Department of Justice represents the Park Service in court. The history, composition, and purposes of the Water Rights Branch presented here were drawn from Williams, Owen (1994) and McGlothlin (1994).

166. McGlothlin (1994). An example of a private interest group that has been active in bringing resource threats to the attention of the Park Service is the National

Parks and Conservation Association (NPCA). For example, the Park Service moved to claim water rights for a wild and scenic river in Tennessee after the NPCA brought to the agency's attention several threats to the river (Owen Williams, 1994).

167. The material that follows is drawn from Williams, Owen (1994).

168. Williams (1996).

169. Haas (1994).

170. See U.S. Bureau of Land Management (November 1992).

171. Ibid.

172. Details of the BLM Riparian-Wetland Initiative can be found in U.S. Bureau of Land Management (1991).

173. U.S. Bureau of Land Management (1991), pp. 23–4.

174. Ibid., table 1, p. 4.

175. The description of the BLM interdisciplinary process that follows is drawn primarily from Jackson et al. (1989) and Muller (1994). Information about the Rio Chama is drawn from the U.S. Bureau of Land Management (December 1992) and Muller (1994).

176. For example, the most recent application of BLM's interdisciplinary study process, on the Arkansas River in Colorado, was delayed for over two years while agency personnel worked out cooperative agreements with the numerous water users and managers that might be affected by the project. The BLM has completed studies for the San Pedro River Properties in southern Arizona (1987), the Beaver Creek National Wild River in Alaska (1987), the Gulkana National Wild River in Alaska (1990), the Dolores River in southwestern Colorado (1990), and the Rio Chama in northern New Mexico (1992).

177. The description of the interdisciplinary process given here is derived in part from Jackson et al. (1989), who described the process as consisting of six basic steps.

178. This emphasis on value definition has sometimes led to the BLM's entire study process being referred to as BLM's "value-driven" process, though the term is not currently in much use.

179. The management plan and other basic information about the Rio Chama are described in more detail in chapter 3.

180. Reed and Drabelle (1984) at p. 10.

181. As quoted by MacDonnell (1985), p. 102.

182. Cooperative Instream Flow Service Group (undated), p. 11.

183. Stalnaker and Lamb (1994).

184. Their name was later changed to National Biological Service.

185. Smith (1994). Some of the BLM state offices that have been particularly active in instream flow matters include Alaska, Arizona, Colorado, and Idaho. The BLM was the first successful applicant for an instream flow right in Nevada. There has been very little instream flow protection activity in some of the other BLM state offices.

186. Reed and Drabelle (1984, p. 25) reported that there is no official in Wash-

ington—aside from the director and his or her associates—specifically in charge of refuges to whom the regional directors report.

187. Willis (December 1993).

188. Information about Region 1 is derived from Oser (1993). Instream flow claims are being made for wildlife refuges in the general adjudication of the Snake River in Idaho.

189. The status of instream flow protection activities in Region 6 was derived from Willis (1993).

190. Information about Region 2 was derived from Cullinan (1993).

191. Information about instream flow protection on the National Wildlife Refuges in Alaska was derived from Bayha (1993).

192. Instream flow reservations in Alaska may be used to protect lakes at a specified level of elevation. Region 7 personnel intend to verify lake levels in remote areas such as the Arctic National Wildlife Refuge by using state-of-the-art satellite observation technology.

193. Bayha (1993).

194. Tarlock (1987) argued that "the Park Service now lacks a consistent park management philosophy. Buffeted by strong constituencies, comparatively little attention has been given to less visible problems such as water flow protection" (p. 38). He noted a similar lack of focus in the popular and scholarly literature on the parks, where "water flow management is given comparatively little attention compared to external threats, land acquisition, human-wildlife conflicts and visitor management" (p. 29).

195. Coggins and Wilkinson (1987), p. 162.

196. U.S. Bureau of Land Management (November 1992), p. 12 and table 2 (p. 16).

197. U.S. Fish and Wildlife Service (1993), pp. 1–17.

198. National Audubon Society (1994).

199. U.S. Fish and Wildlife Service (1993), pp. 1–18.

200. Ibid.

Chapter Nine

1. The origins and activities of the Bureau of Reclamation were also described in chapter 2.

2. 1993 reclamation data from U.S. Bureau of Reclamation (1993).

3. The only major reclamation project still pending is the Animas–La Plata Project in southwestern Colorado. Envisioned over 90 years ago, and now expected to cost over $650 million, the project involves construction of dams on the Animas River near Durango, Colorado, and the La Plata River near the New Mexico state line and facilities to deliver water to agricultural, municipal, and industrial users. Now promoted as a way to settle water claims of the Southern Ute and Ute Moun-

tain tribes, the project would deliver roughly one-third of the annual yield of nearly 200,000 acre-feet to those tribes plus the Navajo nation. The Bureau of Reclamation was expected to release a final environmental assessment in late 1995. The project has been widely opposed on both economic and environmental grounds and will certainly face legal challenges.

4. U.S. Bureau of Reclamation (1992).

5. As of January 1995, there were no ongoing efforts at the Bureau to define a specific instream flow policy, but the draft implementation plan for instream flows had been distributed to all Bureau of Reclamation offices for use as a reference on the topic (Whittington, 1995; Whittington was head of the Bureau's Instream Flow team that developed the draft implementation plan).

6. The four strategies specifically identified in the draft plan are: 1) to determine, in cooperation with others, instream flow needs, conduct studies to identify flexibility in reservoir operations to help meet needs, and modify operations as appropriate; 2) to operate Reclamation projects in coordination with other projects to help meet instream flow objectives; 3) to develop partnerships and joint-venture projects with interest groups to increase the benefits associated with instream flows; and 4) to conduct research on cost-effective measures that could be implemented at Reclamation facilities to improve the quality of instream flows.

The draft's authors explicitly acknowledge that the needs of society have changed substantially since the Reclamation program was initiated, stating, "The challenge is for Reclamation to adapt past decisions and actions to today's standards and public values. . . ." The value of instream flows for fish, wildlife, water quality, stream channel integrity, recreation, and groundwater recharge are all described in the draft, with particular attention being paid to fish and wildlife needs. The draft includes a discussion of planning processes (establishing objectives, quantifying flows, consideration of legal, financial, and physical constraints), multi-objective planning, monitoring and evaluation, and flow quantification studies that must be considered when pursuing instream flow opportunities.

7. The following examples are taken from U.S. Bureau of Reclamation (1992).

8. The draft instream flow implementation plan does, however, note that "enhancing the sport fishery could be at the expense of native species of fish" (U.S. Bureau of Reclamation, 1992, p. 16).

9. The welfare of swans on the Henry's Fork was also discussed in chapter 5, in conjunction with the use of water banking procedures to protect instream values.

10. In the years immediately preceding his appointment as secretary of the interior, Babbitt often suggested that the Bureau of Reclamation had outlived its usefulness and should be abolished. Dan Beard, who formerly worked for reform-minded Congressman George Miller of California, chairman of the House committee in charge of western water affairs, repeatedly stated his intent to change the way that the Bureau operates.

A letter sent out to all Bureau employees in November 1993 by Commissioner Beard (Beard, 1993) provides evidence of the new leadership's commitment to change. Commissioner Beard wrote of Secretary Babbitt's personal commitment to "the transforming of the Bureau of Reclamation from a civil works agency into a preeminent water management agency that is cost effective in serving its customers."

The letter emphasized that the Bureau's mission statement is "To manage, develop, and protect water and related resources in an environmentally sound manner in the interest of the American public" and provided several principles to guide the Bureau in its accomplishment of this mission.

11. Beard (1993).

12. Whittington (1994, 1995). One example of programmatic innovation at the regional level affecting instream flows occurred in the Great Plains region based in Billings, Montana. There, the Investigation of Existing Projects Initiative (IEP) was established to "evaluate our existing projects from the perspective of meeting contemporary public values, including wetlands, recreation, water quality improvements, and fish and wildlife benefits" (U.S. Bureau of Reclamation, 1993). The IEP was designed to enable field offices, on an informal basis, to identify and evaluate possible improvements in project operations which would have the potential to yield additional benefits, particularly with regard to those benefits identified in the commissioner's "Blueprint for Reform" (Beard, 1993).

The process proceeds through several stages. First, a three-to-five page Project Review Report (PRR) is completed for each project, in order to make an initial survey and assessment of potential opportunities and constraints. A Ranking and Selection Criteria Form is then used by a Regional Office team to rank and prioritize PRRs, focusing on the identification of projects "which will improve operating efficiency and conserve and fully utilize resources for expanded uses and new demands, with particular emphasis on non-consumptive environmental benefits." Projects selected through this process are then evaluated more fully, through the use of 1-year appraisal studies. If potential for creating new benefits is found to exist, a report is prepared that recommends items for immediate implementation, more in-depth planning, action under the provisions of the Small Reclamation Projects Act, or local construction to achieve the benefits.

Eighteen PRRs for projects in the Great Plains Region were completed in 1993, and 17 more were scheduled to be completed in 1994. In February 1994, three PRRs were selected for appraisal-level studies.

13. P.L. 102–575, October 30, 1992; 106 Stat. 4600.

14. See *High Country News* (July 15, 1991) for a more detailed accounting of the history, issues, and prospects of the Central Utah Project.

15. Title III, Sec. 301(a)(1), P.L. 102–575.

16. Title III, Sec. 302, P.L. 102–575.

17. Section 303 states:

> The District shall acquire . . . all of the Strawberry basin water rights being diverted to the Heber Valley through the Daniels Creek drainage and shall apply such rights to increase minimum stream flows—
>
> (A) in the upper Strawberry River and other tributaries to the Strawberry Reservoir;
>
> (B) in the lower Strawberry River from the base of Soldier Creek Dam to Starvation Reservoir; and
>
> (C) in other streams within the Uinta basin affected by the Strawberry Collection System in such a manner as deemed by the Commission in consultation

with the United States Fish and Wildlife Service and the Utah State Division of Wildlife Resources to be in the best interest of fish and wildlife.

18. Weland (1995).

19. Section 307.

20. Section 309.

21. Section 313.

22. Title IV, Section 402 of P.L. 102–575. All of the contributions are to be increased proportionally each year based on rises in the Consumer Price Index.

23. Water Education Foundation (1993).

24. Some of the dams constructed as part of the CVP completely blocked salmon from large areas of their historic spawning grounds, and many rivers were dewatered. For example, 22 miles of the San Joaquin below Friant Dam have been completely dewatered, except in flood years. See Dunning (1993b), especially Part IV.

25. Two other major features of the CVPIA enacted reforms in water pricing and made it easier to transfer CVP water from historic uses to new uses, but these reforms are not directly related to the Act's environmental goals.

26. Money for the habitat restoration fund comes from several sources. The federal share, expected to approximate 75 percent of the total, comes from environmental charges of up to $6 per acre-foot for agriculture and $12 per acre-foot for municipalities levied on all deliveries of CVP water, a $25 per acre-foot charge for all water transferred out of the CVP project area, and increased revenues resulting from a new block-rate pricing schedule for water deliveries. The state share will be taken from a $60 million mitigation fund established by the California State Water Resources Control Board as part of a plan for managing water resources in the bay-delta area. Once all habitat projects mandated in the CVPIA are completed, the annual fund will drop to $35 million, largely as a result of reduced environmental fees levied on CVP water deliveries to agricultural and municipal users (Water Education Foundation, 1993).

27. Loomis (1994).

28. In a dry year, the allocation can be reduced as low as 600,000 acre-feet. If conditions are sufficiently wet that water-quality standards are met without any additional allocation from the CVP, then the Bureau can store the water or use it for other purposes.

29. Because of differences between types of contractors for CVP water, and because certain water users were exempted from provisions of the Act, reductions in previous water allocations necessary to free up the 800,000 acre-feet of water for environmental purposes will not be shared equally among those currently using CVP water. The burden will instead be placed on a category of water users known as water service contractors, who have previously received about half (4 million acre-feet) of the CVP's annual yield. These contractors will be required to give up as much as 20 percent of their current allocations—depending on exactly how much water needs to be allocated for environmental purposes to meet water-quality standards in any given year—whereas the recipients of the other half of the CVP's annual yield—primarily the "exchange contractors" and "water rights settlement contractors" discussed in

chapter 6—will not be required to give up any water (Water Education Foundation, 1993).

30. Environmentalists and others have recently questioned the continued existence of a few small dams, most notably in the Pacific Northwest. The removal of two dams on Washington's Elwha River and three dams on Oregon's Rogue River has been proposed in order to promote the return of salmon runs that were decimated when the dams were constructed. All five of these dams are relatively small, however, and the conditions justifying their removal are not likely to exist for a very large proportion of the rest of the West's dams and reservoirs. People are also discovering that the cost of removing dams, including expenditures necessary to rehabilitate natural stream channels, is substantial. In particular, the huge volumes of sediment trapped behind dams has the potential to damage rivers for many years after dams are removed. The total cost of removing the two Elwha River dams was recently estimated to be as high as $209 million, including the cost of buying the dams ($29.5 million), engineering their removal, and disposing of the accumulated sediment (Dellios, 1994).

31. Reclamation facilities might be of benefit to proponents of instream water uses in another way. Uncontracted storage capacity (storage capacity in excess of the amount of water already appropriated and contracted for delivery downstream) exists in several Reclamation facilities around the West and might be used to facilitate negotiated settlements between offstream and instream water users—essentially by enlarging the pie that is being divided rather than forcing one party's gain to come at the other's expense (Glaser, 1993).

32. For a more complete explanation of the Bureau's accounting methods and the role of hydroelectric generation in justifying reclamation projects, see Reisner (1986) and Wahl (1989).

33. BPA was originally given the goals of promoting "the widest possible use of electricity" and selling federal power "at the lowest possible rates consistent with repaying investment obligations over a reasonable period of time" (Bonneville Power Administration, April 1994).

34. Bonneville Power Administration (January 1994a).

35. WAPA's mission, as originally established for the Bureau by the Flood Control Act of 1944, is to market federal hydroelectric resources "in such a manner as to encourage the most widespread use thereof at the lowest possible rates to consumers consistent with sound business principles. . . ." (Western Area Power Administration, 1994).

36. Ibid.

37. Quoted by U.S. Bureau of Reclamation (March 1995b), p. iv.

38. Ibid., p. 17. This is the stated objective of the "no action" alternative, which represents historic operations of the dam.

39. According to the Bureau of Reclamation's draft EIS for the Glen Canyon Dam, annual releases from 1966 to 1989 ranged from 8.23 million acre-feet to 20.4 million acre-feet (in 1984). The minimum release has occurred in about half of the years since the dam was closed in 1963. U.S. Bureau of Reclamation, March 1995b, p. 18.

40. Roughly 1.7 million end-use customers—residential, commercial, and industrial—are served with electricity by the entities—municipal utilities, rural electric cooperatives, irrigation districts, etc.—to which WAPA sells electricity produced at Glen Canyon Dam. U.S. Bureau of Reclamation, March, 1995a, p. 297.

41. U.S. Bureau of Reclamation (March 1995b), p. 19.

42. The EIS process is described in chapter 10.

43. U.S. Bureau of Reclamation (March 1995b), pp. iv and 1, respectively.

44. Grand Canyon Protection Act of 1992, P.L. 102–575, signed into law October 31, 1992.

45. The intent of the Bureau in preparing the EIS was to give significant consideration to input from all affected groups, and the Grand Canyon Protection Act mandates a broad interest group discussion of future dam operations. Among those Congress directed to be consulted by the Bureau of Reclamation are the secretary of energy, the governors of the Colorado River basin states, Indian tribes, and the general public, "including representatives of academic and scientific communities, environmental organizations, the recreation industry, and contractors for the purchase of Federal power produced at Glen Canyon Dam." Grand Canyon Protection Act of 1992, section 1805(c).

46. The preferred alternative identified in the draft EIS called for a maximum flow rate of 20,000 cfs and an upramp rate of 2,500 cfs/hour. After receiving comments on the draft EIS, and after discussions with the U.S. Fish and Wildlife Service and other agencies, the Bureau decided to increase both of these rates, to levels discussed below in the main text. Scientists from the Fish and Wildlife Service and other agencies assured the Bureau that these changes would not have an adverse impact on downstream resources (U.S. Bureau of Reclamation, Fall 1994).

47. U.S. Bureau of Reclamation (Fall 1994).

48. The sequence of flows in the beach/habitat-building test flow is described in more detail in U.S. Bureau of Reclamation (1996). In preparation for the week-long controlled flood, scientists photographed and measured the beaches, moved most of the endangered Kanab ambersnails to higher ground, and even implanted 10 endangered humpback chubs with microchips to facilitate tracking them during the flood. Geologists, hydrologists, geomorphologists, biologists, botanists, and other scientists gathered to monitor what was certainly the most carefully observed flood in history.

49. See Wuethrich (1996) or Collier, Webb, and Andrews (1997) for a more complete description of the test flood.

50. U.S. Bureau of Reclamation (1996).

51. Wuethrich (1996).

52. U.S. Bureau of Reclamation (August 1994).

53. Wuethrich (1996).

54. The opportunity cost of energy produced at hydropower plants is the cost of replacing it at the more expensive thermal plants. Because shifting some energy production at Glen Canyon Dam from peak times to base-load times will displace production at coal-fired plants producing base-load energy but will require more pro-

duction at gas-turbine peaking plants, a rough idea of the economic effect of the shift is obtained by observing the operation and maintenance costs of production at the two kinds of thermal plants. Average O&M costs are currently about 33 mills per kilowatt hour at gas-turbine plants and 19 mills per kilowatt hour at coal-fired plants (Gibbons, 1986; Energy Information Administration, 1985, 1996). (The differences largely reflect the difference in fuel cost.) Thus, the short-run economic cost of each kilowatt hour of shift is the difference between these two rates, or about 14 mills. This cost will essentially be handed on to electricity consumers.

55. U.S. Bureau of Reclamation (March 1995b), p. 62.

56. The two dollar figures reflect two alternative electric energy marketing assumptions (U.S. Bureau of Reclamation, March 1995a, p. 300).

57. The act authorizes the secretary of the interior to use funds from the sale of electric power from the Colorado River Storage Project to pay for the expenses, but states that "such funds will be treated as having been repaid and returned to the general fund of the Treasury as costs assigned to power for repayment under section 5 of the [Colorado River Storage Project] Act of April 11, 1956" (Grand Canyon Protection Act of 1992, Section 1807).

58. Grand Canyon Protection Act of 1992, Section 1809.

59. Wilkinson and Conner (1987).

60. Bates et al. (1993), p. 98.

61. Pacific Northwest Power Planning and Conservation Act, sec. 2.(6). The NPA actually went well beyond these two features, because it also called for increased reliance on energy conservation, regional planning for meeting both electricity and environmental needs, and enhanced opportunities for participation by the public in decision-making processes, but these features are less relevant to the discussion of instream flows.

62. Creation of the Council was also motivated by a desire to balance NPA's expansion of federal power with an expanded role for the states, acting through the new regional authority. However, "rather than being a reaction against federal interests, then, the Act created a way for the region to weave federal interests into the region's plans" (Volkman and Lee, 1988, p. 565).

63. Bonneville Power Administration (March 1994).

64. The Columbia River Inter-Tribal Fish Commission (CRITFC) is composed of the four tribes with treaty fishing rights to the Columbia River—the Warm Springs, Umatilla, Yakima, and Nez Perce tribes. CRITFC has been a leader in salmon conservation efforts since its establishment in 1977.

65. Hirt and Gregg (1991).

66. Mortensen, Craig (September 1994).

67. Ibid.

68. Ibid.

69. Bonneville Power Administration (March 1994). Naiman et al. (1995), p. 585, cited the Power Planning Council's expenditure of funds for recovering degraded salmon and steelhead runs in the Columbia as an example of "monies far in

excess of what is needed for a comprehensive freshwater program already being spent on ineffective and contradictory programs." Their primary objection seems to be that monitoring programs that would enable the measurement of the major sources of mortality at key points in the river and ocean do not exist.

70. Bonneville Power Administration (March 1994), Mortensen (1994).

71. Naiman et al. (1995), p. 585.

72. Mortensen, Craig (September 1994).

73. 16 U.S.C. sections 791(a)-793, 796–818, and 820–825(r) (1988). Currently as amended, especially by the Electric Consumers Protection Act, 16 U.S.C. sections 791a-825s. More detailed information on the history of the FPA can be found in the introductory material by Walston (1990a).

74. For a more detailed discussion of the impact of the 1986 amendments on FERC licensing, see Bearzi (1991).

75. As held by the court in *Escondido Mutual Water Company v. La Jolla Band of Mission Indians*, 466 U.S. 765 (1984).

76. Excludes pumped-storage plants.

77. The data on hydroelectric facilities and licensing presented in this paragraph have been drawn from U.S. Department of Energy (1994), chapters 1 and 2, and are accurate as of 1993.

78. Facilities owned by the Corps of Engineers and Bureau of Reclamation account for 27 percent and 18 percent of United States capacity, respectively.

79. Data derived from U.S. Department of Energy (1994).

80. See Bearzi (1991), p. 328, for a list of court cases and congressional actions designed to force FERC to more fully accommodate environmental and other nonpower interests in its licensing proceedings.

81. 328 U.S. 152 (1946).

82. See Walston (1990a), pp. 91–92.

83. 438 U.S. 645 (1978).

84. 877 F.2d 743 (9th Cir. 1989), aff'd, 110 S.Ct. 2024, 109 L. Ed.2d 491 (1990).

85. The description of the Rock Creek case presented here is drawn from Robie (1990), Walston (1990a), and Bearzi (1991).

86. *California v. Federal Energy Regulatory Commission*, 877 F.2d 749 (9th Circuit, 1989), as cited by Walston (1990a), p. 96 and note 65.

87. From 109 L.Ed.2d at 491, as quoted and cited by Bearzi (1991), p. 329.

88. Bearzi (1991).

89. Smith (1990).

90. Robie (1990).

91. For example, consider the following comments of the conservation group American Rivers, in its fall 1993 newsletter:

The new Federal Energy Regulatory Commission is listening to American Rivers' repeated calls for reform of its hydropower policies. In mid-September, FERC Chair Betsy Moler announced new agency initiatives that will encourage greater public input into the hydropower licensing process, address environmental problems caused by multiple hydropower projects on the same river, and establish a clear policy for dams that should be decommissioned or removed. These new policy announcements directly address issues raised by members of the National Hydropower Relicensing Coalition at FERC's hydropower roundtable in June.

[American Rivers' director of hydropower programs] said that FERC's announcements "show the commission has heard our concerns. We're greatly encouraged. Now our job is to make sure that FERC 'walks its talk' and carries out these new initiatives."

Chapter 10

1. See Tarlock (1985a).

2. Codified at 42 U.S.C. 4321, 4331–4335, 4341–4347. For a comprehensive description of the Act, see Fogleman (1990).

3. In what is sometimes claimed to be the most expensive preparation of an Environmental Impact Statement ever, the Army Corps of Engineers, using contributions from applicants, spent over $40 million preparing an EIS for the proposed Two Forks water development project on the South Platte River near Denver, Colorado. This case is discussed in more detail in the following section on the Clean Water Act.

4. Fogleman (1990), p. 77.

5. Futrell (1988).

6. Ibid.

7. The "Clean Water Act" is actually a compilation of several pieces of legislation. Among these are the Federal Water Pollution Control Act of 1948 and amendments made to that act in 1952, 1956, 1961, 1972, 1977, and 1987 and the Water Quality Act of 1965. Of these, the 1972 and succeeding amendments were by far the most substantive, forming the core of today's Clean Water Act. The Clean Water Act is codified at 33 U.S.C.A. sec. 1251 et seq.

8. Sec. 101(a), 33 U.S.C. sec. 1251(a).

9. Tarlock (1985a), p. 13.

10. 40 C.F.R. sec. 231.2(e), as cited by Sax, Abrams, and Thompson (1991), p. 599, note 2.

11. 758 F.2d 508, 1985.

12. As excerpted by Sax, Abrams, and Thompson (1991), pp. 603–4. Citation omitted.

13. As discussed above, a nationwide permit may be issued for certain kinds of projects determined to have only a minimal impact on the environment. Because the application and approval process for such permits is much quicker than for individual permits—at least for projects that qualify—the Riverside District first pursued this

option. Failure to receive a nationwide permit does not preclude an applicant from subsequently seeking an individual permit.

14. Rhodes, Miller, and MacDonnell (1992).

15. As explained in greater detail in the Endangered Species Act section of this chapter, a "no jeopardy" opinion essentially means that construction and operation of the project are not likely to threaten the existence of any species on the list of threatened or endangered species maintained by the Fish and Wildlife Service. Though this particular opinion was issued subject to the future implementation of specified mitigation measures by the permit applicants, environmental groups generally considered this decision to be inappropriate (*High Country News*, May 9, 1988).

16. *High Country News*, May 9, 1988.

17. This was the lowest rating that the EIS could have received from the EPA (Luecke, 1990).

18. Sax, Abrams, and Thompson (1991), p. 599.

19. For example, see Ransel and Myers (1988) or Birnbaum (1991).

20. 511 U.S. 700, 128 L.Ed.2d 25, 114 S.Ct. 1900 (1994). The following material about the case is drawn from a variety of sources, including the case report; *Water Intelligence Monthly*, June 1994, pp. 11–12; *Federal Parks and Recreation*, vol. 12, no. 12, June 16, 1994, pp. 7–8; and the *Sacramento Bee*, "High Court Backs States on Water Flows," p. A6, Wednesday, June 1, 1994.

21. WAC 173–201–045(1), cited by the court at L.Ed.2d 724, n. 1.

22. 128 L.Ed.2d 716, 732 (1994).

23. Justices Scalia and Thomas dissented.

24. For example, see Adler, Landman, and Cameron (1993), p. 211.

25. Ibid., p. 215.

26. Adler, Landman, and Cameron (1993).

27. Ransel and Myers (1988), Birnbaum (1991), Ransel (1994).

28. Adler, Landman, and Cameron (1993), p. 201.

29. Adler, Landman, and Cameron (1993).

30. Ransel (1994).

31. Ibid.

32. The Endangered Species Act of 1973 is codified at 16 U.S.C. secs. 1531–1543. According to the Supreme Court: "As it was finally passed, the Endangered Species Act of 1973 represented the most comprehensive legislation for the preservation of endangered species ever enacted by any nation" (*Tennessee Valley Authority v. Hill*, 437 U.S. 153, 98 S.Ct. 2279, 57 L.Ed.2d 117, 1978, as excerpted by Bonine and McGarity, 1992, p. 10).

Passage of the Endangered Species Act was preceded by a handful of other actions and pieces of legislation. In 1964 the Committee on Rare and Endangered Species of the Interior Department's Bureau of Sport Fisheries and Wildlife (now the U.S. Fish and Wildlife Service) published a "Redbook" that listed 63 vertebrate species that it considered to be endangered. Immediate precursors of the Endangered Species Act included the Endangered Species Preservation Act of 1966—"a vague policy directive that served primarily as a symbolic statement of congressional support for endangered species protection" (Kohm, 1991, p. 12)—and its updated version, the Endangered Species Protection Act of 1969.

33. The Endangered Species Act also authorizes the two agencies to list species that are threatened or endangered with extinction outside the United States; to date, another 500 or so of these species have been listed.

34. One thousand of these potential additions are listed in "Category 1," which includes species for which sufficient information exists to propose listing, and 2,000 are listed in "Category 2," which are believed threatened or endangered but require more data to make a determination.

35. 46 Fed. Reg. 54748, 54750 (Nov. 4, 1981), as cited by Ernst (1991).

36. 16 U.S.C. sec. 1536 (1976 ed.), as cited by Bonine and McGarity, 1992, p. 8.

37. *Tennessee Valley Authority v. Hill*, 437 U.S. 153, 98 S.Ct. 2279, 57 L.Ed.2d 117, Supreme Court of the United States, 1978.

38. H.R. Rep. No. 93–412, p. 14 (1973), as cited by the Supreme Court in the case of *TVA v. Hill*, as excerpted by Bonine and McGarity (1992), p. 10. Emphasis by the court.

39. *Carson-Truckee Water Conservancy District v. Watt*, 549 F.Supp.704 (D.Nev. 1982), aff'd, 741 F.2d 257 (9th Cir. 1984), cert. denied, 470 U.S. 1083 (1985).

40. See MacDonnell (1985), p. 17.

41. The exemption process was created in response to the controversy surrounding the Supreme Court's decision in *TVA v. Hill.* Agencies wishing to exempt a particular action from the requirements of Section 7 apply to a high-ranking committee: the secretary of agriculture, secretary of the army, the chairman of the Council of Economic Advisors, the administrator of the Environmental Protection Agency, the administrator of the National Oceanic and Atmospheric Administration, and one individual from each affected state. This committee is usually referred to as the God Committee because of its power to authorize the extinction of an endangered species.

To be successful, an agency must show 1) that there are no reasonable and prudent alternatives to an agency's proposed species-threatening action; 2) that the benefits of such action clearly outweigh the benefits of alternative courses of action consistent with conserving the species or its critical habitat, and that such action is in the public interest; 3) that the action is of regional or national significance; 4) that the agency applying for an exemption had not already made an irreversible commitment of resources toward the species-threatening action so it could hold the species hostage to these already-spent costs.

Though an exemption for the Tellico Dam project in Tennessee was denied by the committee, Congress later passed legislation specifically excluding the dam from the provisions of the ESA. The dam was eventually completed. Construction of the dam did not lead to the immediate extinction of the snail darter, as small populations of the species were later found in other streams in the region.

42. In practice, the Fish and Wildlife Service or the Marine Fisheries Service does the consulting.

43. According to Barker (1993, pp. 137, 158): "Nowhere else in this law or any federal law has wording been so clear or so protected from legal misinterpretation. Yet while the legal mandate has been clear, the actual implementation has been murkier than mining runoff." And, "The language of Section 7 may be airtight, but the actual enforcement is flexible to the point of breakage."

44. Campbell (1991).

45. Barker (1993), p. 139. Barker preceded the comment by noting that "the overwhelming majority of consultations are informal. Biological opinions are as infrequent as caribou sightings in the Lower 48 states. Jeopardy opinions are as rare as Snake River sockeye."

46. Lynn Greenwalt (1991, p. 34), director of the U.S. Fish and Wildlife Service from 1973 to 1981.

47. Comment of Hank Fischer, former regional director of the conservation group Defenders of Wildlife, as quoted by Barker (1993), p. 140.

48. Yaffee (1991).

49. Coggins (1991), p. 66.

50. Yaffee (1991), p. 88.

51. Ibid., p. 92.

52. For example, Sections 8 and 9 of the ESA contain provisions related to international trade in endangered species and implementation of the Convention on International Trade in Endangered Species (CITES). Section 6 establishes a system of cooperative programs with states, Section 5 deals with the acquisition of habitat to support listed species, and Section 4 focuses on the development and implementation of species recovery plans.

53. Barker (1993), pp. 157–8.

54. Ibid.

55. Tarlock (1985a).

56. *High Country News*, May 30, 1994, p. 3. As a testament to the highly controversial nature of actions under the ESA, it is interesting to note that the events surrounding this particular application caused critics on all sides to deplore the effects of political influence on the implementing agencies. Environmentalists believed the original "no jeopardy" decision to be ludicrous and thought that it could be explained only as one more instance in which the Fish and Wildlife Service or Marine Fisheries Service exhibited an unwillingness to list species or issue jeopardy opinions in situations in which political pressure was brought to bear by developers. On the other side, Senator Larry Craig of Idaho argued that the Fisheries Service action on behalf of the salmon in May 1994 was actually a result of political pressure from the White House rather than a result of scientifically justified measures required to protect the salmon. It should be no surprise, therefore, that both the Fish and Wildlife Service and the Marine Fisheries Service find it difficult to base their decisions solely on "biological evidence and the best scientific and commercial data available" when implementing provisions of the Endangered Species Act.

57. Environmentalists have also acted to protect the Snake River salmon species through modification of current land uses in the upstream watersheds where the

salmon spawn. In a court case closely watched by a variety of interested parties, a federal district court ruled in July 1994 that the Forest Service was required to have the National Marine Fisheries Service review its land-use activities in the Wallowa-Whitman and Umatilla National Forests for compatibility with protection of listed species. Those bringing the lawsuit successfully sought to use the ESA to control grazing, logging, and road building in the national forests in order to better protect spawning habitat of the Snake River chinook salmon. *High Country News*, August 8, 1994, p. 5.

58. Quoted in *Water Intelligence Monthly*, April 1996, p. 8.

59. *Water Intelligence Monthly*, April 1996.

60. Ibid., p. 8.

61. For example, Yaffee (1991, p. 93) stated that the Windy Gap strategy went too far: "Simply allowing projects to proceed if they pay tribute without knowing what will happen is wrong."

62. *Water Intelligence Monthly*, April 1994, p. 7.

63. *High Country News*, May 16, 1994, p. 7.

64. The cranes most commonly associated with the Platte River are the sandhill cranes, which stop over at the river for up to a month during the course of their migration from Mexico and Latin America to points farther north in the United States and Canada. Their numbers at the Platte River in March often total several hundred thousand, accounting for the vast majority of all sandhill cranes in the world. The endangered whooping crane, one of the highest-profile species protected under the ESA, also uses the river during migration, though probably less regularly than does the sandhill crane.

65. The potential impact of limitations on existing diversions in the national forest headwaters of the South Platte in order to accommodate endangered species hundreds of miles downstream on the Platte River in Nebraska was discussed in greater detail in the special-use permits unit of the national forests section of chapter 8.

66. Wild and Scenic Rivers Act, 16 U.S.C. sec. 1271–87.

67. Tarlock and Tippy (1970) reported that after many years of fighting dam construction on a case-by-case basis, river conservationists first began to attempt to change river development institutions in the 1940s and 1950s, when dam opponents sought to force the use of small headwaters impoundments for controlling dams rather than construction of large multipurpose reservoirs. After being thwarted in this attempt, free-flowing river advocates then turned to the "park and recreation concept" as a means of protecting flows in the late 1950s and early 1960s. Both of these methods preceded the series of events leading to passage of the Wild and Scenic Rivers Act in 1968. For a more detailed account of this history, see also chapter 2 in Palmer (1993).

Tarlock and Tippy (1970, p. 710, note 17) noted that Congress had, on occasion, acted previously to protect the free-flowing nature of rivers directly. For example, the Act of June 29, 1906, ch. 3621 gave the secretary of the army the power to issue revocable permits for the diversion of water from the Niagara River as long as it did not interfere with the navigability of the river or the scenic grandeur of Niagara Falls.

68. There is some variation in the methods by which the number of rivers in the Wild and Scenic Rivers system is calculated and reported. The number reported here (165) is taken from the November 1996 "Rivers Mileage Classifications" table for the national Wild and Scenic Rivers Program maintained by the National Park Service (Haubert, 1997) and represents the number of designations made to the system by Congress or the secretary of the interior rather than the number of rivers in the system. The distinction between rivers and designations is significant for two reasons. First, 8 rivers have been designated more than once, with later designations adding either contiguous or completely separate segments of a river to the system. Combining separate designations of the same river—as versions of the "Rivers Mileage Classifications" table released since about 1993 have done—reduces the total number of rivers, as of November 1996, to 154 (Haubert, 1997). Second, tributaries included within designation legislation are not counted at all, even though they often represent substantial additions to the system. Palmer (1993) made an attempt to count the actual number of rivers protected within the national Wild and Scenic Rivers Program and reported that as of August 1992—when official figures reflected a total of 151 rivers in the system—a total of 314 rivers and tributaries had actually been designated by Congress for inclusion in the system. The reported mileage of protected rivers is the same no matter which set of figures is used, since the mileage of designated tributaries is also counted in the official data.

69. Figure 10.1 depicts designated Wild and Scenic River stretches in the West as of December 1992.

70. The Wild and Scenic Rivers Act contains the following definitions (16 U.S.C. sec. 1273):

1. Wild river areas—Those rivers or sections of rivers that are free of impoundments and generally inaccessible except by trail, with watersheds or shorelines essentially primitive and waters unpolluted. These represent vestiges of primitive America.
2. Scenic river areas—Those rivers or sections of rivers that are free of impoundments, with shorelines or watersheds still largely primitive and shorelines largely undeveloped but accessible in places by roads.
3. Recreational river areas—Those rivers or sections of rivers that are readily accessible by road or railroad and that may have some development along their shorelines and may have undergone some impoundment or diversion in the past.

71. Miller (1994). This method of designation seems to be especially popular for rivers running through private lands, where substantial efforts have recently been made to increase designations under the Act. This process was used, for example, to evaluate an 11 mile stretch of the Klamath River in Oregon, which was designated by the secretary of the interior in September 1994.

72. The recommendation of a river by a federal agency for consideration by Congress is the last step in a long process. The specific process and terms presented below are those of the United States Forest Service, but the general outline is similar to those used by some of the other federal land management agencies.

Usually identification of rivers for study is preceded by some form of either formal or informal inventory of land and water resources that are being managed. Once a river has been brought to the agency's attention, the agency first tries to determine whether or not a river is "eligible" for designation, i.e., possesses any of the "outstandingly remarkable" characteristics called for by the Wild and Scenic Rivers Act. A river does not need to have multiple such characteristics to be considered for designation in the system, but it must have at least one. The determination of what constitutes an outstandingly remarkable feature is subjective, and the agency will often attempt to include interested members of the public in helping to make this determination.

If the river is determined to be eligible, the agency starts to manage the river in a way that will prevent the outstandingly remarkable characteristics from being damaged or lost while further studies occur. The next step is a "suitability" study, in which the agency not only studies the physical characteristics of the river in greater detail but also evaluates other features such as river corridor land-use activities, the river's ecology, the management preferences of local property owners and other interest groups, the water rights situation on the river, and the relative merits of alternative management schemes for the river. The agency's WSRA study activities generally occur within the context of its general forest planning process, and the agency's conclusions—as well as the data used as a basis for the conclusion and a consideration of alternatives—are usually presented within the Environmental Impact Statement (EIS) that accompanies each forest plan. This stage of the process is also open to public participation, as required by the National Environmental Protection Act (NEPA).

If at the end of this extended study process a river is determined not only "eligible" but "suitable" for designation under WSRA, then the Forest Service, through the secretary of agriculture, may recommend the river to Congress for designation.

Though a relatively straightforward process, as outlined above, in practice, the matter can be much more complex. Federal agencies undertake studies only when financial and technical resources are available. Land management agencies have a great many other responsibilities that are not related to the Wild and Scenic Rivers Program, and in many cases these other responsibilities receive the agency's highest priority. For many rivers a great deal of time passes between the date that it is determined to be eligible and the subsequent initiation of a suitability study. Political considerations also play a large role in the process. An agency is unlikely to forge ahead on the eligibility-suitability-recommendation process in the face of significant hostility from local citizens, state agencies, or the state's congressional representatives. Nor is the agency likely to proceed when the national political climate is not conducive to further river designations. Nonetheless, recommendations to Congress from the federal land management agencies continue to be the primary source of new river designations contemplated by Congress.

The material in this note was drawn from a variety of sources, but particularly from Lundeen (1993), Glasser (1993), and Chambers (1993).

73. Twenty-five rivers in Alaska were simultaneously designated as part of the Wild and Scenic Rivers system when Congress passed the Alaska National Interest Lands Conservation Act (ANILCA) in 1980. The 3,284 river miles designated in ANILCA represent the single largest addition ever made to the system. However, ANILCA was not entirely—or even primarily—a rivers conservation bill. The Oregon Omnibus Rivers Bill of 1988, which designated 39 rivers as part of the Wild and

Scenic Rivers system, was the first of the true omnibus Wild and Scenic Rivers bills. Most of the land bordering these rivers is part of the National Forest system, and the U.S. Forest Service was instrumental in putting together the bill (Palmer, 1993).

74. The Wild and Scenic Rivers Act is an important part of this nation's conservation history and environmental legislation and has attracted a great deal of attention from other authors. Readers interested in a more in-depth discussion of other features of the Act than we have presented here are encouraged to refer to those other publications. For readers interested in a discussion of the basic legal features of the Act, a good place to start would be Tarlock and Tippy (1970). It remains an oft-cited reference on the Act in the scholarly literature. Another oft-cited source of information about legal aspects of the Act is Fairfax, Andrews, and Buchshaum (1984), which contains a substantial amount of information about both the federal and California state wild and scenic rivers programs and the interaction between them. Doppelt et al. (1993) provide a shorter summary of the WSRA, pp. 136–52. Readers wishing a more comprehensive and yet highly readable discussion of the Act's history and features, as well as a thorough description of WSRA rivers and the status of the Wild and Scenic Rivers Program as of 1993, will find Palmer (1993) to be a valuable reference. The conservation group American Rivers (Washington, D.C.) is intimately tied to issues surrounding the Wild and Scenic Rivers Act, and their quarterly newsletter *American Rivers* is a good source of information about current events affecting the wild and scenic rivers, as well as about other river conservation issues.

75. 16 U.S.C. sec 1278(a).

76. For example, Palmer (1994, p. 83) lists the Tuolumne and Hells Canyon of the Snake as two circumstances in which the rivers were designated to prevent the further development of hydropower.

77. 16 U.S.C. sec. 1278(a).

78. 16 U.S.C. sec. 1278(a).

79. 16 U.S.C. sec. 1284(c).

80. The assertion of federal water rights in the Wild and Scenic Rivers Act, as quoted in the text, has been variously described by legal analysts as "backhanded," "confusing," "convoluted," "elliptical," "lefthanded," and "negatively stated." However, there is widespread agreement that the Act does indeed assert a federal water right for a quantity of water sufficient to achieve the purposes of the Act. See, for example, Tarlock and Tippy (1970), Baldwin (1990), and Gray (1988). As stated by Baldwin (1990, pp. 4–6), "The critical factor is the intent of Congress, either express or implied, that such rights be created. . . . Considering the purpose of the Act, it seems an inescapable conclusion that Congress intended to create a federal right to some or all of the instream flows of designated rivers or segments. . . . The choice of terminology used should not affect the otherwise clear intent to achieve certain purposes."

Tarlock and Tippy (1970, p. 734) noted that the Wild and Scenic Rivers Act included a "roundabout" assertion of federal supremacy, thus permitting the federal government to "administer wild and scenic rivers unfettered by inconsistent state law." The reference is to section 13(d) of the Act, which reads: "The jurisdiction of the States over waters of any stream included in a national wild, scenic or recreational river area shall be unaffected by this chapter to the extent that such jurisdiction may be exercised without impairing the purposes of this chapter or its administration," 16 U.S.C. sec. 1284(d).

81. This argument has been proposed by Gray (1988). Gray also notes that, even though the language in some portions of the Wild and Scenic Rivers Act refers to "primary" purposes of the Act, this language could not have been specifically designed to address the New Mexico court's primary-secondary purposes distinction, because the Wild and Scenic Rivers Act was passed 10 years before the New Mexico case was decided.

82. For example, the chairman of the House Committee on Interior and Insular Affairs is quoted by Baldwin (1990, p. 8) as saying:

> Enactment of the bill would not in any way affect or impair any valid or existing water rights perfected under state law. In addition, further appropriations could be made and water rights perfected under state law so long as the subsequent appropriations would not adversely affect the designated rivers.

83. 16 U.S.C. 1284(b).

84. Federal agency personnel contacted by the authors, without exception, have stated that they have never heard of any attempt to condemn a private water right under authority of the Wild and Scenic Rivers Act. For example, Glasser (1993), Lundeen (1993), Huntsinger (1994), Williams, Owen (1994). Most doubt that this would ever occur, as it would likely stir up more controversy and opposition than it would be worth. Furthermore, many agency personnel doubt that Congress intended for condemnation to occur. One Bureau of Land Management employee stated that if Congress wanted the agency to condemn a private water right, it would have to explicitly require the agency to do so and thereby absolve the agency from responsibility for the action.

85. Gray (1988).

86. The river protection activities of the National Park Service are described at greater length in chapter 8.

87. For example, federal agencies were able to obtain the rights to virtually the entire flow of the North and Middle Forks of the Flathead River in Montana during negotiations with the Montana Federal Reserved Rights Settlement Commission. Both rivers are components of the Wild and Scenic Rivers system, but WSRA considerations never came into play—the rights were actually granted for other purposes asserted by the National Park Service and U.S. Forest Service, and the amount of water granted to these federal agencies was thought sufficient to protect the values that had led to the rivers' designations under WSRA (Williams, Owen 1994).

88. For example, the Forest Service has occasionally been able to sign Memoranda of Understanding (MOUs) with the Bureau of Reclamation or Corps of Engineers to provide flow regimes below federal dams that will maintain or enhance instream values (Lundeen, 1993).

89. The priority dates of a water right cannot precede the date at which a river is designated as part of the system, so even those rivers that were designated for immediate inclusion in the system at the time the Wild and Scenic Rivers Act was passed

have priority dates no earlier than 1968. A right that junior in the West may be sufficient for indirect protection (through objection to transfers and changes in use) but is not likely to generate actual water for instream purposes if the water is already being used for other purposes.

90. *Forest Watch* (1993), *American Rivers* (1993). The current state administration has asserted ownership of the beds of 205 rivers and lakes in the state, including the beds of 12 rivers in the national Wild and Scenic Rivers system. State ownership claims are based on the asserted navigability of these rivers and lakes at the time that statehood was granted. The issue of state ownership of—and responsibility for—the riverbeds of certain navigable waterways is discussed in greater detail under the "Public Trust" section of chapter 5. The Bureau of Land Management, which manages the Fortymile River, has taken some steps to limit damage from mining on the river by regulating the miners' access and occupation on the surrounding land (Sterin, 1993).

91. For a general description of the efforts necessary to quantify instream flow needs, see the methods of quantification section in chapter 3. For an example of the efforts needed to quantify the instream needs of a wild and scenic river, see Garn (1986).

92. Congress usually is quite sympathetic to claims of state sovereignty over water allocation, but when debating the Wild and Scenic Rivers Act in 1968 Congress rejected an amendment that would have required river managers to use state law to obtain water rights for the wild and scenic rivers. The amendment was rejected on the grounds that it would have frustrated the purposes of the Act (Baldwin, 1990). The rejection of this amendment is further evidence that Congress did indeed intend to create a federal water right for the wild and scenic rivers.

93. Though river managers have never yet condemned a private water right under WSRA, the language in the Act that seems to authorize such condemnations keeps other rights holders on edge and contributes to an atmosphere of uncertainty and conflict.

94. Section 102, P.L. 99–590, 100 Stat. 3330.

95. Section 102, P.L. 99–590, 100 Stat. 3330.

96. "Findings of Fact, Conclusions of Law and Decree," District Court, Water Division No. 1, State of Colorado, Case No. 86CW367, March 17, 1993.

97. Section 1302(C), Title XIII, P.L. 101–628, Nov. 28, 1990.

98. Bevenger (1993).

99. At least one group of conservation writers has claimed that instream flow matters in general are seriously overlooked by federal agencies (Doppelt et al., 1993).

100. Methods not appearing in WSRA that have been employed by river managers to protect the wild and scenic rivers include, among others, negotiated settlements with private citizens, memoranda of understanding with other federal agencies or with state or local authorities, exercise of land-use management authorities, objections to new and/or changed water rights applications before state courts or state administrative agencies, and the establishment of water rights under other authorities. The degree to which these and other methods have been used by the federal

agencies with primary responsibility for management of the wild and scenic rivers is discussed below.

101. For example, the conservation group American Rivers (*American Rivers,* Winter 1994), reporting on the proceedings of a 1993 conference celebrating the 25th anniversary of the Wild and Scenic Rivers Act, states that: 1) the U.S. Forest Service met its 1988 goal of recommending 200 rivers for inclusion in the system by 1993, has committed itself to doubling the number of wild and scenic rivers it oversees in the next five years, and in the next year will study whether 350 of the more than 600 additional rivers are suitable additions to the system; 2) the National Park Service is going to finalize studies on 120 eligible rivers in the National Parks; 3) the Bureau of Land Management is moving ahead with plans to study the wild and scenic rivers potential on its lands throughout Alaska (this study had previously been stalled by political pressure from members of the Alaska congressional delegation); and 4) the U.S. Fish and Wildlife Service will propose adding more than 1,000 miles to the system over the next 10 years.

102. Palmer (1993), p. 63.

103. Doppelt et al. (1993), Palmer (1993).

104. Quoted by Wilkinson (1992).

105. The Wilderness Act of September 3, 1964, P.L. 88–577, 78 Stat. 890 as amended, is codified at 16 U.S.C. secs. 1131–36 (1982).

106. Wilderness Act, Section 4(d)(4), 16 U.S.C. 1133.

107. The figures that follow are based on data obtained from the National Geographic Society (1994), with additions resulting from passage of the California Desert Protection Act made by the authors.

108. All four of the federal land management agencies manage significant tracts of wilderness land. Following passage of the California Desert Protection Act in October 1994—which added almost 8 million acres of southeastern California desert lands to the national wilderness protection system—approximately 40 percent of the nation's wilderness acreage is managed by the Park Service, 35 percent by the Forest Service, 20 percent by the Fish and Wildlife Service, and 5 percent by the BLM. Almost all of the wilderness managed by the Park Service and Fish and Wildlife Service is in Alaska, though the Park Service also manages large tracts of wilderness in California and Washington. Most of the wilderness in the western states outside of Alaska is managed by the Forest Service. BLM holdings were relatively minor—just 1.6 million acres, much of it in Arizona—prior to passage of the California Desert Protection Act, which added over 3.5 million acres of BLM land to the wilderness preservation system.

Outside of California, other recent additions to the national wilderness preservation system adopted by Congress include Fish and Wildlife Service and BLM lands in Arizona and Forest Service lands in Nevada and Colorado. Legislation to designate additional wilderness areas on national forest land in Idaho and Montana is pending in Congress. The land management agencies may continue to make additional recommendations for lands identified during regular land planning processes, such as the preparation of national forest plans by the Forest Service and resource management plans by BLM. With the exception of BLM lands in Alaska, where the BLM was instructed to recommend new wilderness designations solely through land planning processes, these recommendations are generally expected to cover much smaller tracts of land than have already been identified and recommended.

109. See Brown (1991).

110. Office of the Solicitor (1979).

111. Office of the Solicitor (1988). The opinion was based in large part on Section 1133(d)(6) of the Wilderness Act, which includes some standard language about not claiming any exemptions from state water laws: "Nothing in this Chapter shall constitute an express or implied claim or denial on the part of the Federal Government as to exemption from State water laws." The meaning of such provisions is unclear, as evidenced by seemingly contradictory rulings by the U.S. Supreme Court about similar provisions in the Reclamation Act and Federal Power Act, described in the chapter on federal hydropower development and operations above. Many scholars and analysts have asserted that the effect of such statements is merely to ratify the status quo, which includes federal water rights in situations in which they are necessary to accomplish the primary purposes of federal reservations.

112. Letter from Attorney General Edwin Meese III to Donald P. Hodel, Secretary of the Department of the Interior, July 28, 1988, as cited by Tristani (1989).

113. *Water Intelligence Monthly*, June 1994, p. 9.

114. Corrigall (1994).

115. 622 F. Supp. 842 (1985), enforced and modified sub nom. *Sierra Club v. Lyng*, 661 F. Supp. 1490 (D. Colo. 1987).

116. At least one scholar (Abrams, 1986) has argued that reserved water rights are fixed at the time they are created, and the federal official to whom they are entrusted has the obligation to assert the rights to their fullest extent.

117. Judge Kane, in decision for *Sierra Club v. Block*, as quoted and cited by Sax, Abrams, and Thompson (1991), pp. 852–53.

118. *Sierra Club v. Yeutter*, 911 F.2d 1405, Tenth Circuit, 1990.

119. Quoted by Sax, Abrams, and Thompson (1991), p. 859, note 5.

120. Brown (1991).

121. Sec. 8(a)(1–2), P.L. 103–77, August 13, 1993.

122. Sec. 8(b)(1), P.L. 103–77, August 13, 1993.

123. Sec. 8(b)(2)(D), P.L. 103–77, August 13, 1993.

124. As quoted in U.S. Forest Service (1992, p. ii):

> Section 3(a)(3) of the S. 1029, the Colorado Wilderness Bill of 1991, states that "while the Piedra Wilderness designated by section 2(a)(10) of this [bill] is located downstream of numerous State-granted conditional and absolute water rights, the Forest Service can adequately protect the water-related resources of this wilderness area by working in coordination with the Colorado Water Conservation Board through a contractual agreement between the Secretary [of Agriculture] and the Board to protect and enforce instream flow filings established pursuant to the pro-

visions of section 37–92–102(3) of the Colorado Revised Statutes by the Colorado Water Court for Division 7."

125. See U.S. Forest Service (1992).

126. U.S. Forest Service (1992), pp. 1, 2.

127. This argument ignored a major difference between the two water rights situations. The water right at issue in the channel maintenance claims was a federal reserved right, whereas the right proposed for the Piedra would have been issued under state law and held by the Colorado Water Conservation Board, a state agency.

128. Sec. 9(b)(3), P.L. 103–77, August 13, 1993. The three units are referred to in the bill simply as "Areas" rather than as "Wilderness Areas."

129. Sec. 9(c), P.L. 103–77, August 13, 1993.

130. Leshy (1988), pp. 397–98. Leshy's cogent examination of the controversies surrounding the issue of water rights for wilderness areas supports Trelease's observations that "the 'fears' that federal water rights for non-Indian federal land designations would destroy valuable water rights perfected under state law . . . have shown themselves to be 'groundless,' mere 'bogies we have been conjuring up'" (p. 389). Leshy was appointed Interior Department solicitor in 1993.

131. Leshy (1988).

132. It is important to keep in mind that the preclusion of future upstream development due to the existence of prior rights downstream is a feature of the water rights system that is not unique to wilderness areas. This is a basic feature not only of federal reserved water rights, but also of the prior appropriation system in general (Leshy, 1988).

Chapter Eleven

1. For 1900, the population estimates are from the U.S. Bureau of the Census (1904, table 35) and include the part of current-day Oklahoma that was then known as the Indian Territory. For 1990, the estimates are from the U.S. Bureau of the Census (1995, table 27).

2. For 1900, a large city is defined as one of at least 100,000 people (U.S. Bureau of the Census, 1904, table 37). Only four cities were that large: San Francisco, Los Angeles, Denver, and Omaha (U.S. Bureau of the Census, 1904, table 91). For 1990, we have defined "large city" as a consolidated metropolitan area or a metropolitan statistical area, as specified by the U.S. Bureau of the Census (1995, tables 41 and 42).

3. The 1900 estimate is from the U.S. Bureau of the Census (1904, table 64) and includes all persons of at least 10 years of age engaged in gainful occupations. The 1990 estimate is from the U.S. Bureau of the Census (1992, tables 645 and 1073) and is for the entire U.S. The 1990 estimate for the West alone is unlikely to be greater than that for the entire country, as the West is actually more urbanized than the rest of the country. Excluding the 18 western states, 78 percent of the U.S. population lived in large cities in 1990 (U.S. Bureau of the Census, 1995, table 42), to be compared with 82 percent for the western states.

4. See figures 2.1 and 2.2.

5. Solley, Pierce, and Perlman (1993, table 2); this figure includes fresh surface water only.

6. The explanation for the increased marginal value of instream flows can be made quite simply in economic terms; changes in the social and economic character of the West have increased the demand for instream flows even as our offstream uses of water have steadily reduced the supply. All else being equal, reductions in the supply of a desired good increase the good's marginal value. Water, in fact, has long been used to illustrate the basic "law" of supply and demand. Adam Smith, writing in Scotland in the late 18th century, described the paradox of diamonds and water—diamonds were used mainly for decoration but cost a great deal, whereas water was essential for life but available for free. As Smith explained, it was the abundant supply of water that kept down its price and the scarcity of diamonds that helped make them so dear. Instream flow in the West is, to be sure, not as scarce as diamonds, but it has become gradually more scarce, and the increasing scarcity has occurred as social and economic conditions have increased the demand for instream flow.

7. The state methods are described in chapter 6, and the federal methods are described in chapters 8–10.

8. As demonstrated by the debate over federal reserved rights for headwaters wilderness areas (described in chapter 10), even if the protection has no effect on existing water diversions, use of a method is controversial if it is perceived as setting a precedent for use elsewhere.

9. Specific changes are often viewed as symbolic of a larger issue. For example, it is not uncommon for a small amount of instream flow protection to become a symbol of instream flow protection in general and thus of a threat to established consumptive water uses. When one focuses on the larger issue that is symbolized by a small change, the sensibility of the change can become lost in rhetoric. The debate quickly becomes one of the opposing world views rather than one about the specific change at issue.

This tendency to focus on symbols rather than on the specific change at issue is similar, in economic terms, to focusing on the average value of a good rather than on the marginal value. Because of diminishing marginal returns, the marginal value of a good is less than its average value (the average value being equal to the value of the entire quantity supplied divided by quantity). The average value of water in a given use is often very high because the first increments applied to that use are highly valued. However, in a use that is amply supplied, the marginal value may be relatively low. All transfers among water uses occur at the margin. It is a difference in marginal values that, for example, allows a market trade to occur.

10. Solley, Pierce, and Perlman (1993, tables 2, 4 and 16). On average, 56% of water withdrawals for irrigation are estimated to be consumed (i.e., via evaporation or transpiration) once the water reaches the field; another 20% are estimated to be lost during conveyance to the field, leaving 24% to return to the stream. Adding conveyance losses to the amount of delivered water that is consumed would raise the percentage of water consumption attributable to irrigation above the 90% reported in the text.

11. Of the irrigated land in the West, roughly 10% is used to grow cotton, 32%

to grow hay and pasture, and 42% to grow grains (18% in corn, 9% in wheat, 7% in sorghum, and 8% in other grains) (Frederick and Hanson, 1982, table 2.9). Of the grains, about 75% of the corn and other grains, 30% of the wheat, and nearly all of the sorghum is used to feed livestock (U.S. Department of Agriculture, Economic Research Service, 1987). Thus, according to these somewhat dated estimates, over 60% of the West's irrigation water is used to grow food for livestock and another 10% is used to grow cotton.

12. For example, Palmer (1991) notes the widespread recognition among farmers on the Snake River plain of the need to keep at least some water in the river to sustain instream resources.

13. For an extended discussion of agricultural water use and the possibility for conservation to provide substantial additions to instream flows in the Snake River, see Palmer (1991). In chapters 4 and 5 Palmer raises the interesting possibility that agricultural water conservation may even improve farm yields. He notes several instances in which more efficient water use on farms along the Snake River has reduced soil erosion and resulted in greater crop productivity.

References

Abrams, Robert H., (1986), "Water in the Western Wilderness: The Duty to Assert Reserved Water Rights," *University of Illinois Law Review*, vol. 1986, no. 2, pp. 387–405.

——— (1990), "The Big Horn Indian Water Rights Adjudication: A Battle for the Legal Imagination," *Oklahoma Law Review*, vol. 43, pp. 71–86.

Adler, Robert W.; Jessica C. Landman; and Diane M. Cameron, *The Clean Water Act 20 Years Later* (1993), Island Press, Washington, D.C.

Ahadi, Rassool, (April 1995), Water Rights Division, South Dakota Department of Environment and Natural Resources, Pierre, South Dakota, personal communication.

Aiken, J. David, (July 1990), "First Nebraska Instream Appropriation Granted," *Rivers*, vol. 1, no. 3, pp. 231–35.

——— (1993), "Nebraska Instream Appropriation Law and Administration," in Lawrence J. MacDonnell and Teresa A. Rice, eds., *Instream Flow Protection in the West*, revised edition, Natural Resources Law Center, University of Colorado School of Law, Boulder.

Ambler, Marjane, (1987), "A Tale of Two Irrigation Districts," pp. 147–52 in High Country News, eds., *Western Water Made Simple*, Island Press, Washington, D.C.

American Rivers, (Spring 1993, Winter 1994) (American Rivers Inc., Washington, D.C.).

Anderson, Robert T., (1991), "Alaska Legislature Considers Innovative Instream Flow Law," *Rivers*, vol. 2, no. 3, pp. 255–61.

Annear, Tom, (February 1993), "Keeping the Water Where It Flows," *Wyoming Wildlife*, (Wyoming Game and Fish Department), pp. 39–41.

——— (June 1995), instream flow supervisor, Wyoming Game and Fish Department, Cheyenne, Wyoming, personal communication.

Ausness, Richard, (1986), "Water Rights, the Public Trust Doctrine, and the Protection of Instream Uses," *University of Illinois Law Review*, vol. 1986, no. 2, pp. 407–37.

Baldwin, Pamela, (1990), "Water Rights and the Wild and Scenic Rivers Act," CRS Report for Congress, Congressional Research Service, Library of Congress, Washington, D.C. March 30.

Barker, Rocky, (1993), *Saving All the Parts: Reconciling Economics and the Endangered Species Act*, Island Press, Washington, D.C.

Barwin, Robert F. and Kenneth O. Slattery, (1993), "Protecting Instream Resources in Washington State," in Lawrence J. MacDonnell and Teresa A. Rice, eds., *Instream Flow Protection in the West*, revised edition, Natural Resources Law Center, University of Colorado School of Law, Boulder.

Bates, Sarah F.; David H. Getches; Lawrence J. MacDonnell; and Charles F. Wilkinson, (1993), *Searching Out the Headwaters: Change and Rediscovery in Western Water Policy*, Island Press, Washington, D.C.

Bayha, Keith, (December 1993), U.S. Fish and Wildlife Service Region 7, Anchorage, Alaska, personal communication.

Beard, Daniel P., (1993), Commissioner, Bureau of Reclamation, "Blueprint for Reform," internal 12-page document with cover letter addressed to Bureau employees, November 1.

Bearzi, Judith A., (October 1991), "The Delicate Balance of Power and Nonpower Interests in the Nation's Rivers," *Rivers*, vol. 2, no. 4, pp. 326–32.

Becker, Herbert A., (January 1991), "Aboriginal Water Rights," *Rivers*, vol. 2, no. 1, pp. 66–73.

Beecher, Hal A., (April 1990), "Standards for Instream Flows," *Rivers*, vol. 1, no. 2, pp. 97–109.

——— (May 1995), Washington State Department of Fish and Wildlife, Olympia, Washington, personal communication.

Beeman, Josephine P., (1993), "Instream Flows in Idaho," in Lawrence J. MacDonnell and Teresa A. Rice, eds., *Instream Flow Protection in the West*, revised edition, Natural Resources Law Center, University of Colorado School of Law, Boulder.

Beuscher, J. H., (1967), *Water Rights*, College Printing and Publishing, Madison, Wisconsin.

Bevenger, Greg, (June 1993), Shoshone National Forest, Cody, Wyoming, personal communication.

Birnbaum, Elizabeth S., (1991), "Clean Water Act Section 401 Provides the Key to Stream Protection in Hydropower Licensing," *Rivers*, vol. 2, no. 2, pp. 148–53.

Bleed, Anne Salomon, (December 1987), "Limitations of Concepts Used to Determine Instream Flow Requirements for Habitat Maintenance," *Water Resources Bulletin*, vol. 23, no. 6.

Blumm, Michael C., (1992), "Unconventional Waters: The Quiet Revolution in

Federal and Tribal Minimum Streamflows," *Ecology Law Quarterly*, vol. 19, pp. 445–80.

Bodi, Lorraine F., (1988), "Hydropower, Dams, and the National Parks," pp. 449–64 in David J. Simon, ed., *Our Common Lands: Defending the National Parks*, Island Press, Washington, D.C.

Bolton, Barry, (May 1995), assistant chief of fisheries, Oklahoma Department of Wildlife and Conservation, Oklahoma City, personal communication.

Bonine, John E., and Thomas O. McGarity, (1992), *The Law of Environmental Protection: Cases—Legislation—Policies*, 2nd ed., West Publishing, St. Paul, Minnesota.

Bonneville Power Administration, (December 1991), "Pacific Northwest Electric Power Planning and Conservation Act with Index," 8th printing, Department of Energy, Portland, Oregon.

——— (January 1994a), "1993 Fast Facts," brochure, Department of Energy, Portland, Oregon.

——— (January 1994b), "1993 Annual Report," Department of Energy, Portland, Oregon.

——— (March 1994), "Fish and Wildlife," brochure, 2nd printing, Department of Energy, Portland, Oregon.

——— (April 1994), "High Powered Energy: The BPA Story," brochure, 5th printing, Department of Energy, Portland, Oregon.

Bradshaw, Stan, (1993), "The River Through the Recreationist Lens," pp. 43–45 in *Riparian Management: Common Threads and Shared Interests*, U.S. Department of Agriculture, Forest Service, Rocky Mountain Forest and Range Experiment Station, Fort Collins, Colorado, General Technical Report RM-226.

Brooks, Heidi Topp, (1979), "Reserved Water Rights and Our National Forests," *Natural Resources Journal*, vol. 19, no. 2, pp. 433–43.

Brough, Charles Hillman, (1898), *Irrigation in Utah*, Johns Hopkins Press, Baltimore.

Brown, Thomas C., (October 1991), "Water for Wilderness Areas: Instream Flow Needs, Protection, and Economic Value," *Rivers*, vol. 2, no. 4.

Brown, Thomas C., (1993), "Measuring Nonuse Value: A Comparison of Recent Contingent Valuation Studies," pp. 163–203 in John C. Bergstrom, compiler, *W-133, Benefits and Costs in Natural Resource Planning, Sixth Interim Report*, Western Regional Research Publication, University of Georgia, pp. 163–203.

Brown, Thomas C., and Terry C. Daniel, (August 1991), "Landscape Aesthetics of Riparian Environments: Relationship of Flow Quantity to Scenic Quality Along a Wild and Scenic River," *Water Resources Research*, vol. 27, no. 8, pp. 1787–95.

Brown, Thomas C.; Jonathan G. Taylor; and Bo Shelby, (December 1991), "Assessing the Direct Effects of Streamflow on Recreation: A Literature Review," *Water Resources Bulletin*, vol. 27, no. 6.

Butcher, Walter R.; Philip R. Wandschneider; and Norman K. Whittlesey, (1986), "Competition between Irrigation and Hydropower in the Pacific Northwest," pp. 25–66 in K. D. Frederick, ed., *Scarce Water and Institutional Change*, Resources for the Future, Washington, D.C.

Cahoy, Chuck, (June 1995), Office of Legal Counsel, Arizona Department of Water Resources, Phoenix, personal communication.

California State Lands Commission, (1993), *California's Rivers: A Public Trust Report*, California State Lands Commission, Sacramento, California.

Campbell, Faith, (1991), "The Appropriations History," pp. 134–46 in Kathryn A. Kohm, ed., *Balancing on the Brink of Extinction: The Endangered Species Act and Lessons for the Future*, Island Press, Washington, D.C.

Campbell, Thomas A., (1994), "The Public Trust, What's It Worth?" *Natural Resources Journal*, vol. 34, pp. 73–92.

Carlson, John U., and Paula C. Phillips, (1988), "Accommodating Interests in a Shared Resource Between States and the Federal Government," pp. 109–25 in David H. Getches, ed., *Water and the American West: Essays in Honor of Raphael J. Moses*, Natural Resources Law Center, University of Colorado School of Law, Boulder.

Carlson, Wes, (July 1994), U.S. Forest Service, retired, personal communication.

Chambers, Carl, (June 1993), Rocky Mountain Forest and Range Experiment Station, U.S. Forest Service, Fort Collins, Colorado, personal communication.

Clark, Ken, (June 1995), river guide and former official of the St. Vrain Anglers, a Boulder County, Colorado, chapter of Trout Unlimited, personal communication.

Coggins, George Cameron, (1991), "Snail Darters and Pork Barrels Revisited: Reflections on Endangered Species and Land Use in America," pp. 62–74 in Kathryn A. Kohm, ed., *Balancing on the Brink of Extinction: The Endangered Species Act and Lessons for the Future*, Island Press, Washington, D.C.

Coggins, George Cameron, and Charles F. Wilkinson, (1987), *Federal Public Land and Resources Law*, 2nd ed., Foundation Press, Mineola, New York.

Colby, Bonnie G., (June 1990), "Enhancing Instream Flow Benefits in an Era of Water Marketing," *Water Resources Research*, vol. 26, no. 6, pp. 1113–20.

——— (1993), "Benefits, Costs and Water Acquisition Strategies: Economic Considerations in Instream Flow Protection," in Lawrence J. MacDonnell and Teresa A. Rice, eds., *Instream Flow Protection in the West*, revised edition, Natural Resources Law Center, University of Colorado School of Law, Boulder.

Collier, Michael P.; Robert H. Webb; and Edmund D. Andrews, (1997), "Experimental Flooding in Grand Canyon," *Scientific American,* vol. 276, no. 1, pp. 82–89

Collins, Katharine, (1992), "Fear of Supreme Court Leads Tribes to Accept an Adverse Decision," *High Country News*, vol. 24, no. 19, pp. 1, 15, October 19.

Collins, Richard B., (1985), "The Future Course of the Winters Doctrine," *University of Colorado Law Review*, vol. 56, pp. 481–94.

Colorado District Court, Water Division No. 1, "Memorandum of Decision and Order," Case No. W–8439–76, February 12, 1993.

Condon, Austin, (undated), "Briefing Paper for By-Pass Flows as a Condition of

Land Use: Arapaho and Roosevelt NF's," Arapaho and Roosevelt National Forests, Fort Collins, Colorado.

——— (undated), (Colorado), "Water Special Use Permitting History," Lands, Minerals, and Watershed, Arapaho and Roosevelt National Forests, Fort Collins, Colorado.

Cook, Gregory F., (1993), "The Public Trust Doctrine in Alaska," *Journal of Environmental Law & Litigation*, vol. 8, pp. 1–49.

Cooperative Instream Flow Service Group, (undated), "Three-Year Plan (FY '77 through FY '79)," U.S. Fish and Wildlife Service, Washington, D.C.

Corbett, Roger, (1990), *A Method for Determining Minimum Instream Flow for Recreational Boating*, SAIC Special Report 1–239–91–01, Science Applications International Corporation, McLean, Virginia.

Cordell, H. Ken; John C. Bergstrom; Gregory A. Ashley; and John Karish, (February 1990), "Economic Effects of River Recreation on Local Economies," *Water Resources Bulletin*, vol. 26, no. 1.

Corrigall, Keith, (November and December 1994), Bureau of Land Management, Washington, D.C., personal communications.

Cowan, Michael S., (September 1994), power operations and resources director, Western Area Power Administration, Golden, Colorado, personal communication.

Crandall, Kristine B.; Bonnie G. Colby; and Ken A. Rait, (1992), "Valuing Riparian Areas: A Southwestern Case Study," *Rivers*, vol. 3, no. 2, pp. 88–98.

Crow, Margaret, (October 1991), "Federal Reserved Rights for Instream Flow," *Rivers*, vol. 2, no. 4, pp. 333–41.

Cullinan, Steve, (December 1993), U.S. Fish and Wildlife Service Region 2, Albuquerque, New Mexico, personal communication.

Danielson, Jeris A., (1989), "Colorado's Instream Flow Policy," in Suzanne P. Van Gytenbeek, ed., *Proceedings of the Western Regional Instream Flow Conference*, sponsored by Trout Unlimited, Jackson Hole, Wyoming, October 20-21.

Davidson, John, (April 1995), University of South Dakota School of Law, personal communication.

Davis, Ray Jay, (April 1991), "Utah Instream Flow Protection," *Rivers*, vol. 2, no. 2, pp. 154–60.

Debo, Angie, (1970), *A History of the Indians of the United States*, University of Oklahoma Press, Norman.

Dellios, Hugh, (1994), "Dreams May Replace Dams," *Denver Post*, pp. 21A, 24A, March 18.

De Young, Tim, (1993), "Protecting New Mexico's Instream Flows," in Lawrence J. MacDonnell and Teresa A. Rice, eds., *Instream Flow Protection in the West*, revised edition, Natural Resources Law Center, University of Colorado School of Law, Boulder.

Dishlip, Herb, (1993), "Instream Flow Water Rights: Arizona's Approach," in Lawrence J. MacDonnell and Teresa A. Rice, eds., *Instream Flow Protection in the West*, revised edition, Natural Resources Law Center, University of Colorado School of Law, Boulder.

Doerksen, Harvey R., (1991), "Two Decades of Instream Flow: A Memoir," in *Rivers*, vol. 2, no. 2, pp. 99–104.

Doppelt, Bob; Mary Scurlock; Chris Frissell; and James Karr, (1993), *Entering the Watershed: A New Approach to Save America's River Ecosystems*, Island Press, Washington, D.C.

Duerr, Naomi, (June 1995), Division of Water Planning, State of Nevada, personal communication.

Duffield, John W.; Thomas C. Brown; and Stewart D. Allen, (September 1994), "Economic Value of Instream Flow in Montana's Big Hole and Bitterroot River," USDA Forest Service Research Paper RM-317, Rocky Mountain Forest and Range Experiment Station, Fort Collins, Colorado, 64 pages.

Duffield, John W.; Christopher J. Neher; and Thomas C. Brown, (September 1992), "Recreation Benefits of Instream Flow: Application to Montana's Big Hole and Bitterroot Rivers," *Water Resources Research*, vol. 28, no. 9, pp. 2169–81.

Dunbar, Robert G., (1983), *Forging New Rights in Western Waters*, University of Nebraska Press, Lincoln.

——— (January 1985), "The Adaptability of Water Law to the Aridity of the West," *Journal of the West*, vol. 24, no. 1, pp. 57–65.

Dunning, Harrison C., (July 1990), "Dam Fights and Water Policy in California: 1969–1989," *Journal of the West*, vol. 29, no. 3.

——— (1993a), "Instream Flows and the Public Trust," chapter 4 in Lawrence J. MacDonnell and Teresa A. Rice, eds., *Instream Flow Protection in the West*, revised edition, Natural Resources Law Center, University of Colorado School of Law, Boulder.

——— (1993b), "Confronting the Environmental Legacy of Irrigated Agriculture in the West: The Case of the Central Valley Project," *Environmental Law*, vol. 23, pp. 943–69.

Dyke, Steve, (May 1995), North Dakota Game and Fish, Bismarck, North Dakota, personal communication.

El-Ashry, Mohamed T., (1980), "Ground Water Salinity Problems Related to Irrigation in the Colorado River Basin," *Ground Water*, vol. 18, no. 1, pp. 37–45.

Emery, Stephenson D., (1984), "The Limits of Federal Reserved Water Rights in National Forests: *United States v. City and County of Denver,* 656 P.2d 1 (Colo. 1983)," Law Note, *Land and Water Law Review*, vol. 19, pp. 71–82.

Energy Information Administration, (April 1985), "Historical Plant Cost and Annual Production Expenses for Selected Plants," DOE/EIA-0455, Washington, D.C.

Energy Information Administration, (September 1996), "Monthly Energy Review," DOE/EIA-0035, Washington, D.C.

Ernst, John P., (1991), "Federalism and the Act," pp. 98–113 in Kathryn A. Kohm, ed., *Balancing on the Brink of Extinction: The Endangered Species Act and Lessons for the Future*, Island Press, Washington, D.C.

Estes, Christopher, (April 1985), Alaska Department of Fish and Game, Anchorage, personal communication.

Fairfax, Sally K., and Barbara T. Andrews, (1979), "National Forests and Reserved Water Rights in the Western United States," *Journal of Forestry*, vol. 77, pp. 648–51.

Fairfax, Sally K.; Barbara T. Andrews; and Andrew P. Buchsbaum, (1984), "Federalism and the Wild and Scenic Rivers Act: Now You See It, Now You Don't," *Washington Law Review*, vol. 59, pp. 417–70.

Fairfax, Sally K., and A. Dan Tarlock, (1979), "No Water for the Woods: A Critical Analysis of *United States v. New Mexico*," *Idaho Law Review*, vol. 15, pp. 509–54.

Fassett, Gordon W., (1993), "Wyoming's Instream Flow Law," in Lawrence J. MacDonnell and Teresa A. Rice, eds., *Instream Flow Protection in the West*, revised edition, Natural Resources Law Center, University of Colorado School of Law, Boulder.

Fogleman, Valerie M., (1990), *Guide to the National Environmental Policy Act: Interpretations, Applications, and Compliance*, Quorum Books, Westport, Connecticut.

Forest Watch, (January 1993), (Cascade Holistic Economic Consultants, Portland, Oregon), vol. 13, no. 6.

Fradkin, Philip L., (1984), *A River No More: The Colorado River and the West*, University of Arizona Press, Tucson.

Frakt, Arthur N., and Janna S. Rankin, (1982), *The Law of Parks, Recreation Resources, and Leisure*, Brighton Publishing, Salt Lake City, Utah.

Frederick, Kenneth K., with James C. Hanson, (1982), *Water for Western Agriculture*, Resources for the Future, Washington, D.C.

Fritz, Gary, (1989), "State of Montana Instream Flow Policy," in Suzanne P. Van Gytenbeek, ed., *Proceedings of the Western Regional Instream Flow Conference*, sponsored by Trout Unlimited, Jackson Hole, Wyoming, October 20-21.

Fuller, George W., (1946), *A History of the Pacific Northwest—With a Special Emphasis on the Inland Empire*, 2nd ed., Alfred A. Knopf, New York.

Futrell, J. William, (1988), "NEPA and the Parks: Use It or Lose It," pp. 107–26 in David J. Simon, ed., *Our Common Lands: Defending the National Parks*, Island Press, Washington, D.C.

"Future of CVP Is Pondered," (February 1993), *U.S. Water News*, (U.S. Water News, and the Freshwater Foundation), Halstead, Kansas.

Gardner, Douglas J., (1989), "Casenotes: Water Law—The United States Supreme Court Expands the Public Trust Doctrine," *Land and Water Law Review*, vol. 24, pp. 347–56.

Garn, Herbert S., (1986), "Quantification of Instream Flow Needs of a Wild and Scenic River for Water Rights Litigation," *Water Resources Bulletin*, vol. 22, no. 5, pp. 745–51.

Getches, David H., (1988a), "Pressures for Change in Western Water Policy," pp. 143–64 in David H. Getches, ed., *Water and the American West: Essays in Honor of Raphael J. Moses*, Natural Resources Law Center, University of Colorado School of Law, Boulder.

——— (1988b), "Water Planning in the West," *Journal of Energy Law and Policy*, vol. 9, pp. 1–45.

——— (1990), *Water Law in a Nutshell*, 2nd ed., West Publishing, St. Paul, Minnesota.

Gibbons, Diana C., (1986), *The Economic Value of Water*, Resources for the Future, Washington, D.C.

Gillilan, David, (1992), "Innovative Approaches to Water Resource Management," report prepared for the U.S. Congress Office of Technology Assessment, Washington, D.C., as part of its assessment "Systems at Risk from Climate Change."

Glaser, Donald R., (1993), deputy commissioner, U.S. Bureau of Reclamation, comments made at the 1993 annual water law and water policy conference, "Water Reallocation and the Public Interest in Colorado," University of Denver College of Law, Denver, November 13.

Glasser, Steve, (June 1993), Water Rights Program, Watershed and Air Management, Washington Office, U.S. Forest Service, personal communication.

Goldfarb, William, (1988), *Water Law*, 2nd ed., Lewis Publishers, Chelsea, Michigan.

Gordon, Nancy, (September 1995), *Summary of Technical Testimony in the Colorado Water Division 1 Trial*, USDA Forest Service General Technical Report RM-270, Rocky Mountain Forest and Range Experiment Station, Fort Collins, Colorado, 140 pages.

Gordon, Nancy D.; Thomas A. McMahon; and Brian L. Finlayson, (1992), *Stream Hydrology: An Introduction for Ecologists*, John Wiley and Sons, West Sussex, England.

Grant, Douglas L., (1987), "Public Interest Review of Water Right Allocation and Transfer in the West: Recognition of Public Values," *Arizona State Law Journal*, vol. 19, pp. 681–718.

Gray, Brian E., (1988), "No Holier Temples: Protecting the National Parks Through Wild and Scenic River Designation," *University of Colorado Law Review*, vol. 58, pp. 551–98.

——— (1993), "A Reconsideration of Instream Appropriative Water Rights in California," in Lawrence J. MacDonnell and Teresa A. Rice, eds., *Instream Flow Protection in the West*, revised edition, Natural Resources Law Center, University of Colorado School of Law, Boulder.

Greenwalt, Lynn A., (1991), "The Power and Potential of the Act," pp. 31-36 in Kathryn A. Kohm, ed., *Balancing on the Brink of Extinction: The Endangered Species Act and Lessons for the Future*, Island Press, Washington, D.C.

Haas, Dan, (May 1994), National Park Service Rivers, Trails, and Conservation Assistance, Pacific Northwest Office, Seattle, Washington, personal communication.

Hamilton, Joel R.; Norman K. Whittlesey; and Philip Halverson, (February 1989), "Interruptible Water Markets in the Pacific Northwest," *American Journal of Agricultural Economics*, vol. 71, no. 1, pp. 63–75.

Hannum, Eric, (Summer 1992), "Administration of Reserved and Non-Reserved Water Rights on an Indian Reservation: Post-Adjudication Questions on the Big Horn River," *Natural Resources Journal*, vol. 32, pp. 681–704.

Hansen, LeRoy T., and Arne Hallam, (June 1990), "Water Allocation Tradeoffs: Irrigation and Recreation," Resources and Technology Division, Economic Research Service, U.S. Department of Agriculture, Agricultural Economic Report Number 634.

Harbour, Tom, (June 1995), Hydrology Division, Arizona Department of Water Resources, Phoenix, Arizona, personal communication.

Harle, Mary Lu, and Christopher C. Estes, (1993), "An Assessment of Instream Flow Protection in Alaska," in Lawrence J. MacDonnell and Teresa A. Rice, eds., *Instream Flow Protection in the West*, revised edition, Natural Resources Law Center, University of Colorado School of Law, Boulder.

Hatch, Michael D., (1994), "Ecologic vs. Social Welfare in Water Management: A Plea for Balance," unpublished paper, New Mexico Game and Fish Department, Santa Fe.

——— (May 1995), New Mexico Game and Fish, Santa Fe, New Mexico, personal communication.

Haubert, John, (March 1997), Division of Park Planning and Special Studies, National Park Service, Washington, D.C., personal communication.

Haury, Emil W., (September 1945), "Arizona's Ancient Irrigation Builders," *Natural History*, vol. 54, no. 7, pp. 300–10, 335.

——— (May 1967), "The Hohokam: First Masters of the American Desert," *National Geographic Magazine*, vol. 131, no. 5, pp. 670–95.

Heede, Burchard H., (September 1992), *Stream Dynamics: An Overview for Land Managers*, USDA Forest Service General Technical Report RM-72, Rocky Mountain Forest and Range Experiment Station, Fort Collins, Colorado, 26 pages.

Higginson, R. Keith, (1989), "Instream Flow Protection for Fish Resources," in Suzanne P. Van Gytenbeek, ed., *Proceedings of the Western Regional Instream Flow Conference*, sponsored by Trout Unlimited, Jackson Hole, Wyoming, October 20–21.

High Country News, eds., (1987), *Western Water Made Simple*, Island Press, Washington, D.C.

High Country News, (1988), "Push Comes to Shove" and other articles, vol. 20, no. 9, May 9.

——— (1991), "Why Utah Wants 'the Bureau' Out" and other articles, vol. 23, no. 13, July 15.

——— (1994), "Cities Fight to Keep Water Out of the Platte," written by Bob Kretschman, May 16.

Hill, John, (June 1994), former attorney with the U.S. Department of Justice, Denver, Colorado, personal communication.

Hill, Mark T.; William S. Platts; and Robert L. Beschta, (July 1991), "Ecological and Geomorphological Concepts for Instream and Out-of-Channel Flow Requirements," *Rivers*, vol. 2, no. 3, pp. 198–210.

Hirt, Paul, and Frank Gregg, (December 1991), "Response to Changing Values in a Developed River System: Fisheries and Hydropower in the Columbia River Basin," Appendix D of *Institutional Response to a Changing Water Policy Environment*, Water Resources Research Center, University of Arizona, Tucson.

Hobbs, Gregory J., Jr., (1991), "The Reluctant Marriage: The Next Generation (A Response to Charles Wilkinson)," *Environmental Law*, vol. 21, pp. 1082–90.

Holden, Mark, (1993), "Instream Flows in Utah," in Lawrence J. MacDonnell and Teresa A. Rice, eds., *Instream Flow Protection in the West*, revised edition, Natural Resources Law Center, University of Colorado School of Law, Boulder.

Hollon, W. Eugene, (1961), *The Southwest: Old and New*, Alfred A. Knopf, New York.

Howe, C.W., and W.A. Ahrens, (1988), "Water Resources of the Upper Colorado River Basin: Problems and Policy Alternatives," pp. 169–232 in Mohamed T. El-Ashry and Diana C. Gibbons, eds., *Water and Arid Lands of the Western United States*, Cambridge University Press, New York.

Huffman, James L., (1991), "Clear the Air," *Environmental Law*, vol. 21, pp. 2253–57.

Huntsinger, Ron, (February 1994), Soil, Water, and Air Program, Bureau of Land Management, Washington, D.C., personal communication.

Hutchins, Wells A., (1928), "The Community Acequia: Its Origin and Development," *Southwestern Historical Quarterly* (Texas State Historical Association, Austin, Texas), vol. 31, pp. 261–84.

——— (1971), *Water Rights Laws in the Nineteen Western States*, Miscellaneous Publication No. 1206, U.S. Department of Agriculture, Washington D.C.

Hutchinson, Larry, (May 1995), Nebraska Game and Parks Commission, Lincoln, Nebraska, personal communication.

Hyra, R., (1978), *Methods of Assessing Instream Flows for Recreation*, Instream Flow Information Paper No. 6, FWS/OBS-78/34, U.S. Fish and Wildlife Service, Fort Collins, Colorado.

Jackson, William L.; Bo Shelby; Anthony Martinez; and Bruce P. Van Haveren, (March–April 1989), "An Interdisciplinary Process for Protecting Instream Flows," *Journal of Soil and Water Conservation*, pp. 121–26.

Jahn, Laurence R., (January 1990), "Managing Riverine Values and Uses," *Rivers*, vol. 1, no. 1, pp. 1–2.

Johnson, Norman K., and Charles T. DuMars, (Spring 1989), "A Survey of the Evolution of Western Water Law in Response to Changing Economic and Public Interest Demands," *Natural Resources Journal*, vol. 29, pp. 347–87.

Johnson, Ralph W., (1980), "Public Trust Protection for Stream Flows and Lake Levels," *University of California, Davis, Law Review*, vol. 14, pp. 233–67.

——— (1988), "Water Quality Control by the Public Trust Doctrine," pp. 127–42 in David H. Getches, ed., *Water and the American West: Essays in Honor of Raphael J. Moses*, Natural Resources Law Center, University of Colorado School of Law, Boulder.

Just, Rinda, (October 1990), "Recreational Instream Flows in Idaho: Instream Flows—They're Not Just for Fish Anymore," *Rivers*, vol. 1, no. 4, pp. 307–12.

Kahrl, William L., (1982), *Water and Power*, University of California Press, Berkeley.

Kaiser, Ronald A., (November 1994), *Legal and Institutional Barriers to Water Marketing in Texas*, Texas Water Resources Institute, Texas A&M University, College Station, Texas.

Kaiser, Ronald A., and Sharon Kelly, (1987), "Water Rights for Texas Estuaries," *Texas Tech Law Review*, vol. 18, no. 4, pp. 1121–56.

Kaplan, Rachel, (1977), "Down by the Riverside: Informational Factors in Waterscape Preference," pp. 285–89 in *Proceedings: River Recreation Management and Research Symposium*, USDA Forest Service General Technical Report NC-28, North Central Forest Experiment Station, 455 pages.

Karim, Khalid; Maureen E. Gubbels; and Ian C. Goulter, (1995), "Review of Determination of Instream Flow Requirements with Special Application to Australia," *Water Resources Bulletin*, vol. 31, pp. 1063–77.

Keiter, Robert B., (1988), "National Park Protection: Putting The Organic Act to Work," pp. 75–86 in David J. Simon, ed., *Our Common Lands: Defending the National Parks*, Island Press, Washington, D.C.

Kinney, Tom, (Summer 1993), "Chasing the Wind: Wyoming Supreme Court Decision in *Big Horn III* Denies Beneficial Use for Instream Flow Protection, But Empowers State to Administer Federal Indian Reserved Water Right Awarded to the Wind River Tribes," *Natural Resources Journal*, vol. 33, pp. 841–71.

Kirk, Peggy Sue, (1993), "Cowboys, Indians and Reserved Water Rights: May a State Court Limit How Indian Tribes Use Their Water?" *Land and Water Law Review*, vol. 28, pp. 467–88.

Kohm, Kathryn A., ed., (1991), *Balancing on the Brink of Extinction: The Endangered Species Act and Lessons for the Future*, Island Press, Washington, D.C.

Krutilla, John V., (September 1967), "Conservation Reconsidered," *American Economic Review*, vol. 57, no. 4.

Kulakowski, Lois, and Barbara Tellman, (May 1994), *Instream Flow Rights: A Strategy to Protect Arizona's Streams*, revised edition, Water Resources Research Center, University of Arizona, Tucson.

Kunz, John, (April 1995), U.S. Interior Department Solicitor's Office, Denver, Colorado, personal communication.

Kury, Channing R., (1973), "The Prerequisite of a Man-Made Diversion in the Appropriation of Water Rights," *Natural Resources Journal*, vol. 13, pp. 170–75.

Lamb, Berton L., (1993), "Quantifying Instream Flows: Matching Policy and Technology," in Lawrence J. MacDonnell and Teresa A. Rice, eds., *Instream Flow Protection in the West*, revised edition, Natural Resources Law Center, University of Colorado School of Law, Boulder.

Lamb, Berton L., and Eric Lord, (April 1992), "Legal Mechanisms for Protecting Riparian Resource Values," *Water Resources Research*, vol. 28, no. 4, pp. 965–977.

Lee, Heather Blomfield, (July 1990), "Forcing the Federal Hand: Reserved Water Rights v. States' Rights for Instream Protection," *Hastings Law Journal*, vol. 41, no. 6, pp. 1271–1300.

Leopold, Luna B.; M. Gordon Wolman; and John P. Miller, (1964), *Fluvial Processes in Geomorphology*, Freeman, San Francisco. (Reprinted in 1995 by Dover Publications, Mineola, N.Y.)

Leshy, John D., (1988), "Water and Wilderness/Law and Politics," *Land and Water Law Review*, vol. 23, no. 2, pp. 398–417.

——— (1990), "The Prior Appropriation Doctrine of Water Law in the West: An Emperor with Few Clothes," *Journal of the West*, vol. 29, no. 3, pp. 5–13.

Lewis, Ricki, (January 1991), "Can Salmon Make a Comeback?" *BioScience*, vol. 41, no. 1, pp. 6–10.

Litton, R. Burton, Jr., (1977), "River Landscape Quality and Its Assessment," pp. 46–54 in *Proceedings: River Recreation Management and Research Symposium*, General Technical Report NC-28, U.S. Department of Agriculture, Forest Service, North Central Forest Experiment Station, St. Paul, Minnesota, 455 pages.

——— (1984), "Visual Fluctuations in River Landscape Quality," pp. 369–81 in Joseph S. Popedic, Dorothy I. Butterfield, Dorothy A. Anderson, and Mary R. Popadic, eds., *1984 National River Recreation Symposium Proceedings*, Louisiana State University, Baton Rouge, 740 pages.

Livingston, Marie Leigh, and Thomas A. Miller, (August 1986), "A Framework for Analyzing the Impact of Western Instream Water Rights on Choice Domains: Transferability, Externalities, and Consumptive Use," *Land Economics*, vol. 62, no. 3, pp. 269–77.

Loomis, John B., (1987a), "Balancing Public Trust Resources of Mono Lake and Los Angeles' Water Right: An Economic Approach," *Water Resources Research*, vol. 23, no. 8, pp. 1449–56.

——— (1987b), "The Economic Value of Instream Flow: Methodology and Benefit Estimates for Optimum Fows," *Journal of Environmental Management*, vol. 24, pp. 169–79.

——— (June 1994), "Water Transfer and Major Environmental Provisions of the Central Valley Project Improvement Act: A Preliminary Economic Evaluation," *Water Resources Research*, vol. 30, no. 6, pp. 1865–71.

Luecke, Daniel F., (May 1990), "Controversy over Two Forks Dam," *Environment*, vol. 32, no. 4, pp. 42–45.

Lundeen, Deen, (June 1993), Wild and Scenic Rivers Program leader, U.S. Forest Service, Washington, D.C., personal communication.

MacDonnell, Lawrence J., (1985), *The Endangered Species Act and Water Development Within the South Platte Basin*, Completion Report No. 137, Colorado Water Resources Research Institute, Fort Collins, Colorado.

——— (Spring 1989), "Federal Interests in Western Water Resources: Conflict and Accommodation," *Natural Resources Journal*, vol. 29, pp. 389–411.

——— (October 1991), "Water Rights for Wetlands Protection," *Rivers*, vol. 2, no. 4, pp. 277–87.

MacDonnell, Lawrence J., and Teresa A. Rice, (1989), "National Interests in Instream Flows," pp. 69–86 in Lawrence J. MacDonnell, Teresa A. Rice, and Steven J. Shupe, eds., *Instream Flow Protection in the West*, Natural Resources Law Center, University of Colorado School of Law, Boulder.

Marcus, Michael D.; Michael K. Young; Lynn E. Noel; and Beth A. Mullan, (February 1990), "Salmonid-Habitat Relationships in the Western United States: A Review and Indexed Bibliography," General Technical Report RM-188, U.S. Department of Agriculture, Forest Service, Rocky Mountain Forest and Range Experiment Station, Fort Collins, Colorado.

Marsh, Gary, (February 1994), Bureau of Land Management, Recreation Program, Washington, D.C., personal communication.

Marston, Ed, (1987), "The Missouri River: Developed, But For What?" pp. 111–17 in High Country News, eds., *Western Water Made Simple*, Island Press, Washington, D.C.

Martin, Michael; Joel R. Hamilton; and Ken Casavant, (August 1992), "Implications of a Drawdown of the Snake-Columbia River on Barge Transportation," in *Water Resources Bulletin*, vol. 28, no. 4, pp. 673–80.

Matthews, Olen Paul, (March 1995), University Center for Water Research, Oklahoma State University, Stillwater, Oklahoma, personal communication.

Mattick, Michael J., (1993), "Instream Flow Protection in Oregon," in Lawrence J. MacDonnell and Teresa A. Rice, eds., *Instream Flow Protection in the West*, revised edition, Natural Resources Law Center, University of Colorado School of Law, Boulder.

——— (March 1995), Department of Water Resources, Salem, Oregon, personal communication.

McClurg, Sue, (January–February 1993), "Changes in the Central Valley Project," *Western Water* (Water Education Foundation).

McGlothlin, Dan, (March 1994), Water Rights Division, National Park Service, Fort Collins, Colorado, personal communication.

McKinney, Matthew J., (July 1991), "Leasing Water for Instream Flows: The Montana Experience," *Rivers*, vol. 2, no. 3, pp. 247–54.

——— (1993), "Instream Flow Policy in Montana: A History and Blueprint for the Future," in Lawrence J. MacDonnell and Teresa A. Rice, eds., *Instream Flow Protection in the West*, revised edition, Natural Resources Law Center, University of Colorado School of Law, Boulder.

McKinney, Matthew J., and Jonathan G. Taylor, (October 1988), *Western State Instream Flow Programs: A Comparative Assessment*, Instream Flow Information Paper No. 18, Biological Report 89(2), U.S. Department of the Interior, Fish and Wildlife Service, Research and Development, Washington, D.C.

McLelland, Leroy, (May 1995), Nevada Division of Wildlife, Carson City, Nevada, personal communication.

Mead, Katherine Lamere, (1986), "Wyoming's Experience With Federal Non-Indian Reserved Rights: The Big Horn Adjudication," *Land and Water Law Review*, vol. 21, pp. 433–53.

Megahan, Walt, (June 1994), former regional hydrologist, U.S. Forest Service, personal communication.

Metzger, Philip C., (July–August 1986), "Commentary: Wildlife Needs Instream Flows for Survival," *CF Letter* (Conservation Foundation, Washington, D.C.).

Meyer, Christopher H., (1989), "Instream Flows: Coming of Age in America," pp. 135–41 in Suzanne P. Van Gytenbeek, ed., *Proceedings of the Western Regional Instream Flow Conference*, sponsored by Wyoming Trout Unlimited, Jackson Hole, October 20–21.

——— (1993), "Instream Flows: Integrating New Uses and New Players Into the Prior Appropriation System," in Lawrence J. MacDonnell and Teresa A. Rice, eds., *Instream Flow Protection in the West*, revised edition, Natural Resources Law Center, University of Colorado School of Law, Boulder.

Meyer, Michael C., (1984), *Water in the Hispanic Southwest: A Social and Legal History 1550–1850*, University of Arizona Press, Tucson.

Meyers, Charles J., (1971), *A Historical and Functional Analysis of the Appropriation System*, National Water Commission, Arlington, Virginia.

Miller, Kathleen A., (1990), "Water, Electricity, and Institutional Innovation," pp. 367–93 in Waggoner, Paul E., ed., *Climate Change and U.S. Water Resources*, John Wiley and Sons, New York.

Miller, Tracy, (March 1994), National Park Service Rivers, Trails, and Conservation Assistance, Washington, D.C., personal communication.

Mirati, Al, (April 1995), Oregon Department of Fish and Wildlife, Salem, Oregon, personal communication.

Moffett, Ben, and Vickie Carson, eds., (April 1994), *National Park Service Almanac*, published by the Public Affairs Office of the Rocky Mountain Region, National Park Service, Denver, Colorado.

Morgan, Robert, (1989), "Regulating Instream Flows in Utah," in Suzanne P. Van Gytenbeek, ed., *Proceedings of the Western Regional Instream Flow Conference*, sponsored by Wyoming Trout Unlimited, Jackson Hole, October 20–21.

Mortensen, Craig, (September 1994), Hydrothermal Operations Group, Bonneville Power Administration, Vancouver Office, personal communication.

Mortensen, Janet, (1994), "The Upstream Battle in the Protection of Utah's Instream Flows," *Journal of Energy, Natural Resources, and Environmental Law*, vol. 14, pp. 113–38.

Moss, Randy, (May 1995), Texas Parks and Wildlife, Austin, Texas, personal communication.

Muller, Dan, (September and November, 1994), Division of Resource Services, Service Center, Bureau of Land Management, Denver, Colorado, personal communications.

Naiman, Robert J. et al., (1995), "Freshwater Ecosystems and Their Management: A National Initiative," *Science,* vol. 270, 584–85.

National Audubon Society, (March 1994), "Troubled Landscape: The National Wildlife Refuge System," National Audubon Society, Washington, D.C.

National Geographic Society, (July 1994), "The United States: Federal Lands and Wilderness Areas," map, National Geographic Society, Washington, D.C.

National Parks and Conservation Association, (May–June 1995), "Rivers at Risk," in *National Parks: The Magazine of the National Parks and Conservation Association*, pp. 26–31.

National Research Council, (1992), *Water Transfers in the West: Efficiency, Equity, and the Environment*, National Academy Press, Washington, D.C.

National Water Commission, (1973), *New Directions in U.S. Water Policy*, National Water Commission.

Nelson, Cynthia, (May 1995), Washington State Department of Ecology, Olympia, Washington, personal communication.

Norcross, Beth, (June 1993), director of legislative programs, American Rivers, Washington D.C., personal communication.

Odenbach, (April 1995), office of the North Dakota state engineer, Bismarck, North Dakota, personal communication.

Office of the Solicitor, Department of the Interior, (1979), "Federal Water Rights of the National Park Service, Fish and Wildlife Service, Bureau of Reclamation and Bureau of Land Management," 86 Interior Dec. 553.

Office of the Solicitor, Department of the Interior, (1988), M-36914, Supplement III.

Olson, Timothy, (May 1995), Division of Wildlife, Game, Fish and Parks, State of South Dakota, personal communication.

Oser, Bob, (December 1993), U.S. Fish and Wildlife Service Region 1, Portland, Oregon, personal communication.

Palmer, Tim, (1986), *Endangered Rivers and the Conservation Movement*, University of California Press, Berkeley.

——— (1991), *The Snake River: Window to the West*, Island Press, Washington, D.C.

——— (1993), *The Wild and Scenic Rivers of America*, Island Press, Washington, D.C.

——— (1994), *Lifelines: The Case for River Conservation*, Island Press, Washington, D.C.

Pisani, Donald J., (1992), *To Reclaim a Divided West—Water, Law, and Public Policy 1848–1902*, University of New Mexico Press, Albuquerque.

Potter, Lori, (1990), "*Nevada v. Morros*: Instream Flow Rights for Nevada," *Rivers*, vol. 1, no. 1, pp. 65–67.

——— (1993), "People Preserving Rivers: The Public and Its Changing Role in Protecting Instream Flows," in Lawrence J. MacDonnell and Teresa A. Rice, eds., *Instream Flow Protection in the West*, revised edition, Natural Resources Law Center, University of Colorado School of Law, Boulder.

Purkey, Andrew, (April 1995), Director, Oregon Water Trust, Portland, Oregon, personal communication.

Randall, Alan, (1983), "The Problem of Market Failure," *Natural Resources Journal*, vol. 23, no. 1, pp. 131–48.

Randall, Mike, (April 1995), Office of the Nevada State Engineer, Carson City, Nevada, personal communication.

Ransel, K., and E. Myers, (1988), "State Water Quality and Wetland Protection: A Call to Awaken the Sleeping Giant," *Virginia Journal of Natural Resources Law*, vol. 7, p. 339.

Ransel, Katherine, (September 1994), co-director, Northwest Regional Office, American Rivers, personal communication.

Reed, Nathaniel P., and Dennis Drabelle, (1984), *The United States Fish and Wildlife Service*, Westview Press, Boulder, Colorado.

Reed, Scott W., (1990), "Conserved Water in Oregon," *Rivers*, vol. 1, no. 2, pp. 148–49.

——— (1992), "Fish Gotta Swim: Establishing Legal Rights to Instream Flows Through the Endangered Species Act and the Public Trust Doctrine," *Idaho Law Review*, vol. 28, pp. 645–66.

Reiser, Dudley W.; Thomas A. Wesche; and Christopher Estes, (March–April 1989), "Status of Instream Flow Legislation and Practices in North America," *Fisheries*, vol. 14, no. 2, pp. 22–29.

Reisner, Marc, (1986), *Cadillac Desert*, Viking Penguin, New York.

Rhodes, Steven L.; Kathleen Miller; and Lawrence J. MacDonnell, (January 1992), "Institutional Response to Climate Change: Water Provider Organizations in the Denver Metropolitan Region," *Water Resources Research*, vol. 28, no. 1, pp. 11–18.

Robie, Ronald B., (1977), "The Public Interest in Water Rights Administration," Rocky Mountain Mineral Law Institute, vol. 23, pp. 917–40.

——— (January 1990), "State Control of Instream Flows Rejected by Court," *Rivers*, vol. 1, no. 1, pp. 62–65.

Rolfs, Leland E., (1993), "Minimum Desirable Stream Flows in Kansas," in Lawrence J. MacDonnell and Teresa A. Rice, eds., *Instream Flow Protection in the West*, revised edition, Natural Resources Law Center, University of Colorado School of Law, Boulder.

——— (June 1995), Division of Water Resources, Kansas State Board of Agriculture, Topeka, Kansas, personal communication.

Romm, Jeff, and Kathleen Bartoloni, (1985), "New Rules for National Forest Water," *Journal of Forestry* , vol. 83, no. 6, pp. 362–67.

Roos-Collins, Richard, (October 1993), "The Mono Lake Cases," *Rivers*, vol. 4, no. 4, pp. 328–36.

Rosgen, David L.; Hilton L. Silvey; and John P. Potyondy, (March 1986), "The Use of Channel Maintenance Flow Concepts in the Forest Service," *Hydrological Science and Technology: Short Papers*, vol. 2, no. 1, pp. 19–26.

Rusinek, Walter, (1990), "A Preview of Coming Attractions? *Wyoming v. United States* and the Reserved Rights Doctrine," *Ecology Law Quarterly*, vol. 17, pp. 355–412.

Saliba, Bonnie Colby; David B. Bush; and William E. Martin, (September 1987a),

Water Marketing in the Southwest—Can Market Prices Be Used to Evaluate Water Supply Augmentation Projects? USDA Forest Service General Technical Report RM-144, Rocky Mountain Forest and Range Experiment Station, Fort Collins, Colorado, 44 pages.

Saliba, Bonnie Colby; David B. Bush; William E. Martin; and Thomas C. Brown, (Summer 1987b), "Do Water Market Prices Appropriately Measure Water Values?" *Natural Resources Journal*, vol. 27, no. 3, pp. 617–51.

Sanders, Larry D.; Richard G. Walsh; and John B. Loomis, (July 1990), "Toward Empirical Estimation of the Total Value of Protecting Rivers," *Water Resources Research*, vol. 26, no. 7, pp. 1345–57.

Sax, Joseph L., (1970), "The Public Trust Doctrine in Natural Resource Law: Effective Judicial Intervention," *Michigan Law Review*, vol. 68, p. 471.

Sax, Joseph L.; Robert H. Abrams; and Barton H. Thompson, Jr., (1991), *Legal Control of Water Resources: Cases and Materials*, 2nd ed., West Publishing, St. Paul, Minnesota.

Schmidt, Larry, (June 1994), Project Leader, Stream Systems Technology Center, U.S. Forest Service, Fort Collins, Colorado, personal communication.

Shay, Monique C., (1992), "Promises of a Viable Homeland, Reality of Selective Reclamation: A Study of the Relationship Between the Winters Doctrine and Federal Water Development in the Western United States," *Ecology Law Quarterly*, vol. 19, pp. 547–90.

Shelby, Bo; Thomas C. Brown; and Robert Baumgartner, (July 1992), "Effects of Streamflows on River Trips on the Colorado River in Grand Canyon, Arizona," *Rivers*, vol. 3, no. 3, pp. 191–201.

Shelby, Bo; Thomas C. Brown; and Jonathan G. Taylor, (March 1992), *Streamflow and Recreation*, USDA Forest Service General Technical Report RM-209, Rocky Mountain Forest and Range Experiment Station, Fort Collins, Colorado, 26 pages.

Shupe, Steven J., (Spring 1985), "Reserved Instream Flows in the National Forests: Round Two," *Western National Resources Digest*.

——— (1989), "Keeping the Waters Flowing: Stream Flow Protection Programs, Strategies and Issues in the West," pp. 1–22 in Lawrence J. MacDonnell, Teresa A. Rice, and Steven J. Shupe, eds., *Instream Flow Protection in the West*, Natural Resources Law Center, University of Colorado School of Law, Boulder.

Shurts, John, (1984), "FLPMA, Fish and Wildlife, and Federal Water Rights," *Environmental Law*, vol. 15, pp. 115–51.

Sims, Steven O., (1993), "Colorado's Instream Flow Program: Integrating Instream Flow Protection into a Prior Appropriation System," in Lawrence J. MacDonnell and Teresa A. Rice, eds., *Instream Flow Protection in the West*, revised edition, Natural Resources Law Center, University of Colorado School of Law, Boulder.

Skillern, Frank F., (1993), *Texas Water Law,* Sterling Press, San Antonio, Texas.

Skinner, Jay, (May 1995), Colorado Division of Wildlife, Denver, Colorado, personal communication.

Slade, Terry, (April 1995), Water Rights Permitting Section, Texas Natural Resource Conservation Commission, personal communication.

Slattery, Kenneth O., and Robert F. Barwin, (1993), "Protecting Instream Resources in Washington State," in Lawrence J. MacDonnell and Teresa A. Rice, eds., *Instream Flow Protection in the West*, revised edition, Natural Resources Law Center, University of Colorado School of Law, Boulder.

Smith, Gary, (June 1995), Environmental Specialist, California Fish and Game, personal communication.

Smith, Gordon C., (1984), regional attorney, Office of General Counsel, U.S. Forest Service, Denver, Colorado, letter to the Regional Forester (Region 2), Denver, Colorado, April 26.

Smith, Roy, (February 1994), water rights coordinator, Colorado State Office, Bureau of Land Management, personal communication.

Smith, W. B., (January 1990), "The Controversial Effects of Instream Flow Determinations," *Rivers*, vol. 1, no. 1, pp. 3–5.

Soden, Dennis L., (October 1990), "Developing a Framework for Instream Flow Case Studies," *Rivers*, vol. 1, no. 4, pp. 318–21.

Solley, Wayne B.; Robert R. Pierce; and Howard A. Perlman, (1993), "Estimated Use of Water in the United States in 1990," U.S. Geological Survey Circular 1081, U.S. Government Printing Office, Washington, D.C.

Sonorensis, (Summer 1988) (Arizona-Sonora Desert Museum).

Spann, Chic, (February 1993), water resources program manager, Watershed and Air Management, U.S. Forest Service Region 3, Albuquerque, New Mexico, personal communication.

Squillace, Mark, (January 1993), "Transferring Indian Reserved Rights to Instream Flows: Lessons from the Big Horn Adjudication," *Rivers*, vol. 4, no. 1, pp. 48–54.

Staff, George, (April 1995), Texas Natural Resource Conservation Commission, Austin, Texas, personal communication.

Stalnaker, Clair; Berton L. Lamb; Jim Henriksen; Ken Bovee; and John Bartholow, (1995), *The Instream Flow Incremental Methodology: A Primer for IFIM*, Biological Report 29, National Biological Service, U.S. Department of the Interior, Washington, D.C., 45 pages.

Stalnaker, Clair, and Lee Lamb, (October 1994), National Biological Survey, Fort Collins, Colorado, personal communication.

Starlund, Steve, (May 1995), Washington State Department of Parks and Recreation, Olympia, Washington, personal communication.

St. Clair, Jeffrey, (January 1993), "Wild Alaskan Rivers Up for Grabs?" *Forest Watch*, vol. 13, no. 6, pp. 8–9.

Sterin, Bunny, (June 1993), Alaska State Office, Bureau of Land Management, Anchorage, Alaska, personal communication.

Stromberg, Juliet C.; Duncan T. Patten; and Brian D. Richter, (July 1991), "Flood Flows and Dynamics of Sonoran Riparian Forests," *Rivers*, vol. 2, no. 3, pp. 221–35.

Swenson, Lee, and P. Kirt Carpenter, (1989), "Utah's Dammed Stream Fishery Operation of Reclamation Reservoirs in Utah," pp. 22–28 in Suzanne P. Van Gytenbeek, ed., *Proceedings of the Western Regional Instream Flow Conference*, sponsored by Wyoming Trout Unlimited, Jackson Hole, October 20–21.

Tarlock, A. Dan, (1985a), "The Endangered Species Act and Western Water Rights," *Land and Water Law Review*, vol. 20, no.1, pp. 1–30.

——— (1985b), "The Law of Equitable Apportionment Revisited, Updated, and Restated," *University of Colorado Law Review*, vol. 56, pp. 381–411.

——— (1987), "Protection of Water Flows for National Parks," *Land and Water Law Review*, vol. 22, no. 1, pp. 29–48.

——— (1988), "New Commons in Western Waters," pp. 69–89 in David H. Getches, ed., *Water and the American West: Essays in Honor of Raphael J. Moses*, Natural Resources Law Center, University of Colorado School of Law, Boulder.

——— (1989), *Law of Water Rights and Resources*, Release No. 1, Clark Boardman and Company, New York.

——— (1990), "A Decade of Instream Flow Literature in Context," *Rivers*, vol. 1, no. 1, pp. 70–73.

——— (June 1991), "New Water Transfer Restrictions: The West Returns to Riparianism," *Water Resources Research*, vol. 27, no. 6, pp. 987–94.

——— (July 1991), "Global Climate Change and Instream Values: Why Growth Control Must Be Placed on the Western Flow Protection Agenda," *Rivers*, vol. 2, no. 3, pp. 185–89.

Tarlock, A. Dan, and Doris K. Nagel, (1989), "Future Issues in Instream Flow Protection in the West," pp. 137–55 in Lawrence J. MacDonnell, Teresa A. Rice, and Steven J. Shupe, eds., *Instream Flow Protection in the West*, Natural Resources Law Center, University of Colorado School of Law, Boulder.

Tarlock, A. Dan, and Roger Tippy, (1970), "The Wild and Scenic Rivers Act of 1968," *Cornell Law Review*, vol. 55, no. 5, pp. 707–39.

Tellman, Barbara, (April 1995), Water Resources Research Center, University of Arizona, Tucson, personal communication.

Templer, Otis W., (October 1981), "The Evolution of Texas Water Law and the Impact of Adjudication," *Water Resources Bulletin*, vol. 17, no. 5.

Thomas, George, (1920), *The Development of Institutions Under Irrigation*, Macmillan, New York.

Thomas, Jack Ward, (1996), chief, U.S. Forest Service, memo to Regional Foresters, Station Directors, Area Director, IITF Director, and WO Staff, "Implementation of the Federal Agriculture Improvement and Reform Act of 1996," May 17.

Trelease, Frank J., (September 1976), "The Legal Basis for Instream Flows," pp. 1–21 of the *Proceedings of the Symposium and Specialty Conference on Instream Flow Needs*, vol. 2, American Fisheries Society, Bethesda, Maryland.

Tristani, M. Gloria, (Summer 1989), "Comment: Interior Turns off Tap for Wilderness Areas," *Natural Resources Journal*, vol. 29, pp. 877–94.

Underwood, Laura, (1994), "Better Late Than Never: States Regain the Right to

Regulate Stream Flows under the Clean Water Act," *Texas Tech Law Review*, vol. 26, pp. 187–215.

U.S. Army Corps of Engineers, (1976), "Columbia-Snake Inland Waterway," brochure, G.P.O. 697-170.

——— (1982), "Columbia River Projects," brochure, Portland District.

——— (1988), "Waterborne Commerce of the United States: Calendar Year 1986," WRSC-WCUS-86-5, Water Resources Support Center, Navigation Division, May 31.

——— (July 1994), *Missouri River Master Water Control Manual: Review and Update Study: Draft Environmental Impact Statement,* Missouri River Division.

U.S. Bureau of the Census, (1904), *Abstract of the Twelfth Census of the United States 1900, Third Edition*, U.S. Government Printing Office, Washington, D.C.

——— (1995), *Statistical Abstract of the United States 1995*, U.S. Government Printing Office, Washington, D.C.

U.S. Bureau of Land Management, (September 1991), "BLM Riparian-Wetland Initiative for the 1990's."

——— (November 1992), "BLM Water Availability Strategy."

——— (December 1992), "Rio Chama Instream Flow Assessment."

——— (September 1993), *Public Land Statistics 1992*, vol. 177.

U.S. Bureau of Reclamation, (December 1992), "An Implementation Plan for Instream Flows."

——— (1993), "Investigation of Existing Projects Initiative," Memorandum to Interested Parties from the Regional Director, Great Plains Region, Billings, Montana, November 30.

——— (August 1994), "Summary of Cost Operation of Glen Canyon Dam, Glen Canyon Environmental Studies and Environmental Impact Statement," UC-1500, August 8.

——— (Fall 1994), newsletter (Colorado River Studies Office) vol. 8.

——— (March 1995a), "Operation of Glen Canyon Dam: Final Environmental Impact Statement."

——— (March 1995b), "Operation of Glen Canyon Dam: Final Environmental Impact Statement: Summary," 1995–841-505, U.S. Government Printing Office, Washington, D.C.

——— (February 1996), "Glen Canyon Dam: Beach/Habitat-Building Test Flow: Final Environmental Assessment and Finding of No Significant Impact."

U.S. Department of Agriculture, (1987), Economic Research Service, "National-Interregional Agricultural Projections (NIRAP) System," USDA, Washington, D.C.

U.S. Department of Energy, Idaho Operations Office, (January 1994), *Environmental Mitigation at Hydroelectric Projects: Volume II: Benefits and Costs of Fish Passage and Protection.*

U.S. Energy Information Administration, (January 1994), *Electric Power Annual*

1992, USDOE, Office of Coal, Nuclear, Electric and Alternate Fuels, Washington, D.C.

U.S. Department of the Interior, Fish and Wildlife Service, (1975), "Toward a National Program of Substantive Instream Flow Studies and a Legal Strategy for Implementing the Recommendations of Such Studies," contained in memorandum to associate director for environment and research from the chief, Division of Ecological Services, August 21.

——— (January 1993), *Refuges 2003: Draft Environmental Impact Statement: A Plan for the Future of the National Wildlife Refuge System*, issue 5.

U.S. Department of the Interior, National Park Service, (1993), "The National Parks: Index 1993," Washington, D.C.

U.S. Fish and Wildlife Service and U.S. Bureau of the Census, (1993), *1991 National Survey of Fishing, Hunting, and Wildlife-Associated Recreation*, U.S. Government Printing Office, Washington, D.C.

U.S. Forest Service, (December 1976), "The National Forest Management Act of 1976," Current Information Report No. 16.

——— (1992), "Instream Flow Needs Assessment and Recommendations for the Proposed Piedra Wilderness," report prepared by personnel in Forest Service Region 2 (Denver) and in the San Juan National Forest.

——— (February 1994), "Land Areas of the National Forest System: As of September 1993," FS-383.

——— (July 29, 1994), Arapaho and Roosevelt National Forests and Pawnee National Grassland, "Record of Decision: Land-Use Authorization for Joe Wright Dam and Reservoir and Amendment to the Land and Resources Management Plan," signed by Arapaho-Roosevelt Forest Supervisor M.M. Underwood.

U.S. Geological Survey, (1992), "The U.S. Geological Survey Stream-Gaging Program in Alaska," USGS Open File Report 92-106, E.F. Snyder, Anchorage.

U.S. National Park Service, (November 1992), "River Mileage Classifications for Components of the National Wild and Scenic Rivers System," Division of Park Planning and Protection.

U.S. National Park Service, Office of Public Affairs and Division of Publications, (1993), *The National Parks: Index 1993*, U.S. Government Printing Office, Washington, D.C.

U.S. Public Land Law Review Commission, (1970), *One Third of the Nation's Land: A Report to the President and to Congress.*

Vencil, Betsy, (1986), "The Migratory Bird Treaty Act—Protecting Wildlife on our National Refuges—Kesterson Reservoir, a Case in Point," *Natural Resources Law Journal*, vol. 26, no. 3, pp. 609–28.

Volkman, John M., and Kai N. Lee, (1988), "Within the Hundredth Meridian: Western States and Their River Basins in a Time of Transition," *University of Colorado Law Review*, vol. 59, pp. 551–77.

Wahl, Richard W., (1989), *Markets for Federal Water: Subsidies, Property Rights, and the Bureau of Reclamation*, Resources for the Future, Washington, D.C.

——— (July 1990), "Acquisition of Water to Maintain Instream Flows," *Rivers*, vol. 1, no. 3, pp. 195–206.

Wallentine, Kenneth R., (1989), "Note: Wilderness Water Rights: The Status of Reserved Rights after the Tarr Opinion," *B.Y.U. Journal of Public Law*, vol. 4, pp. 181–205.

Walston, Roderick E., (1990a), "State Regulation of Federally-Licensed Hydropower Projects: The Conflict Between *California* and *First Iowa*," *Oklahoma Law Review*, vol. 43, pp. 87–102.

——— (1990b), "The Supreme Court's Changed Perspective of Federal-State Water Relations: A Personal Memoir of the *New Melones* Case," *Journal of the West*, vol. 29, no. 3, pp. 28–39.

Washington State Department of Ecology, (1992), *Trust Water Rights Program: Guidelines*, Publication No. 92–88, September 10, Olympia, Washington.

Water Education Foundation, (January–February 1993), "Changes in the Central Valley Project," *Western Water*, Sacramento, California.

Water Intelligence Monthly, various issues (Stratecon, Inc., Claremont, California).

Weland, Michael C., (May 1995), executive director, Utah Reclamation Mitigation and Conservation Commission, personal communication.

Welder, G. E., (1988), "Hydrologic Effects of Phreatophyte Control, Acme-Artesia Reach of the Pecos River, New Mexico, 1962–82," Water-Resources Investigations Report 87–4148, U.S. Department of the Interior, Geological Survey, Albuquerque, New Mexico.

Western Area Power Administration, (February 1984), "Small-Scale Hydroelectric Power: A Brief Assessment," prepared by Tudor Engineering, Denver, Colorado.

——— (June 1991), "Hydropower: Harnessing Water," WAPA, Denver, Colorado.

——— (1994), "Ideas That Worked: 1993 Annual Report," WAPA, Denver, Colorado.

White, Michael D., (1993), "Colorado Instream Flows," *Rivers*, vol. 4, no. 1 pp. 55–58.

Whittaker, Doug; Bo Shelby; William Jackson; and Robert Beschta, (1993), *Instream Flows for Recreation: A Handbook on Concepts and Research Methods*, U.S. Department of the Interior, National Park Service, Rivers and Trails Conservation Program, Washington, D.C.

Whittington, Mike, (February 1994, January 1995), Bureau of Reclamation area manager for North and South Dakota, personal communications.

Wigington, Robert, (August 1990), "Market Strategies for the Protection of Western Instream Flows and Wetlands," Occasional Papers Series, Natural Resources Law Center, University of Colorado School of Law, Boulder.

——— (1992), "Instream Water Rights: A Protection Tool for Western Rivers," unpublished paper, The Nature Conservancy, Boulder, Colorado.

——— (1993), "Selected Colorado Cases in the Protection of Instream Flows," unpublished paper, The Nature Conservancy, Boulder, Colorado.

Wilkinson, Charles F., (1980), "The Public Trust Doctrine in Public Land Law," *U.C. Davis Law Review*, vol. 14, p. 269.

——— (1985), "Western Water Law in Transition," *University of Colorado Law Review*, vol. 56, no. 3, pp. 317–45.

——— (1988a), "To Settle a New Land: An Historical Essay on Water Law in Colorado and in the American West," pp. 1–17 in David H. Getches, ed., *Water and the American West: Essays in Honor of Raphael J. Moses*, Natural Resources Law Center, University of Colorado School of Law, Boulder.

——— (1988b), "Water Rights and the Duties of the National Park Service: A Call for Action at a Critical Juncture," pp. 261–74 in David J. Simon, ed., *Our Common Lands: Defending the National Parks*, Island Press, Washington, D.C.

——— (1989), "Aldo Leopold and Western Water Law: Thinking Perpendicular to the Prior Appropriation Doctrine," *Land and Water Law Review* vol. 24, no. 1, pp. 14–17.

——— (1990), "Values and Western Water: A History of the Dominant Ideas," Western Water Policy Project Discussion Series Paper No. 1, Natural Resources Law Center, University of Colorado School of Law, Boulder, 10 pages.

——— (January 1990), "The Historical Context for Instream Flows," *Rivers*, vol. 1, no. 1, pp. 6–7.

——— (1991), "Prior Appropriation: 1848–1991," *Environmental Law*, vol. 21, no. 3, part I, pp. v–xviii.

——— (1992), *The Eagle Bird: Mapping a New West*, Vintage Books, New York.

Wilkinson, Charles F., and H. Michael Anderson, (1987), *Land and Resource Planning in the National Forests*, Island Press, Washington, D.C.

Wilkinson, Charles F., and Daniel Keith Conner, (1987), "A Great Loneliness of the Spirit," pp. 54–64 in High Country News, eds., *Western Water Made Simple*, Island Press, Washington, D.C.

Williams, Owen, (February 1994, and January 1996), Water Rights Branch, Water Resources Division, National Park Service, Fort Collins, Colorado, personal communication.

Williams, Wes, (1994), "Changing Water Use for Federally Reserved Indian Water Rights: Wind River Indian Reservation," *U.C. Davis Law Review*, vol. 27, pp. 501–32.

Willis, Cheryl, (June and December 1993), Water Rights Branch, Engineering Division, U.S. Fish and Wildlife Service Region 6, Denver, Colorado, personal communications.

Wolfe, Robert J., and Robert J. Walker, (1987), "Subsistence Economies in Alaska: Productivity, Geography, and Development Impacts," *Arctic Anthropology*, vol. 24, no. 2, pp. 56–81.

Wooster, Dave, (April 1991), "Overview of Corps Water Management Obligations," paper delivered at symposium on management of the least tern and piping plover, Lincoln, Nebraska.

——— (September 1994), Reservoir Control Center, U.S. Army Corps of Engineers, Missouri River Division, Omaha, Nebraska, personal communication.

Wuethrich, Bernice, (1996), "Deliberate Flood Renews Habitats," *Science*, vol. 272, pp. 344–45.

Yaffee, Steven L., (1991), "Avoiding Endangered Species/Development Conflicts Through Interagency Consultation," pp. 86–97 in Kathryn A. Kohm, ed., *Balancing on the Brink of Extinction: The Endangered Species Act and Lessons for the Future*, Island Press, Washington, D.C.

Young, Robert A., and S. Lee Gray, with R.B. Held and R.S. Mack, (1972), "Economic Value of Water: Concepts and Empirical Estimates," Technical Report to the National Water Commission, NTIS no. PB210356, U.S. Department of Commerce, Springfield, Virginia.

Index

Administrative agencies:
 federal:
  protection of instream flow by, 206–23, 301
  see also individual agencies
 state:
  protection of instream flow by, 113–17, 127–28, 140–43
  water resources managed by, 35–39
Aesthetics, 2, 3, 31, 58–60, 119
 quantifying instream flow needs for, 83–85, 89, 90
Aggradation of river channels, 73–74
Agriculture, 304–305
 irrigation for, *see* Irrigation
Air pollution, 243
Alaska, 78
 instream flow protection in, 112, 119, 120, 121, 124, 126, 135, 138, 140, 158, 220–21
 mining in, 21
 public trust doctrine in, 154
 rivers program, 158
 water law in, 26, 27, 37–38
 wild and scenic rivers in, 284, 286
Alaska Department of Fish and Game, 124
Alaska National Interest Lands Conservation Act, 289
Alaska Power Administration, 234
Alaska Water Use Act of 1966, 27
Allocation of water:
 efficiency principle and, 95–97
 equity principle and, 97
 for protection of instream flows, *see* Protection of instream flows
 reaching a balance in, 297–306
American River, 16, 17
Anadromous fish, 47, 244
 salmon, *see* Salmon
Anasazi Indians, 11
Antiquities Act, 184, 185
Aquatic life, *see* Fish and aquatic life
Arapaho-Roosevelt National Forests, 209
Arctic National Wildlife Refuge, 221

Arizona, 21, 206, 287
  instream flow protection in, 112, 113–14, 117, 120, 121, 124, 126, 129–30, 138, 144
  instream water rights in, 144
  water law in, 26, 37
  water rights transfers and, 146
Arizona Department of Water Resources, 113–14, 119, 129–30
Arizona Nature Conservancy, 124
Arizona v. California, 180, 187, 189, 194–95
Arizona Wilderness Act of 1990, 292
Arkansas River, 40, 106
Army Corps of Engineers, 33–34, 71, 87, 216, 217, 234, 245, 246, 260, 261, 266–67, 274, 275
  Two Forks Dam and, 262–63

Babbitt, Bruce, 228
Bald eagle, 89, 90, 92
"Bankfull" discharge, 85–86
Banking of water, 147–48
Base flow:
  defined, 8
  see also Minimum flow
Beard, Daniel, 228
Bequest value, 62, 103
Bighorn Canyon National Recreation Area, 185
Big Horn River, 190
Bill Williams National Wildlife Refuge, 220
Bird watching, 58
Bitterroot River, 104–105
Blue Lake, 203
Boating, 56–57, 69, 88–89, 90, 92
Bonneville Power Administration (BPA), 234–35, 245, 246, 247, 248, 274
Boulder Canyon Act of 1932, 235
Boulder Canyon (Hoover) Dam, *see* Hoover Dam
British settlers, 13
  riparian law and, *see* Riparian water rights
Brown, Melville, 30
Brownlee Dam, 273
Brown trout, 88, 90, 92
Buenos Aires National Wildlife Refuge, 220
Bureau of Indian Affairs, 195
Bureau of Land Management (BLM), 87, 116, 124, 203, 214–18, 219–20, 280
  federal nonreserved water rights and, 199–201
  interdisciplinary study process, 215–18
  reserved water rights and, 186–87
  resources of, 222
  Riparian-Wetland Initiative for the 1990s, 214–15
  state water rights and, 205–206
  wilderness areas and, 289, 290
Bureau of Reclamation, 33, 87, 195, 204, 225–33, 245, 246, 251, 275, 276
  changes at, 226–29
  future of reclamation program, 232–33
  hydropower and, 233–34, 238–44
  instream flow protection and, 226–32
  river-basin accounting methods, 233–34

Cache la Poudre River, 284–85, 287
California, 286
  federal use of state water rights in, 202, 206
  Gold Rush, 16–17, 19
  instream flow protection in, 112, 120, 121, 138, 139–40, 141, 143, 144, 164
  instream water rights in, 144
  public trust doctrine in, 151–52
  rivers program, 158
  water law in, 15, 19–20, 25–26, 27, 37
  water rights transfers and, 146
  Wild and Scenic Rivers program, 138

California Department of Fish and Game, 142, 251, 252
California Fish and Game Code, 142
California State Water Resources Control Board, 141, 152–53, 158, 203, 231, 251–52
California v. Federal Energy Regulatory Commission, 251–52, 253
California v. United States, 251
California Water Code, 141
California Wild and Scenic Rivers Act, 158
"Call" on a river, placing a, 36, 172–73
Canals, 17, 225, 304
  early civilizations building, 10
  settlement of the West and, 11
"Canoeing zero" flow, 83
Cappaert v. United States, 180, 184–85
Cedar River, 251
Central Utah Project Completion Act (CUPCA), 229–30
Central Valley Project, 33, 204, 230–32
Central Valley Project Improvement Act (CVPIA), 230–32
Channel incision, 75
Channel maintenance, 72–76, 190, 193
  quantifying instream flow needs for, 85–87
Cities, diversion of water for, *see* Municipal water supplies; *specific cities and projects*
Citizen initiatives, 112
City of Thornton v. City of Fort Collins, 117–18
Clark's Fork of the Yellowstone River, 285, 287
Clean Water Act, 78–79, 178, 259–67, 277, 301
  effectiveness in protecting instream flows, 266–67
  goal of, 259
  Section 404, 260–63, 266
  Sections 303 and 401, 263–66, 267
Coffin, George, 28–29
Coffin v. Left Hand Ditch Co., 26–27, 28–29
Coldiron, William, 199
Colorado, 287
  federal reserved water rights and, 183, 185, 189–90, 191–93
  instream flow protection in, 112, 117–18, 119, 121, 123, 125–26, 130, 138, 144, 160–61
  instream water rights in, 144
  mining in, 20–21
  public trust doctrine in, 155
  Riverside Irrigation District, 261
  water law in, 26–27, 28–29, 38–39
  water project permits in, 208–11
  water rights transfers and, 146–47
  wilderness bill, 292–94
Colorado-Big Thompson Project, 33, 161
Colorado Division of Wildlife, 130, 135, 160, 294
Colorado River, 77, 100, 167, 238, 275–76
Colorado River Storage Project Act of 1956, 235, 237–38
Colorado Water Conservancy District v. Colorado Water Conservation Board, 117
Colorado Water Conservation Board, 119, 123, 125–26, 130, 131, 135, 146–47, 155, 160, 294
Colorado Wilderness Act of 1993, 292–94
Columbia Basin Project, 33
Columbia River, 9, 71, 72, 227–28, 234, 244
Columbia River Gorge, 137
Columbia River Inter-Tribal Fish Commission, 246
Colville Confederated Tribes v. Walton, 194
Commercial uses of water, 40
Common law, water rights and, *see* Riparian water rights
Compacts, *see* Private agreements
Conservation, 162–64, 305
Constitution, U.S., 203

commerce clause, 178
property clause, 178, 187
supremacy clause, 178–79, 187, 255
Cottonwood trees, 52, 53
Court decisions and instream flow protection:
state courts, *see individual states and decisions*
Supreme Court, *see* Supreme Court, U.S.
Creative Act of 1891, 187

Dakota Territory, 15, 21
Dams, 8, 55, 63, 301, 304
Bureau of Reclamation and, 225, 226, 227, 232
era of construction of, 32–34, 40, 42
erosion and, 74–75
fish movement and migration and, 50, 227–28, 244–48, 273–75
hydropower and, *see* Hydropower
mining and, 16, 17
prohibiting construction of new, 158–59
see also names of specific dams
Definition of terms, 7–8
Degradation of river channels, 74–75
DeKay v. United States Fish and Wildlife Service, 117
Denver Water Department, 261, 262
Departure analysis, 213
Desert Lands Act of 1877, 22
Devil's Hole National Monument, 180
Devil Valley National Monument, 184–85
Dewatering of streams, 8, 40, 106
Diminishing marginal returns, 96–97, 105–106
Dinosaur National Monument, 185
Diversion of water for offstream uses, *see* Offstream uses, diversion for
Doerksen, Harvey, 1, 4, 6
Domestic uses of water, *see* Municipal water supplies
Donation of water rights, 126, 301
Dosewallips River, 265
Downstream senior water rights, *see* Senior water rights downstream
Dry-year option, 147

Economics:
allocation of water and, 96–97, 104–106
recreational uses of water and, 60–61
Efficiency principle and allocation of water, 95–97
Electric Consumers Protection Act of 1986, 249, 250
Electricity, generation of, *see* Hydropower; Thermoelectric plants
Eminent domain, *see* Takings
Empirical observation, 81–82
Endangered American Wilderness Act, 289
Endangered Species Act (ESA), 209, 246, 255, 267–77, 301
consultations under, 270–71, 272
impact on instream flows, 273–77
implementation of, 272–73
listing of species under, 268–69
politics and, 272–73
prohibition of takings under, 269
Section 7 protection, 269–71
England, water law in, 151
riparian, *see* Riparian water rights
Environmental Assessments (EAs), 256, 257
Environmental Impact Statements (EISs), 256, 257–58
costs of preparation of, 242, 243
for Glen Canyon Dam, 239, 240, 242
for Two Forks Dam, 262, 263
Environmental protection, 3, 31, 43, 61–64

federal legislation, 255–96
see also individual acts
increased interest in, 62, 111, 231, 298
Environmental Protection Agency (EPA), 78–79, 231, 260, 263, 264
Equity and allocation of water, 97
Erosion, 74–75
Exchange of water, 204
Existence value, 62, 103
Experts, instream flow needs determined by judgment of, 84, 88

Farm Bill of 1996, 212
Feather River, 17
Federal Energy Regulatory Commission (FERC), 213, 245, 248–53, 267, 280–81, 301
Federal government:
jurisdictional conflicts with states, 128–29, 202–203
nonreserved water rights, 197–202
protection of instream flow and, *see* Federal protection of instream flow
reserved water rights, *see* Reserved water rights, federal
water law and, 21–22
water resources development programs, 33, 225–53
hydropower programs, 233–53
reclamation programs, 225–33
see also names of specific projects
Federal Lands Policy and Management Act of 1976, 197, 199–201, 205–206, 207, 289
Federal Power Act (FPA), 248–49, 250, 251
Federal protection of instream flows, 177–223, 301
additional activities fostering, 212–19
administrative controls used for, 206–12, 301
Constitutional foundations for, 177–79
environmental protection legislation and, 255–96
see also individual acts
nonreserved water rights, 197–202
reserved water rights doctrine, *see* Reserved water rights, federal
resources for, availability of, 221–23
state water rights, federal use of, 202–206
summary, 223
table listing methods for, 299–300
variation with agencies in pursuing, 219–21
water resources development programs and, 33, 225–53
Federal Water Power Act of 1920, 248
First in time, first in right, 12, 18
First Iowa Hydroelectric Cooperative v. Federal Power Commission, 250–51
Fish and aquatic life, 2, 3, 9, 31, 38, 69
conflicts among species, 51
endangered species, 273–77
instream flow and needs of, 45–51, 90, 91, 131–32
quantifying instream flow needs of, 79–83
see also Fishing and fisheries
Fish and Wildlife Act of 1956, 185
Fish and Wildlife Conservation Act of 1934, 218
Fish and Wildlife Coordination Act, 205
Fish and Wildlife Service, U.S., 114–15, 185, 205, 218–19, 220–21, 231, 246, 260, 262, 280
endangered species and, 209, 211, 246, 255, 268, 270, 271–72, 272, 275–76, 276–77
Instream Flow Group (IFG), 218–19

Fish and Wildlife Service (*continued*)
resources of, 223
wilderness areas and, 289, 290
Fishing and fisheries, 9, 10, 49–50, 57–58, 88, 90, 119
federal reserved water rights for, 194
instream flow protection for, 227–28, 245–48
quantification goals, 131–32
see also Fish and aquatic life
Floating, *see* Boating
Flood control, 33, 34, 54, 55, 217, 234, 246
Flood Control Act of 1944, 235
Flow regime of a river, 7
Fluvial geomorphology, 76, 190, 191
Forest Service, U.S., 87, 114, 124, 216, 280, 288
administrative protection of instream flows by, 206–12
federal nonreserved water rights and, 201–202
permit process and, 206–12
reserved water rights and, 181, 187–93, 207–208
state water rights and, 206
wilderness areas and, 289, 293, 294
Fort Belknap Indian Reservation, 179
Fortymile River, 284
Free rider problem, 103
French settlers, 13
Friends of the Earth, 152
Fur trade, 13

General Dam Act of 1906, 33
Geographic factors protecting instream flows, 98, 99–100
Geological Survey, U.S., 87
Gila National Forest, 181, 187
Gila River, 10, 40, 106
Gila Wilderness Reserve, 288
Glen Canyon Dam, 75, 100, 237–44
Gold Rush, 16–17, 19
Grand Canyon Environmental Studies (GCES), 239
Grand Canyon Protection Act of 1992, 235, 239, 243
Grand Coulee Dam, 33, 273
Grazing, 55
Great Plains, 10
Groundwater pumping, 55, 180, 185

Halligan Reservoir, 161
Hell's Canyon Dam, 273
Hetch Hetchy Reservoir, 32
History of water use and water law in the West, 3, 12–43
Hohokam civilization, 10–11
Homestake Mining Company, 114
Hoover Dam, 33
Hydraulic mining, 16–17
Hydrologic records, historical, 85, 86
Hydropower, 33–34, 38, 40, 64–70
federal programs, 233–53
agencies responsible for, 233–35
examples of emerging importance of instream uses in operation of, 237–48
operating rules for hydropower facilities, 235–37
for regulation of nonfederal hydropower facilities, 248–53
see also names of specific projects
instream flows and, 67–70
instream versus offstream facilities, 64
licensing of hydropower facilities, 248–50, 252, 253, 280–81, 301
marketing of, 234–35
power demand and, 68–69
pumped-storage plants, 64
run-of-the-river facilities, 65, 67
significance to the West of, 65–67
storage facilities, 64–65, 67, 69

Idaho, 21, 286
instream flow protection in, 112, 117, 119, 120, 121, 126, 131, 135, 137, 138, 142, 144
instream water rights in, 144

public trust doctrine in, 154
rivers program, 158
water law in, 26, 27
wilderness bill, 293
Idaho Department of Water Resources, 131, 154
Idaho Protected Rivers Act, 158
Idaho Water Resources Board, 158
Illinois River, 158–59
Inaccessible streams, 98, 99–100
Indians, *see* Native Americans
Industrial use of water, 2, 34, 38, 40
In Re Application A-16642, 117
In Re Water of Hallett Creek Stream System, 202–203
Instream flow:
defined, 8
popular understanding of, 4, 132–34, 302
protection of, *see* Protection of instream flows
quantifying necessary, methods for, 79–90, 213
timing and location of, 91–93
Instream Flow Incremental Methodology (IFIM), 82–83, 84, 88, 216, 219
Instream flow water rights, *see* Water rights, instream
Instream reservations, 135, 137, 138–40
Instream uses of water, 45–94
conflicts among, 90–94
as nonconsumptive, 133, 168
quantifying necessary instream flow for, methods for, 79–90, 213
transfer of water rights:
from instream use to offstream use, 149–51
from offstream use to instream use, 146–48, 168–76
see also specific uses
Interior Department, Office of the Solicitor, 198–99, 290–91
Interstate water allocations, 99, 100
compacts for, 99, 100, 167–68
Irrigation, 33, 42, 225, 233, 246–47, 305
by early civilizations, 10–11
as major offstream use, 2, 24, 31, 39, 40, 106, 304
settlement of the West and, 22–24
water rights transfers and, 145
Irwin, Mathew, 20
Irwin v. Phillips, 20
Island Park Dam, 227

Justice Department, U.S., Office of Legal Counsel, 198, 199

Kansas:
instream flow protection in, 112, 120, 121, 138, 139
water law in, 15, 26, 37
water rights transfers in, 147
Klamath River, 186, 193
Knowledge about instream flow issues, limited, 4, 132–34, 302
Kortes Dam, 227
Krulitz, Leo, 198, 290
Krulitz Opinion, 198, 290, 291

LaCreek National Wildlife Refuge, 115
Lake Powell, 237
Law and water usage, *see* Water law
Leases, *see* Water leases
Least tern, 276
Left Hand Ditch Company, 28
Lefthand Water Conservancy District, 160
Legal right to water usage, *see* Water law; Water rights
Legislatures, state:
instream flow protection measures and, 112–13
Leopold, Aldo, 288
Leshy, John, 291
Lewis and Clark expedition, 13
Licensing of hydropower facilities, 248–50, 252, 253, 280–81, 301
Longmont Reservoir, 160

Los Angeles, California, 32, 152
Louisiana Territory, 13

McClellan v. Jantzen, 117
Mandan Indians, 10
Market mechanism, 42
 instream flow protection and, 4, 99, 101, 301–302
Marks v. Whitney, 151–52
Marshall, Bob, 288
Marshall, James, 16, 17
Martz, Clyde, 198–99
Mead, Elwood, 30
Meese, Edwin, 291
Mettler v. Ames, 29
Migratory Bird Treaty Act of 1918, 185
Minimum flow:
 choosing between optimum and, 106–109, 129–32
 defined, 8
 states setting levels of, 138–40
Mining, 16–22, 23
 creation of rules governing, 17–18
 Gold Rush, 16–17, 19
 gravel, 74
 staking of claims, 18
 water diversion for, 11–12, 16–17
 water pollution and, 78
Mining Act of 1866, 21–22, 26
Missouri River, 10, 71, 234, 287
Missouri River Basin, 139
Missouri River Navigation and Bank Stabilization Project, 71–72
Mono Lake, 152–53, 304
Mono Lake Committee, 152
Montana, 21, 286
 instream flow protection in, 112, 117–18, 121, 123, 135, 137, 139, 144, 160
 instream water rights in, 144
 public trust doctrine in, 154
 rivers program, 159
 water law in, 26, 29
 water rights transfers in, 147
 wilderness bill, 293
Montana Board of Natural Resources and Conservation, 139
Montana Division of Fish, Wildlife, and Parks, 117–18, 159
Montana Water Use Act, 139
Mormons, 23–24, 25, 30
Muckleshoot Indian Tribe v. Trans-Canada Enters, 194
Multiple regression method, 81–82
Multiple-Use Sustained Yield Act, 188–91, 197
Municipal sewage treatment plants, 78, 304
Municipal water supplies, 2, 31, 33, 34, 38, 40, 42, 225, 304
 water rights transfers for, 145

National Audubon Society, 104, 152
National Audubon Society v. Superior Court of Alpine County, 152–53, 157, 304
National Biological Survey, 219
National Conservation Areas, 186–87
National Environmental Policy Act (NEPA), 209, 236, 256–59, 277
National Forest Management Act of 1976, 198, 201–202
National Marine Fisheries Service, 246
 endangered species and, 268, 270, 271–72, 274, 275
National monuments, water rights in, 184–85
National Park Service, 118, 280
 reserved water rights and, 182–85, 295
 resources of, 222
 Rivers, Trails, and Conservation Assistance (RTCA) office, 213–14
 Water Rights Branch, 212–13
 wilderness areas and, 289, 290
National Park Service Organic Act of 1916, 182, 183, 184
National Recreation Areas, 185

National Survey of Fishing, Hunting, and Wildlife-Associated Recreation, 60
National Water Commission, 62
National Wilderness Preservation System, 289–96
National Wildlife Refuge Administration Act of 1966, 185
Native Americans, 245
  reserved water rights and, 179, 193–97
  salmon fishing and, 49–50
  water usage by, 10–11
Natural flows, 8, 63
Nature Conservancy, The, 104, 131, 148, 160, 161–62, 301
Navigation, 14, 34, 70–72, 89, 178, 246–47
Nebraska:
  instream flow protection in, 112, 117, 121, 123, 130–31, 144
  instream water rights in, 144
  water law in, 15, 26, 27, 37
  water rights transfers and, 146
Negative externalities, allocation of water to instream uses and, 102
NEPA, *see* National Environmental Policy Act (NEPA)
Nevada:
  instream flow protection in, 112, 116, 117, 120, 121, 124, 144
  instream water rights in, 144
  mining in, 20, 21
  water law in, 26, 37, 203
Nevada v. Morros, 117
Nevada Wilderness Act, 292
New Melones dam project, 251
New Mexico:
  instream flow protection in, 112, 116, 122
  water law in, 26
New Mexico Department of Game and Fish, 216
New Mexico Game and Fish Department, 92
New Mexico Interstate Steam Commission, 167–68
New Mexico Wilderness Act, 292
Niobrara River, 287
Nonconsumptive nature of instream water uses, 133
Nonexcludable goods, 103, 104
Nonreserved water rights, federal, 197–202
Norris, George, 34
North Dakota:
  instream flow protection in, 112, 116–17, 122
  public trust doctrine in, 154
  water law in, 26, 37
North Dakota Game and Fish Department, 116–17
North Dakota Water Conservation Commission, 154
Northern spotted owl, 272
North Platte River, 227
North Poudre Irrigation Company, 161, 162
North St. Vrain Creek, 160
Northwest Planning Act (NPA), 245–48
Northwest Power Planning Council, 245–46

Observed conditions, protection of instream flow based on, 106
O'Connor, Sandra Day, 266
Offstream uses, diversion for, 1–2, 132
  historically, 10–12
  water law and, *see* Water law
  water rights transfers to instream uses, 146–48, 168–76, 298–301
  see also specific uses, e.g., Commercial use of water; Irrigation
Oklahoma:
  instream flow protection in, 112, 116, 122, 142–43
  rivers program, 158–59
  water law in, 26, 27, 37
Oklahoma Department of Wildlife and Conservation, 116

Oklahoma Water Resources Board, 116, 159
Olson, Theodore, 199
Olson Opinion, 199
Optimum flow, choosing between minimum and, 106–109, 129–32
Oregon, 20, 286
  instream flow protection in, 112, 122, 123, 124, 130, 137, 139, 144, 162–63, 164
  instream water rights in, 144
  Scenic Waterways Program, 112, 137, 157–58
  water law in, 15, 25–26, 27, 37
  water rights transfers in, 147
Oregon Water Resources Commission, 162
Organic Act of 1897, 187–91, 207
O'Shauhnessy Dam, 32
Outdoor Recreation Resources Review Commission, 111
Oxbow Dam, 273
Oxygen concentration of streams, dissolved, 48–49

Pacific Northwest Power Planning and Conservation Act, 245–48
Participants in state protection activities, 120–29
Pecos River, 167–68
Permits:
  Clean Water Act, 259–63, 266–67
  from state agencies, 204, 301
PHABSIM (Physical Habitat Simulation System), 81, 82, 83, 88, 216
Phantom Canyon Preserve, 161–62
Phillips, Robert, 20
Piedra River, 293, 294
Platte River, 106, 276–77
Pollution:
  air, 243
  water, *see* Water pollution
Pomeroy, John Norton, 30
Population growth, 3, 16, 39, 56, 297
Positive externalities, allocation of water to instream uses and, 102–104
Poudre River, 161–62, 209–11
Powder River, 106
Powell, Lewis F., Jr., 188
Power River, 40
Practicably irrigable acreage (PIA), 194, 195, 196
Preservation flow defined, 8
Prior appropriation doctrine, 120, 141, 162, 203
  beneficial use requirements, 31, 32, 96, 98, 162
  changes arising in 20th century, 34–38, 40–41
  development of, 17–21
  dominance over riparian law, 24–27
  federal law and, 21–22
  first in time, first in right, 18–19, 31
  key components of, 31
  notice and, 18
  use it or lose it principle, 19
Private agreements for instream flow protection, 118, 161–62, 301
Protection of instream flows, 298–301
  amount of water to be kept instream, determining, 106–109, 129–32, 138–40
    see also Quantification of necessary instream flow
  defined, 8
  effect on other water uses, 165–76
  evolution of water law for, 42–43
  federal government and, *see* Federal protection of instream flow
  need for, reasons for, 1–4
  opponents of, 4, 133, 134, 302–303
  reaching a balance, 297–306
  special measures for, arguments for and against, 97–106
  state measures for, *see* State protection of instream flow
  table listing methods for, 299–300
Public goods, 104
Public interest:
  instream flow protection and, 126–27, 140–41
  permit application approval and, 204

water law and, 36–37, 41
Public Land Law Review Commission, 187
Public trust doctrine, 118, 120–23, 133, 151–57, 301
PUD No. 1 of Jefferson County v. Washington Department of Ecology, 265–66, 267
Pueblo Indians, 11

Quantification of necessary instream flow, 79–90, 213, 284
see also Protection of instream flows, amount of water to be kept instream, determining

Rainfall in the West, 2
Ralph Price Reservoir, 160, 161
Ranching, 31
Reclamation Act of 1902, 33, 202, 225, 235, 251
McCarran Amendment, 181, 189
Reclamation Projects Authorization and Adjustment Act of 1992, 229
Recreational uses of water, 2, 4, 31, 34, 38, 55–58, 69, 75, 119
economic significance of, 60–61
quantifying instream flow needs for, 83–85, 88–89, 90
Regulated flows, 8, 63–64
Rehnquist, William H., 181
"Rental" of water, *see* Water leases
Reservation of instream flows, 135, 137, 138–40
Reserved water rights, federal, 179–97, 207–208, 301
Bureau of Land Management lands and, 186–87
Forest Service and, 187–93
Indian reservations and, 179, 193–97
limits of, 181–82
national monuments and, 184–85
national park lands and, 182–84
national wildlife refuge lands and, 185–86
origin of, 179–80
Wild and Scenic Rivers Act and, 281–85, 287
wilderness areas and, 290–96
see also Nonreserved water rights, federal
Reservoirs, 8, 55, 63, 297, 304
Bureau of Reclamation and, 225, 226, 232
location of, 40, 41
modified operations of, 159–61, 227–28, 232
see also names of individual reservoirs
Retroactive measures, 302
public trust doctrine, application of, 156, 157
Return flow, 77, 170, 171
Rio Chama, study of instream flow needs on the, 89–89, 90, 91–94, 215–17
Rio Grande Compact of 1929, 217
Rio Grande River, 40, 106, 287
Rio Mimbres, New Mexico, 187–89
Riparian vegetation, *see* Vegetation, riparian
Riparian water rights, 13–16, 18
conflicts resulting due to shift from, 28–31
dominance of prior appropriation doctrine over, 24–27
federal claims of, 202–203
natural flow doctrine, 14
reasonable use interpretation of, 14–15
state codification of, 5
Rivers and Harbors Act of 1899, 178, 260
Riverside Irrigation District v. Andrews, 261
River terminology, 7
Rock Creek Hydroelectric Project, 251–52, 253
Rocky Mountain National Park, 183
Rogue River, 137
Roman law, 14, 151
Roosevelt Dam, 33
Russian settlers, 13

Sacramento River, 9, 71, 78
St. Vrain Anglers, 160
St. Vrain Corridor Committee, 160–61
Salmon, 49–50, 137, 227–28, 244, 246, 247, 273–75
  life cycle of chinook, 46–47
Salt River, 10, 40, 106
San Bernadino Wildlife Refuge, 220
San Francisco, California:
  water supply for, 32
San Francisco Bay, 78, 151
San Joaquin River, 40, 71, 78, 106
San Juan River, 11
San Pedro Riparian National Conservation Area, 187
Scarcity of water in the West, 2
Sediment transport, 73, 86–87
Senior water rights downstream, 98, 100
  transfer of, 148–49
Settlement of the West, 11–13
  irrigation and, 22–24
Shipping, commercial, 70–72
Sierra Club, 291
Sierra Club v. Block, 291–92
Snail darter, 269–70
Snake River, 40, 71, 72, 106, 167, 186, 193, 227, 273–75
  water bank, 147–48, 160
Snake River Gorge, 137
South Dakota:
  instream flow protection in, 112, 114–15, 117, 120, 122, 142–43, 144
  instream water rights in, 144
  water law in, 26, 37
  water rights transfers in, 146, 147
South Dakota Water Management Board, 114–15
Southwestern Power Administration, 234
Spanish West, water in the, 11–12, 23
Speculation in water, 150
Stanislaus River, 251
State Department of Parks v. Idaho Department of Water Administration, 117
State protection of instream flow, 111–64
  barriers to further protection, 132–35
  combining methods for, 164
  creation of programs, 112–18
  federal government and, 128–29
  methods for, 137–64, 298–301
  minimum versus optimum instream flow, 129–32
  by modified reservoir operations, 159–61, 227–28
  parties allowed to participate, 120–29
  by private agreement, 118, 161–62, 301
  public trust doctrine and, *see* Public trust doctrine
  purposes, 118–20
  rivers programs, *see* Wild and Scenic Rivers programs, state
  table listing methods for, 299–300
  through conservation, 162–64
  through water rights, *see* Water rights
  timing of flow and level of benefits achieved, 109
  with water rights transfers, *see* Water rights transfers
State water rights:
  federal use of, 202–206
  jurisdictional conflicts with federal government, 128–29, 202–203
Steelhead, 244, 247
Steelhead runs, 228
Stock-Raising and Homestead Act of 1916, 186
Storage reservoirs, *see* Reservoirs
Story, Justice, 14
Superfund sites, 78
Supreme Court, U.S.:
  Clean Water Act and, 265–66, 267
  Federal Energy Regulatory Commission's authority and, 250–52
  federal reserved water rights and, 179–80, 181–83, 184–85, 187, 188–89, 194, 199, 291

snail darter decision, 269–70
state power over water rights and, 22
Surplus flows of water, 99, 100–101
Swimming, 57

Takings, 203–204, 282–83
under Endangered Species Act, 269
Tarr, Ralph, 290–91
Taylor Grazing Act of 1934, 199, 205
Tennant, Donald, 79–80
Tennant (or Montana) method, 79–80
Tennessee Valley Authority, 270
Tennessee Valley Authority v. Hill, 270
Texas, 287
instream flow protection in, 112, 115–16, 119, 122, 142–43
new water development projects for, 100–101
water law in, 15, 26, 37
Texas Natural Resource Conservation Commission, 142, 225
Texas Parks and Wildlife Department, 115
Thermoelectric plants, 34, 38, 64
Thomas, George, 30–31
Transaction costs of water transfer, 101–102
Transfers of water rights, *see* Water rights transfers
Transportation, 9
Trelease, Frank, 3, 140, 295
Trinity River, 78
Trout fisheries, 227
Trout Unlimited, 160
Trumpeter swans, 227
Trumpeter Swan Society, 148, 160
Turbidity of streams, 48
Two Forks Dam, 261–63
Tyler v. Wilkinson, 14

United States v. Adair, 194
United States v. City and County of Denver, 185, 189
United States v. New Mexico, 181–83, 188–89, 199, 202, 291
Upper Clark Fork River Basin, 139
User surveys, 84–85, 88
Utah:
instream flow protection in, 112, 119, 120, 122, 142, 144
instream water rights in, 144
Mormons and, 23–24, 25, 30
public trust doctrine in, 154
water law in, 26, 30, 37
water rights transfers and, 146

Values, 3, 43, 297
Vegetation, riparian, 2, 88, 90
quantifying instream flow needs of, 83
streamflow and, 52–53
wildlife and, 53–54
Verde River, 287

Wade, Chief Justice (Colorado state supreme court), 29, 30
Washington, 20, 286
instream flow protection in, 112, 119, 122, 137–38, 138–39, 144, 163, 164
instream water rights in, 144
rivers program, 158, 159
trust water rights program, 163, 164
water law in, 15, 26, 37
Washington Department of Ecology, 138–39, 163, 265
Washington Water Resources Act, 138
Washington Wilderness Act, 292
Waste decomposition, 77–78
Water allocation, *see* Allocation of water
Water banking, 147–48
Water law:
administrative agencies and, *see* Administrative agencies
changes in, 4, 28–31, 36–39, 40–43, 298

Water law (*continued*):
  conflicts arising from changes in, 28–31
  at end of 19th century, 31–32
  federal role in defining, 21–22
  first in time, first in right, 12, 18–19, 31
  historical development of, 3, 12–43
  instream uses and water allocation process, 101
  nonholders of water rights and, 102
  parties affected by dispute resolution, protection of, 12
  prior appropriation doctrine, *see* Prior appropriation doctrine
  public interest and, 36–37, 41
  riparian law, *see* Riparian water rights
  Spanish, 12
  use versus ownership of water, 36
  see also Water rights
Water leases, 99, 147, 160, 301
Water pollution, 76–77
  fish and aquatic life and, 50
  waste decomposition and, 77–78
Water quality, 76–79, 119
  agriculture's effect on, 304–305
  avoidance of pollution, 76–77
  Clean Water Act requirements, 78–79, 259, 263–66
  dilution of waste and sewage and, 78
  waste decomposition and, 77–78
  see also Water pollution
Water Quality Act of 1965, 263
Water Research Center, 1
Water rights:
  donation of, 126, 301
  instream, 135, 137, 138, 143–45, 204–205, 298
  reserved, federal, *see* Reserved water rights, federal
  states recognizing instream, 144
  transfers of, *see* Water rights transfers
  see also Water law
Water Rights Task Force, 212
Water rights transfers, 145–51, 168–76, 298–301
  historical consumptive use and, 171, 174, 175
  no injury rule, 170, 174–75
  temporary, 147–48
  transaction costs of, 101–102
Water speculation, 150
Water volume terminology, 7
Western Area Power Administration (WAPA), 234, 235, 242
Wetter perimeter method, 80–81, 83
Whitewater, 56–57
Whooping crane, 276
Wild and Scenic Rivers programs:
  federal, 87, 157, 159, 277–87
    administration of, 280
    designation process, 87, 213, 279–80, 287
    general characteristics of wild and scenic rivers, 278–79
    map, 278
    policy statement, 277–78
    protection of instream flows and, 283–87
    water rights and, 281–85, 287
  state, 112, 137, 138, 157–59
Wildcat Reservoir, 261
Wilderness Act of 1964, 288–89, 290, 291
Wilderness areas, 287–96
  instream flows in, 290–94
  water rights in, 290–96
Wildlife, 2, 3, 31, 38, 50, 69, 75, 119, 245
  endangered and threatened, *see* Endangered Species Act (ESA)
  fish and aquatic, *see* Fish and aquatic life; Fishing and fisheries
  national refuges, *see* Wildlife refuges, national
  quantifying instream flow needs of, 89, 90
  riparian areas and, 51–55
  watching, 58
Wildlife refuges, national, 220–21
  reserved water rights and, 185–86
  resources of, 222–23
Williamette River, 71
Willow trees, 52, 53

Wind River Reservation, 195–97
Windy Gap project, 275
Winters v. United States, 179–80
Wyoming, 287
  federal reserved water rights and, 190–91, 196–97
  instream flow protection in, 112, 119, 122, 123, 135, 144
  instream water rights in, 144
  water law in, 26, 37
Wyoming Game and Fish Department, 130

Yahu River, 17
Yakima River, 227–28
Yampa River, 167
Yellowstone River Basin, 139

About the Authors

David M. Gillilan and Thomas C. Brown both received their bachelor's degrees in economics, Gillilan from Swarthmore College and Brown from American University. They both received their master's degrees in water resources administration from the University of Arizona, where Brown also received a Ph.D. in watershed management and economics. Gillilan was a research associate at Colorado State University when working on this book and is now a law student at the University of Colorado. Brown continues to work as an economist for the Rocky Mountain Experiment Station, a division of the research branch of the U.S. Forest Service in Fort Collins. Gillilan co-authored *Divided Waters: Bridging the U.S.–Mexico Border*, released in 1995 by the University of Arizona Press. Brown has authored numerous journal papers and monographs on economic valuation, water management, and related topics.